The Handbook of Environmental Chemistry

Volume 2 Part J Reactions and Processes

O. Hutzinger
Editor-in-Chief

Springer-Verlag Berlin Heidelberg GmbH

Bioaccumulation
New Aspects and Developments

Volume Editor: B. Beek

With contributions by
B. Beek, S. Böhling, U. Bruckmann, C. Franke,
H. J. Geyer, L. G. Hansen, J. L. M. Hermens,
U. Jöhncke, A. Kaune, A. Kettrup,
D. Mackay, D. C. G. Muir, J. Paasivirta,
G. G. Rimkus, I. Scheunert, D. T. H. M. Sijm,
G. Studinger, M. Zeeman

Springer

Environmental chemistry is a rather young and interdisciplinary field of science. Its aim is a complete description of the environment and of transformations occuring on a local or global scale. Environmental chemistry also gives an account of the impact of man's activities on the natural environment by describing observed changes.

"The Handbook of Environmental Chemistry" provides the compilation of today's knowledge. Contributions are written by leading experts with practical experience in their fields. The Handbook will grow with the increase in our scientific understandig and should provide a valuable source not only for scientists, but also for environmental managers and decision makers.

Library of Congress Cataloging-in-Publication Data

The Natural environment and the biogeochemical cycles / with contributions by P. Craig ... [et al.].
v. <A–F > : ill. ; 25 cm. – (The Handbook of environmental chemistry : v. 1)
Includes bibliographical references and indexes.
ISBN 978-3-642-08293-1 ISBN 978-3-540-68091-8 (eBook)
DOI 10.1007/978-3-540-68091-8
1. Biogeochemical cycles. 2. Environmental chemistry. I. Craig, P. J., 1944– . II. Series.
QD31. H335 vol. 1 [QH344] 628.5 s [574.5'222]–DC20

ISSN 1433-6839
ISBN 978-3-642-08293-1

Originally published by Springer-Verlag Berlin Heidelberg New York in 2000
Softcover reprint of the hardcover 1st edition 2000

Production Editor: ProduServ GmbH Verlagsservice, Berlin
Cover design: E. Kirchner, Springer-Verlag
Typesetting: Fotosatz-Service Köhler GmbH, Würzburg
Printed on acid-free paper SPIN: 10503050 52/3020 - 5 4 3 2 1 0

Advisory Board

Editor-in-Chief

Prof. Dr. em. Otto Hutzinger
Universität Bayreuth
Postfach 10 12 51
D-95440 Bayreuth, Germany
E-mail: otto.hutzinger@uni-bayreuth.de

Volume Editor

Prof. Dr. Bernd Beek
Umweltbundesamt
Seecktstraße 6 – 10
D-13581 Berlin, Germany
E-mail: bernd.beek@uba.de

Preface

Environmental Chemistry is a relatively young science. Interest in this subject, however, is growing very rapidly and, although no agreement has been reached as yet about the exact content and limits of this interdisciplinary discipline, there appears to be increasing interest in seeing environmental topics which are based on chemistry embodied in this subject. One of the first objectives of Environmental Chemistry must be the study of the environment and of natural chemical processes which occur in the environment. A major purpose of this series on Environmental Chemistry, therefore, is to present a reasonably uniform view of various aspects of the chemistry of the environment and chemical reactions occurring in the environment.

The industrial activities of man have given a new dimension to Environmental Chemistry. We have now synthesized and described over five million chemical compounds and chemical industry produces about hundred and fifty million tons of synthetic chemicals annually. We ship billions of tons of oil per year and through mining operations and other geophysical modifications, large quantities of inorganic and organic materials are released from their natural deposits. Cities and metropolitan areas of up to 15 million inhabitants produce large quantities of waste in relatively small and confined areas. Much of the chemical products and waste products of modern society are released into the environment either during production, storage, transport, use or ultimate disposal. These released materials participate in natural cycles and reactions and frequently lead to interference and disturbance of natural systems.

Environmental Chemistry is concerned with reactions in the environment. It is about distribution and equilibria between environmental compartments. It is about reactions, pathways, thermodynamics and kinetics. An important purpose of this Handbook, is to aid understanding of the basic distribution and chemical reaction processes which occur in the environment.

Laws regulating toxic substances in various countries are designed to assess and control risk of chemicals to man and his environment. Science can contribute in two areas to this assessment; firstly in the area of toxicology and secondly in the area of chemical exposure. The available concentration ("environmental exposure concentration") depends on the fate of chemical compounds in the environment and thus their distribution and reaction behaviour in the environment. One very important contribution of Environmental Chemistry to the above mentioned toxic substances laws is to develop laboratory test methods, or mathematical correlations and models that predict the environmental fate of new chemical compounds. The third purpose of this Handbook is to help in the basic understanding and development of such test methods and models.

The last explicit purpose of the Handbook is to present, in concise form, the most important properties relating to environmental chemistry and hazard assessment for the most important series of chemical compounds.

At the moment three volumes of the Handbook are planned. Volume 1 deals with the natural environment and the biogeochemical cycles therein, including some background information such as energetics and ecology. Volume 2 is concerned with reactions and processes in the environment and deals with physical factors such as transport and adsorption, and chemical, photochemical and biochemical reactions in the environment, as well as some aspects of pharmacokinetics and metabolism within organisms. Volume 3 deals with anthropogenic compounds, their chemical backgrounds, production methods and information about their use, their environmental behaviour, analytical methodology and some important aspects of their toxic effects. The material for volume 1, 2 and 3 was each more than could easily be fitted into a single volume, and for this reason, as well as for the purpose of rapid publication of available manuscripts, all three volumes were divided in the parts A and B. Part A of all three volumes is now being published and the second part of each of these volumes should appear about six months thereafter. Publisher and editor hope to keep materials of the volumes one to three up to date and to extend coverage in the subject areas by publishing further parts in the future. Plans also exist for volumes dealing with different subject matter such as analysis, chemical technology and toxicology, and readers are encouraged to offer suggestions and advice as to future editions of "The Handbook of Environmental Chemistry".

Most chapters in the Handbook are written to a fairly advanced level and should be of interest to the graduate student and practising scientist. I also hope that the subject matter treated will be of interest to people outside chemistry and to scientists in industry as well as government and regulatory bodies. It would be very satisfying for me to see the books used as a basis for developing graduate courses in Environmental Chemistry.

Due to the breadth of the subject matter, it was not easy to edit this Handbook. Specialists had to be found in quite different areas of science who were willing to contribute a chapter within the prescribed schedule. It is with great satisfaction that I thank all 52 authors from 8 countries for their understanding and for devoting their time to this effort. Special thanks are due to Dr. F. Boschke of Springer for his advice and discussions throughout all stages of preparation of the Handbook. Mrs. A. Heinrich of Springer has significantly contributed to the technical development of the book through her conscientious and efficient work. Finally I like to thank my family, students and colleagues for being so patient with me during several critical phases of preparation for the Handbook, and to some colleagues and the secretaries for technical help.

I consider it a privilege to see my chosen subject grow. My interest in Environmental Chemistry dates back to my early college days in Vienna. I received significant impulses during my postdoctoral period at the University of California and my interest slowly developed during my time with the National Research

Council of Canada, before I could devote my full time of Environmental Chemistry, here in Amsterdam. I hope this Handbook may help deepen the interest of other scientists in this subject.

Amsterdam, May 1980 *O. Hutzinger*

Seventeen years have now passed since the appearance of the first volumes of the Handbook. Although the basic concept has remained the same some changes and adjustments were necessary.

Some years ago publishers and editor agreed to expand the Handbook by two new open-ended volume series: Air Pollution and Water Pollution. These broad topics could not be fitted easily into the headings of the first three volumes. All five volumes series are integrated through the choice of topics and by a system of cross referencing.

The outline of the Handbook is thus as follows:

1. The Natural Environment and the Biochemical Cycles,
2. Reactions and Processes,
3. Anthropogenic Compounds,
4. Air Pollution,
5. Water Pollution.

Rapid developments in Environmental Chemistry and the increasing breadth of the subject matter covered made it necessary to establish volume-editors. Each subject is not supervised by specialists in their respective fields.

A recent development is the 'Super Index', a subject index covering chapters of all published volumes, which will soon be available via the Springer Homepage http://www.springer.de or http://www.springer-ny.com or http://Link.springer.de.

With books in press and in preparation we have now published well over 30 volumes. Authors, volume-editors and editor-in-chief are rewarded by the broad acceptance of the 'Handbook' in the scientific community.

May 1997 *Otto Hutzinger*

Contents

Foreword

Bioaccumulation as an enhancing factor in exposure of organisms to environmental chemicals has become of increasing importance in environmental research and risk analysis during recent years. As a now classical approach, the assessment of environmental hazards due to chemical contaminants is based upon the comparison of external exposure concentrations and toxic concentration levels of a particular substance. As modifiers of exposure and – as a consequence – of toxicity, degradation and accumulation phenomena were included in this approach. During the last decade it has become increasingly clear that Bioaccumulation and Biomagnification of chemicals in biota via the food chain, or better the food web, may be the prerequisite for adverse effects in individuals, species, and ecosystems because environmental concentrations of xenobiotics are very often too low to exert deleterious effects immediately. Furthermore, even sophisticated eco-toxicity testing for chronic effects cannot rule out a possible risk of delayed or long-term effects which may be unknown as yet (as happened recently with the so-called endocrine-disrupting chemicals). This risk is increasing by magnitudes with time if hardly any or no reduction in environmental concentrations of xenobiotics occur due to lack or inhibition of degradation processes (the so-called persistent organic pollutants, POPs). Thus, there is good evidence to assume that bioaccumulating chemicals need particular attention in environmental hazard assessment.

This book gives a state-of-the-art report on reliable determination of Bioaccumulation and an up-dated review of Bioaccumulation of organic compounds, including endocrine-disrupting chemicals and POPs, in fish and other organisms in the first chapter. For a more sophisticated comparison of exposure and toxic (effect) concentrations in hazard assessment of environmental chemicals it will become more and more necessary to compare *internal* exposure concentrations rather than *external* ones with toxic effect levels in organisms. In the second chapter a concept of the Internal Effect Concentration as a link between Bioaccumulation and Ecotoxicity is presented. The internal concentration deals with additivity of mixtures of chemicals, and it may become indeed more meaningful in the future to compare additive internal "matrices" of groups of similar chemicals rather than single-chemical concentrations with endpoints responsible for biological (toxic) effects. Due to coaccumulation of many toxic substances it is difficult to trace back damage in ecosystems to particular chemicals in most cases, but it is certain that Bioaccumulation of xenobiotics has caused long-term adverse effects in ecosystems (third chapter). In the final chapter a review is given of existing concepts for the assessment of Bioaccumulation, and a comprehensive concept for the assessment of Bioaccumulation, Biomagnification via the food web, and Secondary Poisoning due to enriched concentrations of environmental chemicals in food is presented.

Berlin, August 1999 Bernd Beek

Bioaccumulation and Occurrence of Endocrine-Disrupting Chemicals (EDCs), Persistent Organic Pollutants (POPs), and Other Organic Compounds in Fish and Other Organisms Including Humans*

Harald J. Geyer[1,*] · Gerhard G. Rimkus[2] · Irene Scheunert[3] · Andreas Kaune[4] · Karl-Werner Schramm[1] · Antonius Kettrup[1,4] · Maurice Zeeman[5] · Derek C.G. Muir[6] · Larry G. Hansen[7] · Donald Mackay[8]

[1] GSF-National Research Center for Environment and Health GmbH, Munich, Institute of Ecological Chemistry, P.O. Box 1129, D-85758 Neuherberg, Germany
[2] Food and Veterinary Institute (LVUA) Schleswig-Holstein, Department of Residue and Contamination Analysis, P.O. Box 2743, D-24517 Neumünster, Germany
[3] GSF-National Research Center for Environment and Health GmbH, Munich, Institute of Soil Ecology, P.O. Box 1129, D-85758 Neuherberg, Germany
[4] Technical University Munich, Institute of Ecotoxicological Chemistry and Environmental Analysis, D-85350 Freising-Weihenstephan, Germany
[5] U.S. Environmental Protection Agency, Office of Pollution Prevention and Toxics, Risk Assessment Division (7403), 401 M St., S.W., Washington, D.C. 20460, USA
[6] National Water Research Institute, Environment Canada, Burlington, Ontario, Canada L7R 4A6
[7] University of Illinois, 2001 S. Lincoln Avenue, Urbana IL 61302, USA
[8] Trent University, Peterborough, Ontario, Canada K9 J 7B8
* Corresponding author

Bioaccumulation of chemicals by aquatic organisms, especially fish, mussels and *Daphnia*, is an important criterion in risk assessment. Bioconcentration from water must be considered in context with toxicity, biotic and abiotic degradation and other physical-chemical factors in order to protect the freshwater and marine environments with their organisms. Furthermore, it is necessary to prevent human exposure from contaminated aquatic food, such as fish, mussels, and oysters. This review outlines the factors such as toxic effects, bioavailability, chemical concentration in the water, pH of the water, and lipid content of the organisms, which are known to affect the bioconcentration of chemicals in aquatic organisms. Quantitative structure-activity relationships (QSARs) for predicting the bioconcentration potential of chemicals in algae, *Daphnia*, mussels, and fish are presented. Specific classes of organic chemicals, such as endocrine-disrupting chemicals (EDCs), super-hydrophobic persistent organic pollutants (POPs) (2,3,7,8-tetrachlorodibenzo-*p*-dioxin, octachlorodibenzo-*p*-dioxin, Mirex, and Toxaphene), tetrachlorobenzyltoluenes (TCBTs), polybrominated benzenes (PBBz), polybrominated biphenyls (PBBs), polybrominated diphenyl ethers (PBDEs), polychlorinated diphenylethers (PCDEs), nitro musk compounds (NMCs), polycyclic musk fragrances (PMFs), and sun screen agents (SSAs) are critically reviewed and discussed. Furthermore, predictions for some metabolites, especially hydroxylated aromatics, of these chemical classes which may have endocrine-disrupting effects are made. The selected bioconcentration factors on a wet weight basis (BCF_W) and on a lipid basis (BCF_L) in aquatic organisms, such as algae (*Chlorella* sp.), water fleas (*Daphnia* sp.), mussels (*Mytilus edulis*), oysters (*Crassostrea vir-*

* Disclaimer: This document has been reviewed by the Office of Pollution Prevention and Toxics, US Environmental Protection Agency and approved for publication. The views expressed are those of the author and approval does not signify that the contents necessarily reflect the views and policies of the Agency nor does mention of tradenames or commercial products constitute endorsement or recommendation for use.

The Handbook of Environmental Chemistry, Vol. 2 Part J
Bioaccumulation (ed. by B. Beek)

ginica), and different fish species, of these chemicals are presented in tables. Furthermore, the chemical structure, physico-chemical properties, such as selected log K_{OW} values, and other data are compiled. In the cases where no bioconcentration factors (BCFs) were published the BCF values of chemicals in fish and mussels were predicted from QSARs using the n-octanol/water partition coefficient (K_{OW}) as the basic parameter. A new classification scheme for organic chemicals by their hydrophobicity (log K_{OW}) and by their worst-case bioconcentration factors on a lipid basis (BCF_L) is also presented.

Keywords: Bioaccumulation, Bioconcentration, Bioconcentration factor (BCF), Endocrine-disrupting chemicals (EDCs), Persistent organic pollutants (POPs), Xenoestrogens, Xenoantiestrogens, Xenoandrogens, Xenoantiandrogens, Super-hydrophobic compounds, TCDD, OCDD, PCBs, PCDDs, PCDFs, PBDEs, PCDEs.

1
Introduction

Bioaccumulation of pesticides and other chemicals in aquatic organisms first gained public attention in the 1960s. Residues of DDT, DDD, DDE, and methyl mercury were discovered in fish and wildlife. The bioaccumulation potential of a chemical in aquatic organisms, such as fish is, in addition to toxicity, and biotic and abiotic degradation, an important criterion in the assessment of environmental hazards [1–7]. A high bioaccumulation potential of a chemical in biota increases the probability of toxic effects being encountered in aquatic and terrestrial organisms including humans and their environment. Therefore, many proposed and existing regional and international regulatory classification schemes, guidelines, and risk assessments use estimates of bioaccumulation to indicate whether chemicals may be hazardous to aquatic organisms, if their bioconcentration factor (BCF) exceeds designated threshold values [2–7].

In the European Union (EU), any chemical with a bioconcentration factor on a wet wt. basis (BCF_W) > 100 is considered to have the potential to bioaccumulate and is classified as "dangerous to the environment", because it could impair the health of an aquatic organism or of predators feeding on that organism. The administrative directorate of the EU, the European Commission, therefore has recommended a BCF_W value of 100 as a trigger for hazard classification of chemicals [6]. The U.S. EPA uses a BCF_W > 1000 as the trigger for high concern for potential bioaccumulation effects [9]. In Canada chemicals with a BCF_W value >5000 are considered to bioaccumulate and are recommended for "virtual elimination". If a chemical has a BCF_W value > 500 it is considered as hazardous [8]. Chemicals with elevated bioconcentration factors are also of concern for regulators because they are considered capable of biomagnification in the food chain. Bioaccumulation properties of chemicals are one of the triggers of the U.S. EPA and the EU environmental risk assessment process. This may become internationally applicable through intergovernmental mechanisms, e.g. the North Sea Conference in the EU, the United Nations International Marine Convention, the "Great Lakes Water Quality Agreements" in North America, and the International Forum on Chemical Safety.

Aquatic organisms may be contaminated by chemicals by several pathways: directly via uptake through gills or skin as well as indirectly via ingestion of food or contaminated sediment particles [3]. For clarity the terminology associated with such studies should be given.

2 Definitions and Terminology

2.1 Bioconcentration

Bioconcentration is the result of direct uptake of a chemical by an organism only from water. Experimentally, the result of such a process is reported as the bioconcentration factor (BCF). Consequently, the BCF is defined as the ratio of steady state concentration of the chemical in aquatic organisms (C_F) such as fish, mussels, water flea (*Daphnia*), algae etc. and the corresponding **freely dissolved chemical concentration** in the surrounding water (C_W) [2 a, b, c, 4, 10–14]:

$$BCF = \frac{C_F}{C_W} \frac{[\text{ng kg}^{-1}]}{[\text{ng L}^{-1}]} \tag{1}$$

Instead of BCF sometimes the abbreviation K_B is also used, however, for clarity we do not recommend the use of this abbreviation. For aquatic organisms three different bioconcentration factors (BCF) can be given [13]:

(1) on a wet weight basis (BCF_W),
(2) on a lipid basis (BCF_L), and/or
(3) on a dry weight basis (BCF_D).

All three BCF values can be viewed as essentially unitless because 1 l water has a mass of 1 kg; so the dimensions of the chemical concentration in water are equivalent to the dimensions of the chemical concentration in the organisms [13–16].

It was shown by Geyer et al. [17] and others [18] that the BCF_W value of lipophilic organic chemicals is dependent on the lipid content of the organism (see Sect. 5.5). Therefore, for the sake of comparison, the most important BCF value of a lipophilic chemical in an organism is that on the lipid basis (BCF_L). The BCF_L values can easily be calculated from BCF_W values, if the lipid content (L in % on a wet weight basis; L_W (%)) of the organism is known:

$$BCF_L = \frac{BCF_W \cdot 100}{L_W\ (\%)} \tag{2}$$

Sometimes the lipid content of the organisms is given on a dry weight basis (L_D in %). In this case the water content (%) of the organisms must also be measured. But more important is the lipid content on a wet weight basis (L_W in %) of the organisms.

2.2
Biomagnification, Bioaccumulation and Ecological Magnification

The definition of bioconcentration has to be distinguished from the terms of indirect contamination such as biomagnification, bioaccumulation, and ecological magnification [12, 19].

(a) The term **biomagnification** is used for the dietary uptake via contaminated food. The **biomagnification factor** (BMF) of a chemical is the ratio between the concentrations in fish and food at steady state [20a]. Again, the BMFs may be expressed on wet, dry, or lipid basis.
(b) **Bioaccumulation** is defined as the uptake of substances from both food and water.
(c) **Ecological magnification** means increasing chemical concentrations in the food chain [19a].

One of the latest most comprehensive review of trophic transfer and biomagnification potential of chemicals in aquatic ecosystems was published by Suedel et al. [19b]. They summarized literature on trophic transfer of chemicals from field and laboratory experiments. Results were expressed in terms of trophic transfer coefficient (e.g. concentration of a chemical in consumer tissue divided by the concentration of chemical in food). They compared these values and estimates of overall potential chemical trophic transfer through aquatic food webs with data from aquatic food web models. The authors analyzed data on organic chemicals, such as atrazine, dieldrin, DDT, DDE, hexachlorocyclohexane, Kepone, Toxaphene, polychlorinated biphenyls (PCBs), polynuclear aromatic hydrocarbons (PAHs), and tetrachlorodibenzo-*p*-dioxin (TCDD), and on inorganic compounds. From their results some general conclusions can be drawn:

a) The majority of chemicals evaluated do not biomagnify in aquatic food webs;
b) for many of the compounds examined, trophic transfer does occur but does not lead to biomagnification in aquatic food webs;
c) DDT, DDE, Toxaphene and methyl mercury have the potential to biomagnify in aquatic ecosystems;
d) the lipid content of predators directly influences biomagnification potential of lipophilic chemicals;
e) even those compounds for which evidence for biomagnification is strongest show considerable variability and uncertainty regarding the magnitude and existence of food web biomagnification in aquatic ecosystems;
f) the food web model reviewed [19d] provided similar estimates for most of the organic compounds examined (log K_{ow} values between 5 and 7) with model predictions falling within the range of values of all compounds except dieldrin.

These conclusions are in agreement with other literature. Opperhuizen [19c] found that the feeding rate of fish [0.02 g/(g d)] compared to the ventilation rate [2000 ml water/(g d)] is very low. Thus uptake from food contributes significantly if the concentration of the chemical in food is 100,000 times higher than the concentration of the chemical in water.

Because for most chemicals the uptake from water (bioconcentration) is of the greatest importance [20b,c], the following sections deal mainly with bioconcentration. However, for very hydrophobic chemicals with log n-octanol/water partition coefficients (log K_{ow}) >6.3, bioaccumulation is of relevance [20b]. In particular, some of the main factors which are affecting the bioconcentration potential are described. Because it is known that many environmental chemicals and/or especially their metabolites can have endocrinic disrupting or estrogenic properties, this chapter deals with some of these chemicals, including some of their metabolites. Furthermore, selected bioconcentration factors, especially of persistent organic pollutants (POPs) in aquatic organisms, such as algae, water fleas, mussels, oysters, and fish are presented.

3
Theory of Bioconcentration and Elimination of Chemicals in Aquatic Organisms

3.1
Bioconcentration Kinetics

The bioconcentration process of non-degradable chemicals can generally be interpreted as a passive partitioning process between the lipids of the organisms and the surrounding water. This process can be described by the first order two-compartment (water and aquatic organism) model. The conventional equation describing the uptake and elimination of a persistent chemical by aquatic organisms, such as fish, mussels, and *Daphnia*, is given as Eq. (3):

$$\frac{dC_F}{dt} = k_1 \cdot C_W - k_2 \cdot C_F \tag{3}$$

where k_1 is the uptake rate constant (day^{-1}), k_2 is the elimination or depuration rate constant (day^{-1}), C_w is the chemical concentration in water, and C_F the chemical concentration in fish. At steady state, $dC_F/dt = 0$ and the BCF value can be calculated by Eq. (4):

$$BCF = \frac{k_1}{k_2} = \frac{C_F}{C_W} \tag{4}$$

The bioconcentration factor can be estimated by exposing fish or other aquatic organisms, for an appropriate time period, to a constant chemical concentration in water by using a flow-through system until a steady-state concentration in the organism is reached. However, for many chemicals – especially very hydrophobic chemicals – a steady-state cannot be reached in an appropriate time. Therefore, the kinetic approach is the only method which can be used for the determination of a "real" BCF value.

If during the experiment, the fish are growing and the chemical is metabolized, the specific growth rate constant (k_G) and the metabolism rate constant (k_M) must be included in Eq. (3):

$$\frac{dC_F}{dt} = k_1 \cdot C_W - (k_2 + k_G + k_M) \cdot C_F \tag{5}$$

If the concentration reaches steady-state, i.e., $dC_F/dt = 0$, the BCF value is given by equations (6) and (7):

$$k_1 \cdot C_W = (k_2 + k_G + k_M) \cdot C_F \tag{6}$$

$$BCF = \frac{C_F}{C_W} = \frac{k_1}{k_2 + k_G + k_M} \tag{7}$$

It should be noted that the BCF can also be determined solely from the uptake curve of the chemical in the organisms. The method and equations for calculating the BCF values in this way were recently published by Wang et al. [23]. An important paper on different compartment models and the mathematical descriptions of uptake, elimination and bioconcentration of xenobiotics in fish and other aquatic gill-breathing organisms was given by Butte [24].

3.2
Elimination Kinetics and Biological Half-Life ($t_{1/2}$)

The elimination or depuration of chemicals from aquatic and terrestrial organisms often follows first order kinetics and can be described by Eq. (8):

$$C_t = C_0 \cdot e^{-k_2 t} \tag{8}$$

where C_t is the concentration in the organism at time t, C_0 is the concentration in the organism at time t_0 at the start of the depuration or elimination phase if the contaminated organism is put into clean water. The elimination constant k_2 can be calculated after integration of Eq. (9):

$$k_2 \cdot t = \ln \frac{C_0}{C_t} \tag{9}$$

or using base 10 log values:

$$k_2 = \frac{2.303}{t} \cdot \log \frac{C_0}{C_t} \tag{10}$$

An important criterion in hazard assessment of organic chemicals is the biological half-life ($t_{1/2}$). The half-life of a chemical is the time required to reduce the concentration of this chemical by one-half in tissue, organ, or in the whole organism. If the elimination rate k_2 was determined the $t_{1/2}$ can be calculated by Eq. (11):

$$t_{1/2} = \frac{\ln 2}{k_2} = \frac{0.693}{k_2} \tag{11}$$

However, if the elimination phase takes a long time, as is the case for highly superhydrophobic persistent chemicals, the increase in body weight has to be considered [25a]. Compensation for so-called "growth dilution" can be made if the growth rate constant (k_G) during the elimination phase is known by using Eq. (12):

$$t_{1/2} = \frac{0.693}{k_2 + k_G} \tag{12}$$

In case that the k_G is not known, this adjustment can be eliminated by multiplying the chemical concentration by the total weight of the organism. Estimation of $t_{1/2}$ based on body burden provides a better basis for comparisons of $t_{1/2}$ of a chemical among studies with the same organism [25a] (see also Sect. 8.2.3). However, recently it was shown that the half-life of a chemical in different aquatic organisms is dependent on its lipid content [29a, b, 40]. For persistent lipophilic chemicals $t_{1/2}$ increases with the lipid content of the organism (Fig. 1).

3.3
Equations to Predict the Half-Life ($t_{1/2}$) or Elimination Rate Constant (k_2)

The biological half-lives ($t_{1/2}$) of a chemical in organisms have important implications in hazard assessment and can also be used to assess the importance of

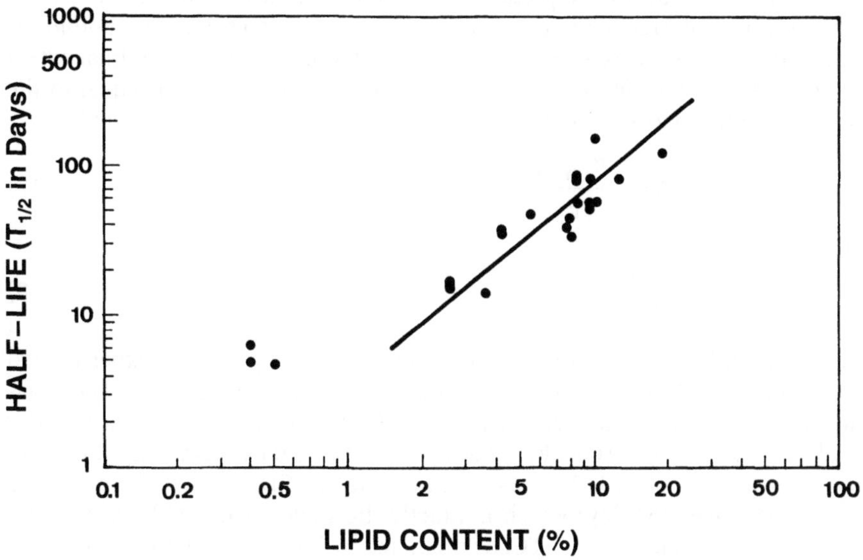

Fig. 1. The relationship between half-lives ($t_{1/2}$) of 2,3,7,8-tetrachlorodibenzo-*p*-dioxin (TCDD) in mussels and fish and their lipid content (L%). The linear regression equation is: $\log t_{1/2} = 1.36 \log L\% + 0.546$; $n = 25$, $r^2 = 0.764$, $p < 0.01$ (two tailed). 76.4% of the half lives variability of TCDD can be explained by the lipid content. Data were taken from Geyer et al. [29a]

the bioconcentration and biomagnification pathways for accumulation of chemicals in fish and other organisms [25a]. Therefore, it is useful if it is possible to predict the $t_{1/2}$ of an organic chemical from its physico-chemical properties in an aquatic gill-breathing organism, such as fish etc. In the following section equations are derived to predict the $t_{1/2}$ of an organic chemical in organisms if the uptake rate (k_1), the BCF or the log K_{OW}, and the lipid content of the gill-breathing organisms is known.

In Eq. (11) the elimination constant (k_2) is substituted by k_1/BCF_W:

$$t_{1/2} = \frac{0.693}{k_2} \quad \text{and} \quad k_2 = \frac{k_1}{BCF_W} \tag{13a, b}$$

It follows

$$t_{1/2} = \frac{0.693}{k_1} \cdot BCF_W \tag{14}$$

Because BCF_W depends on the lipid content (L, in %) of the organisms, it can be replaced by

$$BCF_W = \frac{BCF_L \cdot L}{100} \tag{15}$$

to give

$$t_{1/2} = \frac{0.693}{100} \cdot \frac{BCF_L \cdot L}{k_1} \tag{16}$$

Equation (16) can be used to predict the half-life of organic chemicals in fish and other aquatic gill-breathing organisms if the chemical does not form bound residues. In case that the organic chemical is metabolized only to a minimal extent or not at all the bioconcentration factor on a lipid basis (BCF_L) is equal to the n-octanol/water partition coefficient (K_{OW}) (see Sect. 7) so that Eq. (16) gives

$$t_{1/2} = 0.00693 \cdot \frac{K_{OW} \cdot L}{k_1} \tag{17}$$

or

$$\frac{1}{k_2} = \frac{K_{OW} \cdot L}{k_1 \cdot 100} \tag{18}$$

Equations (16) and (17) were examined on their accuracy by using experimental kinetic data of 2,3,7,8-tetrachlorodibenzo-*p*-dioxin (TCDD) uptake and elimination kinetics in the fish medaka (*Oryzias latipes*) determined by Schmieder et al. [28]: $BCF_L = 5{,}100{,}000$; $k_1 = 2{,}300$ days^{-1}; $k_2 = 0.0045$ days^{-1}; lipid content $L = 10\%$; log k_{OW} of TCDD = 6.64. The half-life of TCDD in medaka predicted by Eq. (16) gives 154 days which is exactly the value measured by Schmieder et al. [28]. However, the $t_{1/2}$ of TCDD predicted by Eq. (17) gives 132 days, which is in satisfactory agreement with the measured $t_{1/2}$ value.

From Eq. (17) it is obvious that the half-life of persistent organic chemicals is increasing with its n-octanol/water partition coefficient and the lipid content of

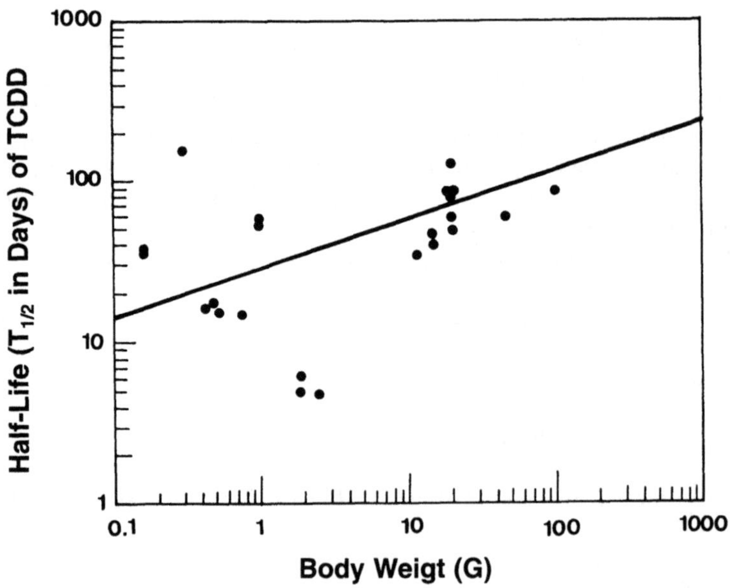

Fig. 2. The relationship between half-lives ($t_{1/2}$) of 2,3,7,8-tetrachlorodibenzo-*p*-dioxin (TCDD) in mussels and fish and their body weight (BW in g). The linear regression equation is: $\log t_{1/2} = 0.306 \log BW + 1.44$; $n = 25$, $r^2 = 0.609$, $p < 0.01$ (two tailed). Only 61% of the half lives variability of TCDD can be attributed to the differences in body size (weight). Data were taken from Geyer et al. [29a]

the organisms. Because the half-life is decreasing with increasing temperature [25a, b], equations (16) and (17) are valid for a given temperature or a small deviation from this value. If the half-life is very long the growth rate constant (k_G) must also be taken into account. Because the $t_{1/2}$ increases with body size (BW) (see Fig. 2 and references [29a, 40]), it could also be necessary to include the BW in equations (17) and (18) if the size of the organism is very great.

3.4
Application of the Half-Life ($t_{1/2}$) or Elimination Rate Constant (k_2)

It is known that for very hydrophobic chemicals it can take a very long time (months to years) to reach steady-state concentrations in fish and other organisms. If the BCF value is calculated by dividing the non-equilibrated chemical level in fish by the chemical concentration in water, the bioconcentration factor is underestimated. However, steady-state residue concentrations can be extrapolated if the half-lives or elimination rate constants are available.

The increase in residue level in fish (C_F) or other organisms as a function of time (t) is given by Eq. (19):

$$C_F = \frac{k_1}{k_2} \cdot C_W \left(1 - e^{-k_2 \cdot t}\right) \tag{19}$$

Replacing $k_1 \cdot C_W$ by C_U, which is the amount of chemical uptake per day, gives Eq. (20):

$$C_F = \frac{C_U}{k_2} \cdot \left(1 - e^{-k_2 \cdot t}\right) \tag{20}$$

These relationships are useful in planning bioconcentration studies. Furthermore, Eq. (20) can be used to estimate the level of a chemical in an organism as percent of steady-state (equilibrium) level reached at time t:

$$\text{Steady-state level (\%)} = 100 \cdot \left(1 - e^{-k_2 \cdot t}\right) \tag{21}$$

or replacing k_2 by $0.693/t_{1/2}$ gives:

$$\text{Steady-state level (\%)} = 100 \cdot \left(1 - e^{-\frac{0.693 \cdot t}{t_{1/2}}}\right) \tag{22}$$

By means of Eq. (22) the percentage of chemical steady-state level in relation to time of chemical uptake in half-lives was calculated and presented in Table 1. Furthermore, as an example the uptake of TCDD in medaka was calculated. 95% of the steady-state TCDD level is reached if the time of uptake is 4.3 × half-life and 98.4% is reached in 6 × half-life. This means that for TCDD in medaka (10% lipid content) 95% of steady-state TCDD level is reached in 1.8 years and 98.4% in 2.5 years, respectively. The experimentally determined half-life of TCDD in medaka was 154 days [28].

Table 1. Level of a chemical in an organism during constant uptake from water or food, respectively, as per cent of steady-state level in relation to time (t in half-lives, $t_{1/2}$), calculated by means of Eq. (22)

Time of chemical uptake in half-lives ($t_{1/2}$)	Steady-state level of chemical attained (%)
$1 \times t_{1/2}$	50.0
$2 \times t_{1/2}$	75.0
$3 \times t_{1/2}$	87.5
$4 \times t_{1/2}$	93.7
$4.3 \times t_{1/2}$	95.0
$5 \times t_{1/2}$	96.9
$6 \times t_{1/2}$	98.4
$7 \times t_{1/2}$	99.2
$8 \times t_{1/2}$	99.6
$9 \times t_{1/2}$	99.8
$10 \times t_{1/2}$	99.9

Note: At the beginning ($t = 0$) of the chemical uptake the level in the organism is nil.

4
Determination of Bioconcentration Factors (BCFs)

Recently it was recommended by the Organization for Economic Cooperation and Development (OECD), Paris, that the existing five standardized and internationally harmonized OECD guidelines for bioaccumulation of chemicals in fish No. 305 A-E [21] should be replaced by the single modified version of the Flow-through Kinetic Fish Test [22]. This method is valid when applied to organic chemicals with log n-octanol/water partition coefficients (log K_{ow}) between 1.5 and 6.0 but may still be applied to super-hydrophobic compounds having log K_{ow} values >6.0. In this kinetic approach the uptake rate constant (k_1) and the elimination rate constant (k_2) are determined in separate experiments. The elimination is usually estimated by placing the contaminated aquatic organisms such as fish, mussels etc., in clean flowing water and measuring the decrease of the concentration in the organism with time. It is important to note that if chlorinated tap water is used in the flow-through system the water has to be dechlorinated; otherwise toxic effects can occur which can modify the bioconcentration factor. The BCF should be determined in an appropriate concentration range, where values are independent of concentration of the test chemical in water and are ecologically meaningful, and where no toxic effects occur. The concentration of the test chemical must be well below its water solubility, otherwise the obtained BCF value is too small (see Sect. 5.4).

For performing the bioconcentration flow-through fish test see [22]. An apparatus for continuously saturating water with hydrophobic organic chemicals was described by Veith and Comstock [49]. However, an exposure system with generator column [26, 27] is recommended for very hydrophobic chemicals [28,

50a, b]. During the uptake phase the concentration of the chemical in the water must be analyzed at appropriate times. For more information on the performance of the Kinetic approach see references [21, 22, 26, 28]. For very lipophilic chemicals it is important to measure the depuration for a relatively long time (some months corrected for growth of course). Otherwise, a great elimination rate constant is calculated, and thus a small BCF value is obtained.

5
Factors Affecting Bioconcentration

The bioconcentration of a chemical by aquatic organisms is dependent on many factors. It is clear that the BCF is dependent on the physico-chemical properties of the tested chemical, such as water solubility and lipophilicity measured as n-octanol/water partition coefficient (log K_{ow}). The higher the K_{ow} value of a chemical, the higher the bioconcentration potential in a specific aquatic organism, if this chemical is not metabolized. However, there are also many other external and internal factors which can influence the BCF value (see Table 2). Therefore, it is important that the variation of temperature is less than ±2°C, the concentration of dissolved oxygen is >60% of saturation, and the concentration of the test chemical is maintained within ±20% of the values measured during the uptake phase. Since the dissolved and particulate organic matter may significantly influence the bioconcentration of organic chemicals in fish and other gill-breathing organisms, the total organic carbon (TOC) present in the water should not exceed 10 mg l^{-1}. It is important that the flow-through test is performed in accordance with the OECD Test Guideline No. 305 [22].

It was also found that the bioconcentration potential is dependent on the age, sex, and species of the aquatic organisms. Many of these factors can be eliminated if the test is performed under identical conditions with the organisms of the same species, strain, sex, age, etc. or if the bioconcentration factor is related to the lipid content of the organism (see Sect. 5.4). Some other important factors which may affect the bioconcentration potential of chemicals in fish and other aquatic organisms are the toxic effects, bioavailability, concentration of the chemicals in water, pH of the water, and especially the lipid content of the organisms. These factors will be discussed in more detail in the following sections.

5.1
Toxic Effects

It was found that adverse effects, disease and mortality in both treated and control fish can influence the kinetics of the chemical in fish. Mortality, therefore, should normally be <10% at the end of the test. Geyer et al. [29] found that the elimination rate of a chemical in aquatic gill-breathing animals is greater, if toxic effects occur and especially if the lipid content is decreasing during the test. That means that the half-life ($t_{1/2}$) and the bioconcentration factor of a chemical is smaller if the concentration in the water is so high that toxic effects occur. Therefore, the concentration of the test chemical in the water has to be so low that

Table 2. Biotic and abiotic factors which can influence the bioconcentration, bioaccumulation and/or biomagnification of chemicals in fish and other aquatic organisms

A. Biotic factors		B. Abiotic factors
(1) Species (2) Strain (3) Sex (male /female) (4) Genetic background (5) Developmental stage a) eyed-egg b) hatching c) swim-up fry d) young e) adult (6) Body composition (7) Body weight (8) Body length (9) Age (young, adult) (10) Spawning (11) Health status a) disease b) parasitism, etc. (12) Hormone status a) L-thyroxine (T_4) b) L-3,5,3′-triiodothyronine (T_3) c) testosterone etc. (13) Intermediary metabolism	Lipid content of the organism	(1) Diet composition (fat, protein, carbohydrate content) (2) Food deprivation, malnutrition, starvation (3) Manipulation of the body composition of the growing organism for some months with e.g. a) anabolic steroid b) thyroxine c) diet etc. (4) Season of the year (summer, fall, winter, spring) when the test is performed with fish from natural environment
(14) Metabolism rate (15) Elimination rate (k_2) (16) Half-life ($t_{1/2}$) of the chemical (17) Toxic effects (18) Liver function (19) "Growth dilution" in the aquatic organism (20) Changing of the lipid content of the organism during the test etc.		(5) Temperature of water (6) Quality of the water a) pH (is important for the BCF of ionizable organic chemicals) b) oxygen content c) hardness d) salinity e) chlorine concentration f) total organic carbon (TOC), humic substances, suspended solids, etc. (7) Ratio of biomass to water volume (g fish/l water) (8) Static or flow-through test system (9) Changing of test-chemical concentration in water during the uptake phase (10) Concentration of the test-chemical in water (11) Purity of the chemical (^{14}C) (12) Bioavailability etc.

Biotic factors which can influence the lipid content of the organisms are numbers A. (1) to A. (13). Abiotic factors which can also influence the lipid content of the organisms are numbers B. (1) to B. (4).

no toxic or only minimal adverse effects in fish and other aquatic organisms occur. Essentially the test organism must not be stressed during the test, otherwise its physiological parameters change, affecting the rates of uptake and elimination.

5.2 Bioavailability

Transport of chemicals into and through biological membranes requires that the compound in the surrounding water be available in a dissolved form. Environmental factors that can reduce the chemical amount in true solution will reduce the uptake rate and bioconcentration and/or bioaccumulation potential. The most important processes which may influence and reduce bioavailability of hydrophobic chemicals are: binding to particulates and dissolved organic matter (DOM); and adsorption to humic acids, sediments, and other suspended macromolecules. It is also important that formation of colloidal suspensions, especially of very hydrophobic chemicals, can reduce the effective water exposure concentration and its bioavailability. Bioavailability is defined as the external availability of a chemical to an aquatic organism, as opposed to the classic pharmacological definition of internal bioavailability after injection or ingestion [51]. In most studies a reduction of the uptake and the bioconcentration factor of the chemicals in the presence of organic materials has been found. In the following part some examples are presented.

Gobas et al. [30] in 1989 investigated the bioconcentration potential of polychlorinated biphenyls (PCBs), polybrominated benzenes (PBBzs) and polybrominated biphenyls (PBBs), and other super-hydrophobic chemicals, such as decachlorobiphenyl and Mirex. These authors also pointed out the importance of bioavailability for bioconcentration of super-hydrophobic chemicals. Their study showed that the bioavailable fraction of the super-hydrophobic chemical decachlorobiphenyl can be as low as 3% and of Mirex can be as low as 2.2%. For decachlorobiphenyl, a BCF was found that was one to two orders of magnitude lower than the true BCF.

Servos and Muir [31] in 1989 investigated the effects of dissolved organic matter from Canadian lakes on the bioavailability of 1,3,6,8-tetrachlorodibenzo-*p*-dioxin to the amphipod *Crangonyx laurentianus*. They found a relationship between the binding of the compounds to organic material and the reduction of the uptake in these organisms. In another paper Servos, Muir, and Webster [32] pointed out the importance of organic matter for the bioavailability and thus for the bioconcentration factor of chlorinated dioxins in aquatic organisms.

The uptake of five chlorinated benzenes and three polychlorinated biphenyls from sediment suspension has been investigated by Schrap and Opperhuizen [33]. In order to examine the availability of these chemicals, the uptake from water was compared with that from sediment suspension. In the two experiments, the total amount of the chemicals was the same. The only difference was the fact that the chemicals were partly sorbed on the suspended sediment in one system, whereas the chemicals were truly dissolved in the water in the other. For all five chlorobenzenes, bioconcentration factors were found to be reduced when the fish were exposed to these chemicals in the sediment suspension. It was obvious that there was a greater reduction with increasing lipophilicity (log K_{ow}) of the chemical (trichlorobenzene < tetrachlorobenzene < pentachlorobenzene < hexachlorobenzene).

For discussion and more examples of bioavailability of chemicals see Hamelink et al. [52a], Gobas and Russell [52b], Schrap [53], Schrap and Opperhuizen [33], and Delbeke et al. [34]. In a critical review, Haitzer et al. [55] came to the conclusion that the bioconcentration factors of most organic chemicals were reduced in the presence of humic substances. An increase of the bioconcentration factors of organic compounds in aquatic organisms, especially of low DOM concentrations, was found in seven out of 27 of the reviewed studies [55]. However, some authors found also a decrease while others found an increase of the BCFs for the same lipophilic chemical. The DOM-caused decrease in bioconcentration were attributed to binding of the chemical to particulate and/or DOM, leading to aggregates which are too large to be taken up via gills by the gill-breathing organisms. However, no explanation can be given at this time for DOM-caused increase in bioconcentration.

BCF data reported for very lipophilic and super-hydrophobic chemicals in many cases have been underestimated from experiments with high content of particulate or dissolved organic matter. Bioconcentration factors must be related to the "bioavailable" chemical concentration in the water, because only the truly dissolved fraction of the chemical is actually bioavailable [5, 13, 30] (see also Sect. 8.2).

5.3
Concentration of the Test Chemical in the Water

The real bioconcentration factor on a lipid basis (BCF_L) of a chemical should be independent of its concentration in the water. In all cases, however, where bioconcentration factors differ by some orders of magnitude for the same chemical, although they have been determined under nearly equal experimental conditions with fish of the same species, strain, sex, age, body weight, and lipid content, it has to be questioned whether a "true" bioconcentration factor was found. Consequently, all other experimental conditions have to be reexamined.

Generally, a chemical must be truly dissolved (each molecule with a hydration shell) in order for it to be transferred through the gills and/or across the absorbing epithelium. Therefore, exposure of a chemical in excess of its water solubility will underestimate the bioconcentration factor. Geyer et al. [35–37] have shown that the BCF values especially of some super-hydrophobic or super-lipophilic chemicals with log K_{ow} values > 6 and with cross sections larger than 9.5 Å have been underestimated and that the real BCF values of these compounds are considerably higher. Examples are the BCF values of octachlorodibenzo-*p*-dioxin (OCDD) and Mirex which in all cases were tested far above their water solubility leading to relative low bioconcentration factors signaling low risk. These chemicals will be presented and discussed in more detail in Sect. 8.2. Therefore, the concentration of the test chemical in the water has to be considered as one important factor influencing the BCF value and should not exceed true water solubility. This is especially important for chemicals with relatively low water solubility [3, 5, 13, 35–37]. This issue is also important for petroleum hydrocarbons which may be tested as a mixture, for example in crude oils, and often are present as a separate liquid phase under experimental conditions.

If the chemical is surface active, for example an alkyl benzene sulfonate used in detergents, it will form micelles above a critical micelle concentration (CMC). This is effectively a solubility limit for such substances and it is essential that the test conditions be below the CMC, otherwise the BCF will be underestimated. Finally it should be noted that actual concentrations in the water may differ considerably from "nominal" concentrations deduced by adding a known mass of chemical to a known volume of water, because much of the chemical may sorb to the walls of the tank and to pumps and filters. Further, substances of relatively high air-water partition coefficients will evaporate appreciably from solution especially as a result of aeration. For these reasons actual concentration measurements are essential, and nominal values should not be trusted.

5.4 pH of the Water

Some chemicals, such as chlorinated phenols, carboxylic acids, sulfonic acids, amino acids, alkaloids, and amines are capable of ionization depending on the pH of the water. Because the n-octanol/water partition coefficient (K_{OW} value) of ionizable organic chemicals depends on the pH of the water the bioconcentration factor of these groups of compounds also depends on the pH of the water. For an ionizable organic chemical the K_{OW} value is largest if this compound is in the non-ionized form. That means for weak acids, such as pentachlorophenol (PCP), other chlorinated phenols, 2,4,5-trichlorophenoxy acetic acid (2,4,5-T), and 2-methyl-4,6-dinitrophenol (DNOC), the n-octanol/water partition coefficient [60, 61, 63] and the bioconcentration factor increase with decreasing pH of the water [57, 64] (see Figs. 3 and 4). However, for weak bases, such as *p*-chloroanilines, methylanilines, benzidine, tributyltin (TBT), and triphenyltin (TPT), the K_{OW} [65–67] and the BCF values increase with increasing pH of the water. This fact has to be considered in quantitative structure-activity relationships (QSARs) for bioconcentration and/or toxicity of ionizable chemicals for which the K_{OW} depends on pH.

This phenomenon may be also important for all natural estrogens and endocrine-disrupting chemicals (EDCs) which are weak acids, such as 17β-estradiol, estriol, ethynylestradiol, diethylstilbestrol, nonylphenol, octylphenol, bisphenol-A (BPA), tetrabromobisphenol-A (TBBA), hydroxy polychlorinated biphenyls, and other compounds with hydroxylated aromatic rings.

5.5 The Lipid Content of the Organisms

The bioconcentration of chemicals is generally considered to be a partitioning process of the chemicals between the lipids of aquatic organisms, such as fish, mussels, oysters etc., and the water. This process is controlled by the relative solubilities or activities of the chemical in the lipids of the aquatic organisms and

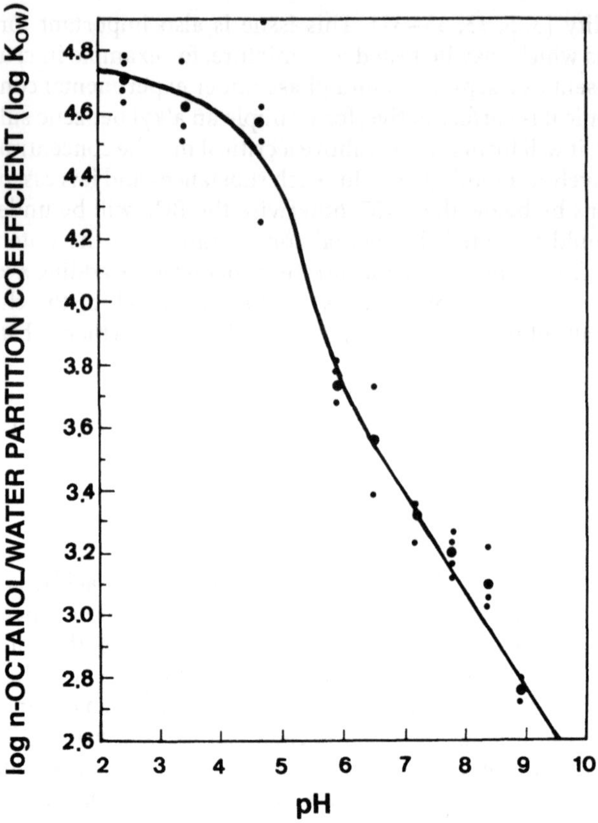

Fig. 3. Relationship between the apparent n-octanol/water partition coefficient (K_{OW}) of pentachlorophenol (PCP) and the pH of the water (data from Kaiser [56])

in water. It was shown by Geyer et al. [38–40] and others [18] that there is a clear relationship between the bioconcentration factor on a wet weight basis BCF_W of a chemical, such as trichlorobenzene [38], lindane (γ-HCH) [40] (see Fig. 5), pentachlorophenol (PCP) [39] (Fig. 6), and chlorinated benzenes etc. in different or the same fish species, and their lipid content. That means that for aquatic organisms in general the greater the lipid content the greater the bioconcentration potential of a chemical. Because, under normal conditions, the lipid content of aquatic organisms increases with body weight (see Fig. 7) and/or age, the concentration of a chemical and/or the bioconcentration factor on a wet weight basis (BCF_W) under steady-state conditions is higher in organisms with higher body weight and/or age. However, during spawning, the aquatic organisms lose a large amount of lipids. Therefore, during this time, the concentration of chemicals and/or the bioconcentration factor (BCF_W) is decreasing in these organisms. For algae the bioconcentration potential of a chemical seems to be mainly dependent on the specific surface of the algae [54]. However, Streit [12] found also a significant positive relationship between bioconcentration factor of a lipophilic organochlorine compound in freshwater diatom algae and the algal lipid content.

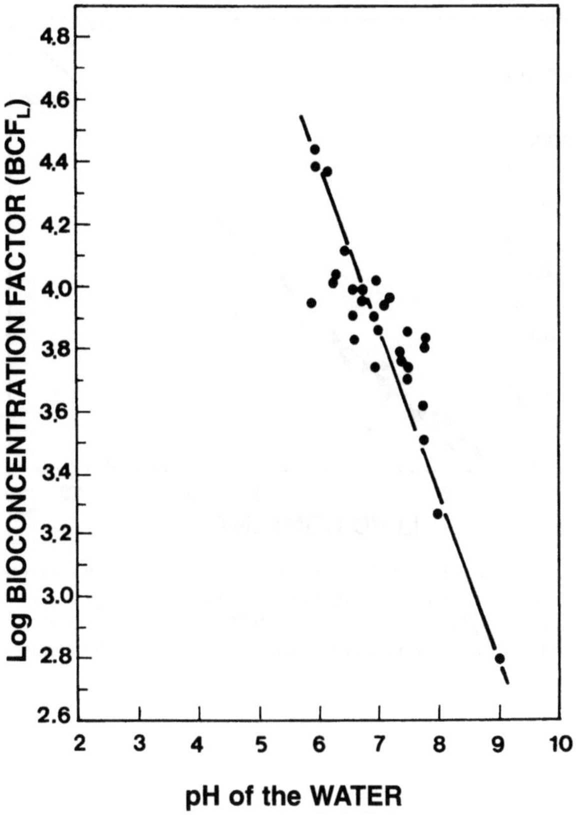

Fig. 4. Relationship between the steady-state bioconcentration factors on a lipid basis (BCF_L) of pentachlorophenol (PCP) in four different fish species and the pH of the water (BCF data of PCP are taken from Stehly and Hayton [57], Bude [58], Veith et al. [62] and McKim et al. [59])

The BCF_W values of trichlorobenzene in eight different fish species compiled by Geyer et al. [38] ranged from 124 in rainbow trout with 1.8% lipid to 2,100 in fathead minnow with 10.5% lipid (see Table 3). The mean BCF_W value was 847 with a coefficient of variation of 57%. The bioconcentration factors on a lipid basis (BCF_L) ranged from 6,890 to 23,790 with a mean value of 15,400. Using a lipid weight basis for calculating bioconcentration factors reduced the coefficient of variation from 57% to 32% of the mean. The reason for the relatively great coefficient of variation of 32% for the mean BCF_L value may be due to the biological variability of the different fish species, analytical problems in the determination of trichlorobenzene, different metabolism rates of TCB in different species of fresh water fish, and/or to the different methods for the determination of the lipid content [38].

In an international ring test with lindane it was found that the relative standard deviation (S.D.) of the bioconcentration factor on a wet weight basis (BCF_W) was 38%, whereas the S.D. was 23% if the BCF was related to the lipid

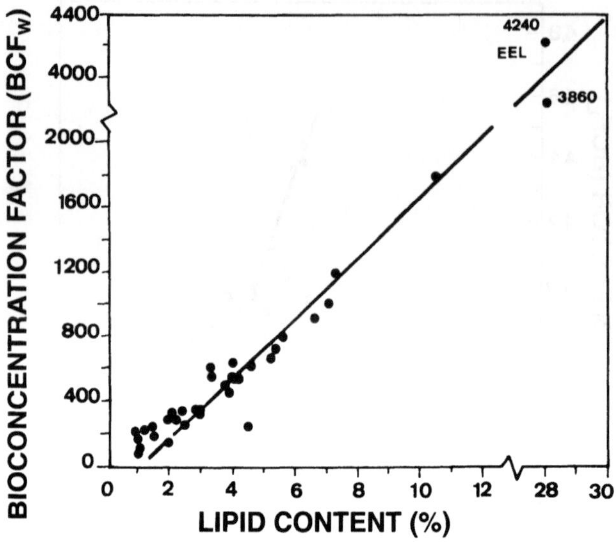

Fig. 5. Relationship between the steady-state bioconcentration factors on a wet weight basis (BCF_W) of lindane (γ-HCH) in mussel, *Daphnia*, and different fish species and their lipid content (L_W in % on a wet weight basis). The highest BCF_W values 3860 and 4240 were calculated for eels from the outdoor environment. From Geyer et al. [40] (with permission)

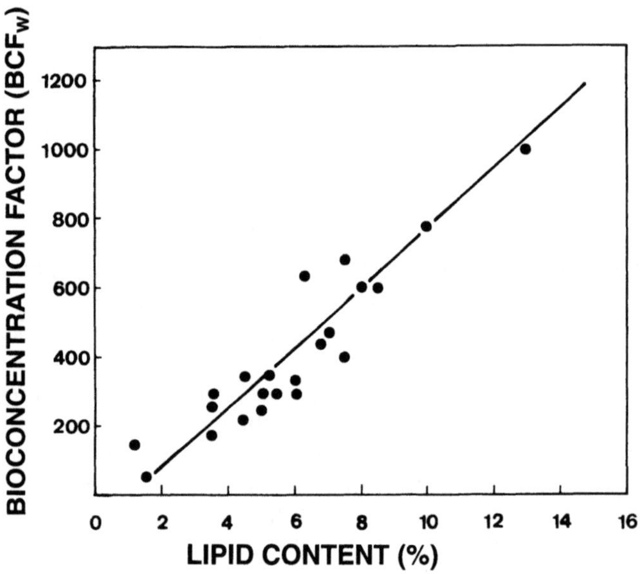

Fig. 6. Relationship between the steady-state bioconcentration factors on a wet weight basis (BCF_W) of pentachlorophenol (PCP) in mussel and different fish species and their lipid content (L_W in % on a wet weight basis). In all experiments the pH of the water was ca. 7 (H. J. Geyer unpublished)

Table 3. Influence of lipid content (%) on the bioconcentration of 1,2,4-trichlorobenzene in fish

Fish Species	Lipid (%)	Bioconcentration Factor (BCF)	
		BCF_W [a]	BCF_L [b]
Rainbow trout (*Oncorhynchus mykiss*) [c]	1.8	124	6,890
Carp (*Cyprinus carpio*)	2.2	190	8,636
	2.2	200	9,090
	2.2	220	10,000
	2.2	455	20,680
Rainbow trout (hatching)	3.2	349	10,906
(*Oncorhynchus mykiss*) [c]	3.2	710 [d]	22,188 [d]
Carp (*Cyprinus carpio*)	4.4	460	10,455
	4.4	540	12,270
Golden ide (*Leuciscus idus*)	5.0	914	18,280
Zebra fish (*Brachidanio rerio*)	5.2	730	14,040
	5.2	810	15,580
Tilapia (*Tilapia nilotica*)	5.2	680	13,080
	5.2	870	16,730
Guppy (female) (*Poecilia reticulata*)	5.4	702 [d]	13,000 [d]
	5.4	756 [e]	14,000 [e]
Bluegill sunfish (*Lepomis macrochirus*)	5.7	960	16,842
	5.7	1,320	23,160
Guppy (*Poecilia reticulata*)	5.8	1,350	23,280
	5.8	1,380	23,790
Rainbow trout (*Oncorhynchus mykiss*) [c]	7.7	1,300	16,880
	7.7	1,600	20,780
Guppy (*Poecilia reticulata*)	8.2	910	11,100
	8.2	1,080	13,170
Rainbow trout (*Oncorhynchus mykiss*) [c]	8.3	1,300	15,660
	8.8	3,200 [f]	36,364 [f]
Fathead minnow (*Pimephales promelas*)	10.5	2,100	20,000
Arithmetic mean ($\bar{x}$)	5.2	846.5	15,403
Standard deviation (± SD)	2.2	485	4945
Coefficient of Variation (CV%) [g]	42	57	32

Source: Taken with permission from Geyer et al. [38].

[a] BCF_W : Bioconcentration factor on a wet weight basis.

[b] BCF_L : Bioconcentration factor on a lipid weight basis $\frac{BCF_W \times 100}{\text{Lipid (\%)}}$.

[c] Formerly named *Salmo gairdneri.*

[d] 1,2,3-Trichlorobenzene.

[e] 1,3,5- Trichlorobenzene.

[f] Outlier (R-Test by Nalimov) not included in statistical analysis.

[g] $CV = \frac{SD \times 100}{\text{Mean } (\bar{x})} (\%)$.

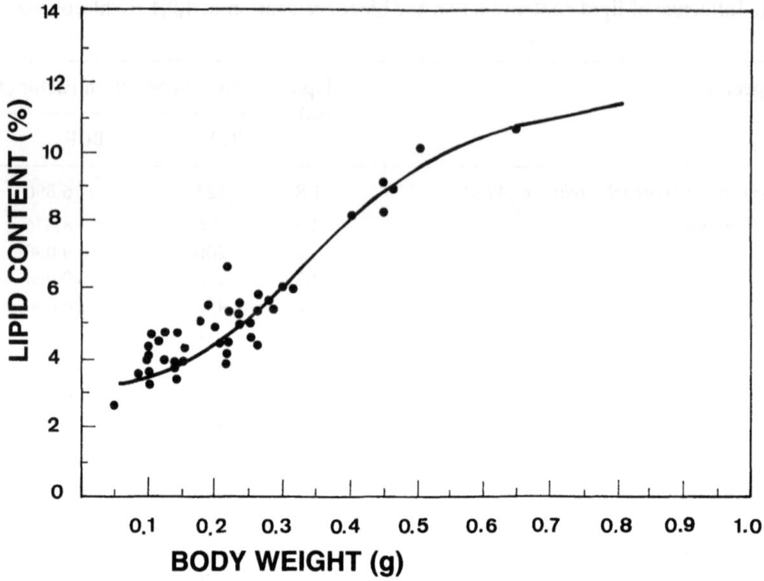

Fig. 7. Relationship between the lipid content (% on a wet wt. basis) and the body weight of fathead minnows (*Pimephales promelas*). Data are from Larry Brooke [395], Daniel Call [396], Gilman Veith [397], and Gregory Lien [398]

weight basis [41]. These and other examples have shown that the deviations of BCF values of a chemical can be significantly reduced if the BCF is based on the total lipid content of the fish and/or other gill-breathing animals. This is also very important if the BCF values of a chemical in different aquatic organisms are compared. Therefore, the method used for the determination of the total lipid content of the aquatic organisms is of great importance. In the following section methods for lipid determinations are presented and discussed.

6
Determination of the Total Lipid Content of Aquatic Organisms

Several methods have been developed for the determination of the total lipid content of aquatic and terrestrial organisms and their tissues. The determination in most cases is performed by extracting the lipids with organic solvents. However, the amount of "extractable organic matter" is dependent on the used organic solvent or the mixture of solvents [42, 43]. In cases where only hydrophobic organic solvent(s), such as diethyl ether, hexane, pentane, benzene, dichloromethane, or petrol ether or a mixture, e.g. hexane + dichloromethane 1:1 are used, the amount of "extractable organic matter" is lower than in cases where a mixture of a hydrophobic and hydrophilic organic solvent such as chloroform-methanol, hexane-acetone, or hexane-isopropanol are used. In the last case, the bioconcentration factors on a lipid basis (BCF_L) are lower. We have to keep this fact in mind because this is important in all cases where the lipid content is included in the result, such as BCF_L, bioaccumulation factor (BAF), or the concentration of chemicals in organisms.

Because in the literature it is not always distinguished between "fat" and "lipid", for clarity the definitions should be given:

(a) The extractable **neutral** organic matter is named **"total fat"** or **"total neutral lipids"**, in case that only hydrophobic organic solvents such as hexane, petrol ether, or benzene are used for extraction.
(b) If the extraction is performed with different organic solvents of different polarity (e.g. chloroform + methanol 1 : 1, or hexane + acetone 2 : 1), the extractable organic matter is called **"total lipid"**.

In the future greater attention should be paid to this aspect. It is also important to give the total lipid content on a wet weight basis (L_W in %) of the investigated organism or tissue. Because the bioconcentration potential is dependent on the **lipid content** of the organism, the method used for lipid determination is of great relevance [42, 43 a, b]. For extraction of lipids only organic solvents of different polarity should be used. In the following paragraph two methods for total lipid determination are recommended.

6.1 The Lipid Determination of Fish by the Modified Blight and Dyer Method

The most popular and generally effective method for lipid determination of fish is the modified Blight and Dyer method [44] see also [42]. The extraction of lipids is performed with a mixture of chloroform and methanol (1 : 1). For the procedure of this method see [42, 44]. Unfortunately, methanol is distinctly toxic, producing headaches if the laboratory is inadequately ventilated, and chloroform has been suspected of being carcinogenic. It is assumed that for these reasons this method was not accepted as an official OECD Guideline, although it was proposed for review panel in 1980. Therefore, this method should be used only if the results of extraction have to be compared with those of other laboratories.

Instead of using a mixture of chloroform and methanol, the extraction of lipids by a mixture of hexane and acetone (2:1) is recommended. This mixture has almost all desirable extraction properties and is superior to the other mixtures with respect to the undesirable properties of these. However, this method for lipid determination is very time consuming. Therefore, in the following section, a fast and easy method for the determination of the lipid content of fish on a fresh weight basis by the modified procedure of Ernst et al. [45], Beck and Mathar [46], and Schmitt et al. [47] is described.

6.2 The Lipid Determination of Fish by the "Cold Extraction Method" [48]

The fish are killed by immersion in liquid nitrogen. Quartz sand (30 g) and 60 g anhydrous sodium sulfate (Na_2SO_4) are mixed in a mortar. The sample of 1 – 10 g of fish is cut into pieces, accurately weighed (accuracy ± 0.1 mg), and added on top of this mixture. The fish homogenate is ground to a dry powder. If the mixture still appears humid, more sand/sodium sulfate is added. The powder is poured into a glass column (diameter, 2 cm; length, 50 cm), fitted with a 200-ml reservoir and removable Teflon stopcock. The column contains glass wool and 1 cm of sand on the bottom. A layer of sand is added on top of the fish mixture. The sample is extracted slowly (overnight) with 300 ml of hexane/acetone (2:1) at an adjusted flow rate of ca. 3 ml/min. The lipid extracts are collected in a tared 250-ml round-bottomed glass flask. After evaporation of the solvent in a rotary evaporator, the flask is dried, cooled to room temperature until constant weight, and weighed (accuracy ± 0.1 mg). The lipid content of fish on a wet weight basis (L_W in percent) is calculated by Eq. (23):

$$\text{Lipid} = \frac{\text{Weight of extract in g} \cdot 100}{\text{Wet weight of sample in g}} \qquad [\%] \tag{23}$$

This cold extraction method is successfully performed in our and other laboratories for more than 20 years. Because hexane is neurotoxic, isohexane or heptane can be used as a hydrophobic solvent. A modified method can also be used as a semi-micro method for the lipid determination in fish, mussels, oysters, *Daphnia*, and other aquatic organisms or tissues.

7
Quantitative Structure-Activity Relationships (QSAR) for Bioconcentration

At the present time between 60,000 and 72,000 industrial chemicals may be in current production and in commercial use throughout the world [68, 70, 72]. A total of 100,000 chemicals is quoted by the OECD [73]. About 3000 chemicals account for 90% of total world-wide production and between 200–1000 new synthetic chemicals enter the market each year [68, 70]. In Europe existing chemicals have been listed in the European Inventory of Existing Chemical Substances (EINECS). The EINECS list covers 100,106 existing chemicals, i.e. those which were on the market of the European Community between January 1971 and September 1981 [71]. Other figures suggest that in the European Community alone, 50,000 chemicals are in use [70]. The systematic evaluation of existing chemicals in Europe began in 1986 when the German Chemical Industry Association (VCI) made a survey of existing chemicals with a production/importation volume in Germany in excess of 10 tons per year. The result of this inventory [71] shows that the number of existing chemicals of 4,600 which are of economic importance is far below the number given in EINECS. Mackay et al. [69] suggest that perhaps 500 compounds are of environmental concern because of their presence in various compartments of the environment, their toxicity, their persistence, or their tendency for bioaccumulation in aquatic and terrestrial organisms.

Since it is impossible to test all these available chemicals and newly introduced substances with long-term testing procedures, it would be useful to be able to predict their bioconcentration potential. Because bioconcentration is defined as an equilibrium partitioning process between aquatic organisms and the surrounding medium (e.g. fish/water partitioning at steady state), modeling efforts are based on analogous partitioning processes such as n-octanol/water partition (K_{ow}). The lipids of aquatic organisms, such as *Daphnia*, mussels, oysters, and fish, is the principal site for bioconcentration. Because octanol is often a satisfactory surrogate for lipids, the n-octanol/water partition coefficient ($\log K_{ow}$) has become one of the most important parameters in studies on the behavior and impact of organic chemicals in the environment [68, 69]. K_{ow} has been particularly useful in the prediction of bioconcentration factors of organic chemicals in aquatic organisms such as algae [74, 75], water flea (*Daphnia*) [75], mussels [75, 76], and fish [62, 77–79]. In general the bioconcentration factors of chemicals are increasing with increase of their K_{ow} values. Usually a linear correlation (Quantitative Structure-Activity Relationships: QSAR) between log BCF_W of different chemicals and their $\log K_{ow}$ is observed [62, 74–79]:

$$\log BCF_W = A \cdot \log K_{ow} + B \tag{24}$$

Examples of such linear Quantitative Structure-Activity Relationships (QSARs) for bioconcentration of different lipophilic chemicals in various aquatic organisms, such as algae, *Daphnia*, poly- and oligochaeters, crustacea, mollusks, and fish were compiled by Connel [78, 79] and Nendza [80] and some are presented in Table 4. It was argued that these linear regressions should be applied to chemicals with $\log K_{ow}$ values smaller than ca. 6. For super-hydrophobic chemicals, such as octachlorodibenzo-*p*-dioxin (OCDD), Mirex, and some organic pigments with $\log K_{ow} > 6$, experimentally determined BCF values were much lower than predicted from their $\log K_{ow}$ values. Therefore numerous non-linear correlations, such as polynominal, $\log K_{ow}$ dependent functions to predict bioconcentration of organic chemicals in fish were derived [80–84 a].

However, Jager and Hamers [84 b] and Schwartz [84 c] in their studies on estimation methods for bioaccumulation in risk assessment of organic chemicals came to the conclusion that the decrease of the polynominal relationship at high K_{OW} is caused only by a few BCF data on polychlorinated dibenzo-*p*-dioxin (PCDDs) congeners. Furthermore, the polynominal approach (Eq. 25) of the Technical Guidance Document (TGD) [84 d] seems to underestimate the BCF values of chemicals in fish at high K_{OW} values (> 6) significantly.

$$\log BCF_W = 2.74 \log K_{OW} - 0.20 (\log K_{OW})^2 - 4.72 \tag{25}$$

Jager and Hamers [84 b] concluded that this equation as advised in TGD is questionable and may result in serious underestimation of the BCF of chemicals which are not metabolized in fish. For the purpose of initial risk assessment they proposed a "BCF + growth" model which can be simplified to the straight line, with a maximum BCF value reached at $\log K_{OW} = 6$ [84 b].

Furthermore, it was argued by Yen et al. [85], Gobas and Schrap [86], Schwartz [84 c] and Geyer et al. [87, 88] that the main reason for the low BCF values of super-hydrophobic chemicals was because they were tested at relatively high concentrations in the water which were some orders of magnitude higher than their water solubility. This indicates that all these super-hydrophobic chemicals in the

Table 4. Summary of regression analysis for bioconcentration of organic chemicals by algae, water flea (*Daphnia*), mussels, and different fish species. Furthermore, the equation to predict the bioaccumulation factor (BAF) of chemicals in human (fat) is presented

Organism	Method[a]	Equation[b]	log K_{OW} range	N[c]	R^{2d}	Reference
Algae (*Chlorella fusca*)	LR	$\log BCF_W = 0.681 \log K_{OW} + 0.164$	0.94–6.40	41	0.803	[74]
	GM	$\log BCF_W = 0.740 \log K_{OW} - 0.050$		41	0.803	[75]
Water flea (*Daphnia magna*)	LR	$\log BCF_W = 0.850 \log K_{OW} - 1.100$	1.65–6.74	52	0.913	[75]
	GM	$\log BCF_W = 0.889 \log K_{OW} - 1.280$		52	0.913	[75]
Mussel (*Mytilus edulis*)	LR	$\log BCF_W = 0.858 \log K_{OW} - 0.808$	1.73–6.19	16	0.914	[75, 76]
	GM	$\log BCF_W = 0.899 \log K_{OW} - 0.970$		16	0.914	[75, 76]
Fish[e]	LR	$\log BCF_W = 1.000 \log K_{OW} - 1.320$	1.00–6.89	71	0.950	[77a]
Fish[e]	GM	$\log BCF_W = 1.000 \log K_{OW} - 1.336$		71	0.950	[77b]
Fish[f]	LR	$\log BCF_L = 0.956 \log K_{OW} + 0.220$	1.87–8.60	69	0.986	this work
Fish[f]	GM	$\log BCF_L = 0.962 \log K_{OW} + 0.190$	1.87–8.60	69	0.986	this work
Human (fat)	LR	$\log BAF_L = 0.745 \log K_{OW} - 1.190$	2.50–5.95	8	0.939	[150, 151]

[a] LR, least-squares regression method; GM, geometric mean functional regression method.

[b] BCF_W, bioconcentration factor on a wet weight basis.

$$BCF_L\text{, bioconcentration factor on a lipid basis} = \frac{BCF_W \cdot 100}{L_W(\%)}$$

L_W, lipid content on a wet weight basis.

$$BAF_L\text{, bioaccumulation factor on a lipid basis} = \frac{\text{concentration in human (fat) [ng} \times \text{kg}^{-1}]}{\text{concentration in total diet [ng} \times \text{kg}^{-1}]}$$

[c] N, number of chemicals.

[d] R^2, regression coefficient.

[e] This equation is valid only for fish with a total lipid content of 4.8% and if the organic chemical is not or only minimal metabolized.

[f] This equation is only valid for organic chemicals which are not or only minimal metabolized in fish and which give no bound residues.

water were present in a sorbed state. Because only "truly dissolved" chemicals are able to be taken up *via* gills [86, 89–91], the use of supersaturated chemical concentrations will clearly underestimate the BCF values [86–88, 91]. That means that the "real" BCF values of the most super-hydrophobic chemicals are some orders of magnitude higher than the BCF values so far experimentally determined in the laboratory. Schmieder et al. [28] tested the bioconcentration of the super-hydrophobic chemical 2,3,7,8-tetrachlorodibenzo-*p*-dioxin in fish with concentration below its water solubility. As a consequence, experimentally achieved BCF_L values of TCDD match those predicted from the K_{ow} value. An extended discussion of this issue can be found in Sect. 8.2 of this chapter and in references [86–91]. Furthermore, in most QSAR equations the bioconcentration factors on a wet weight basis (BCF_W) instead of BCF values on a lipid basis (BCF_L) of chemicals in fish were correlated with their log K_{OW} values. Very often also BCF_W values were used for establishing QSARs of chemicals which were metabolized to a great extent or did not reach steady-state. Therefore, it was necessary to recalculate the correlation between bioconcentration factors on a lipid basis (BCF_L) and measured n-octanol/water partition coefficients. However, it was necessary to select critically these BCF_L values and log K_{OW} data:

(I) It is noted that only steady-state BCF values from flow-through tests with fish for which the lipid content is known, were taken from the literature.

(II) Only organic chemicals which are relatively resistant to metabolism in fish were used for the correlation. If chemicals are metabolized to hydrophilic compounds they are eliminated faster and therefore the BCF_L values are lower than predicted from their log K_{OW} value [92, 93].

(III) The BCF values of chemicals which give bound residues are higher than predicted from their log K_{OW} values. One example is methylmercuric chloride (CH_3HgCl) which has a very low log K_{OW} value of 0.405 [94]. However, methylmercury has a very high bioconcentration factor (BCF_W) between 10,000 and 1,000,000 in fish [95, 97] because this compound is associated with protein sulfhydryl groups in the organism. Therefore methylmercury has also very long half-lives ($t_{1/2}$) between 204 and 348 days in fishes (for reviews see [96, 97]).

(IV) If the concentration of the test chemical in the water was higher than its water solubility, the BCF values are too low and were omitted for establishing the log BCF_L versus log K_{OW} correlation.

(V) BCF_L values of chemicals were also omitted for establishing the QSAR if during the test a high number of fish died because these BCF_L values are also lower than predicted from their K_{OW} value.

(VI) BCF_L values of chemicals in fish were omitted from the correlation if during the bioconcentration test the lipid content of the aquatic organism is changing very fast and substantially. Galassi and Calamari [98] and Galassi et al. [99] measured significant differences between BCF_L values of 1,2,4-trichlorobenzene, 1,2,3-trichlorobenzene, and γ-hexachlorocyclohexane (lindane) in different life-stage of rainbow trout, such as eyed-egg, hatching, half-absorbed yolk, and early juvenile. It is known that during these early life-stages of fish their lipid content and composition is decreasing and changing very fast (see Fig. 8 and references 99 and 100). As a consequence the BCF_L values are significantly different from BCF_L values predicted from their log K_{OW} value.

(VII) The "best" or "right" n-octanol/water partition coefficients which were experimentally determined e.g. by the slow-stirring method or the HPLC method (in agreement with the OECD guideline) were used for establishing the QSAR.

(VIII) As shown in Sect. 5.4 the log K_{OW} value of ionizable organic chemicals depends on the pH. Therefore, the BCF_L values of these chemicals also depend on the pH. Consequently, the BCF_L data of these ionizable organic compounds have to be correlated with their log K_{OW} values at the pH (normally about 7) of the water, which prevailed during the bioconcentration test. In most, if not all QSARs this fact was so far not considered.

Using linear regression analysis the following regression Eq. (26) was obtained:

$$\log BCF_L = 0.956 \log K_{OW} + 0.22 \tag{26}$$

The number of chemicals included in the regression was $n = 69$, the coefficient of determination $r^2 = 0.986$, and the significance level $p < 0.0001$. The graphic expression of Eq. (26) is presented in Fig. 9. To include the errors in both de-

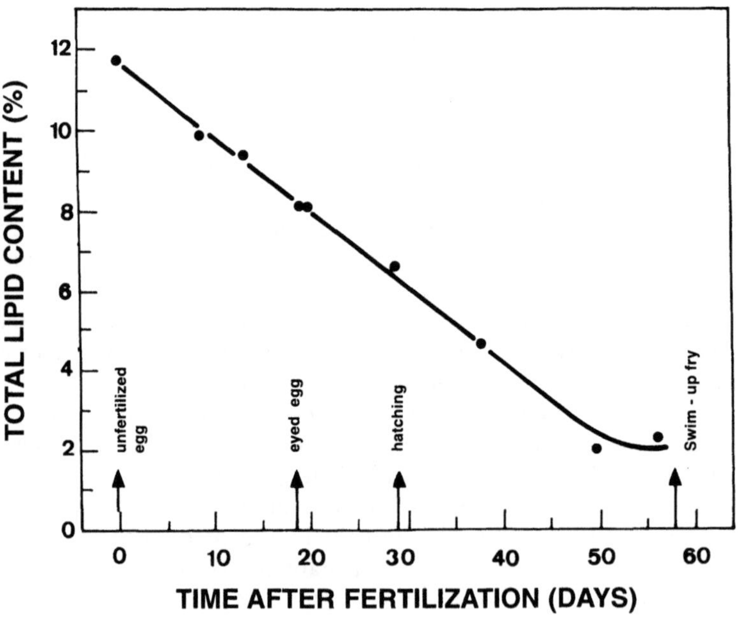

Fig. 8. The total lipid content (% on a wet weight basis) of rainbow trout (*Oncorhynchus mykiss*) eggs during development. The lipid data of unfertilized egg, fertilized egg (9 days after fertilization), and just before eye stage (13 days) are from reference [100], all other data are from reference [99]

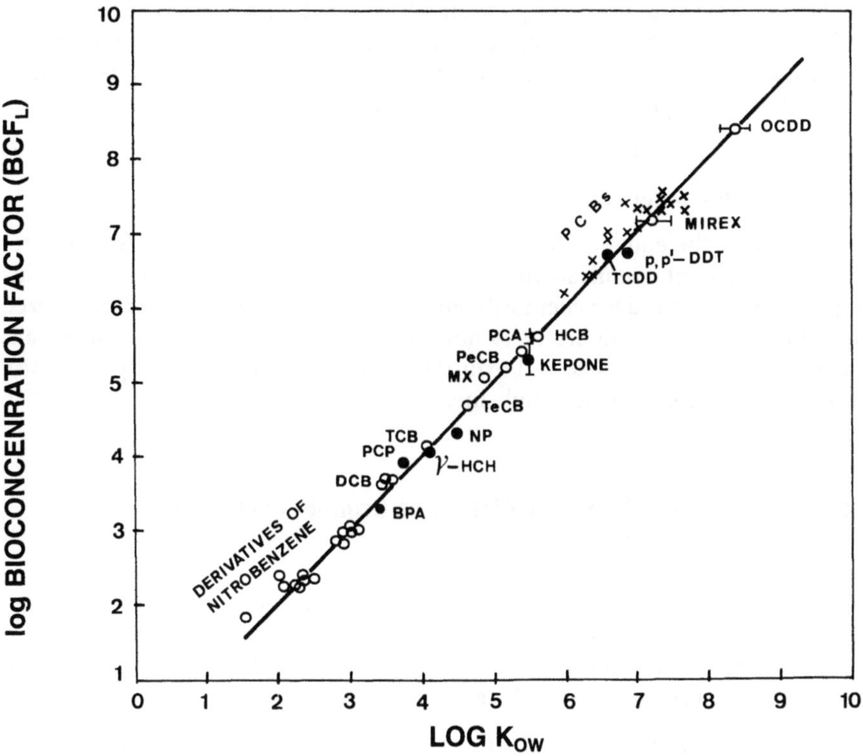

Fig. 9. Relationship between the steady-state bioconcentration factors on a lipid basis (BCF_L) of chemicals in different fish species and the n-octanol/water partition coefficient (K_{OW}) (log/log scale). (●) Solid circles are chemicals with known endocrine-disrupting properties. Abbreviations of the chemicals: *p,p′*-DDT; 2,2-bis-(*p*-chlorophenyl)-1,1,1-trichloroethane. OCDD; octachlorodibenzo-*p*-dioxin. TCDD; tetrachlorodibenzo-*p*-dioxin. HCB; hexachlorobenzene. PCA; pentachloroanisole. PeCB; pentachlorobenzene. MX; musk xylene. TeCB; tetrachlorobenzene. NP; nonylphenol. TCB; tetrachlorobenzene. *γ*-HCH; *γ*-hexachlorocyclohexane (Lindane). PCP; pentachlorophenol. DCB; dichlorobenzene. BPA; bisphenol-A. PCBs; polychlorinated biphenyls

pendent and independent variables the geometric mean (GM) functional regression method published by Halfon [77b] was also used (see equation in Table 4). Equation (26) can be used for prediction of BCF_L values of relatively persistent organic chemicals in fish and other aquatic gill-breathing organisms such as *Daphnia*, mussels, and oysters if their lipid content is known.

Equation (26) is essentially the correlation suggested in 1982 by Mackay [77a] which was $BCF_W = 0.048\ K_{OW}$ if the lipid content of fish is 4.8%. Equation (26) implies that the coefficient A in Eq. (24) is 1.0. Often values less than 1.0 are observed, probably because of bioavailability considerations and failure to achieve equilibrium during the limited test time. It is recommended that Eq. (26) be used to estimate BCF_L and hence BCF_W using measured lipid contents.

A final point is that when K_{OW} is small, i.e. less than 20, much of the chemical may be present in the fish, the mussel or the gill-breathing organism in aqueous solution and BCF may be underestimated. From a theoretical viewpoint it can be argued that BCF_W should be correlated as Eq. (27):

$$BCF_W = L \cdot K_{OW} + W \tag{27}$$

where W is the water content and L the lipid content of the organism.

In Table 4, the equation to predict the bioaccumulation factor (BAF_L) of relatively persistent chemicals in human (fat) is also presented [191, 192]. This equation is only valid for chemicals which are not or only minimal metabolized in human. It is also important to note that for super-hydrophobic chemicals, such as octachlorodibenzo-*p*-dioxin (OCDD) and Mirex, no steady-state BAF value is reached during the whole life.

8 Bioconcentration of Specific Classes of Organic Chemicals in Aquatic Organisms

In this section, the physico-chemical properties, especially the n-octanol/water partition coefficients (log K_{OW}), and the measured or predicted bioconcentration factors (BCF_W and BCF_L values) of the following classes of environmental chemicals are presented and critically discussed:

(1) natural hormones, synthetic hormones, and endocrine-disrupting chemicals (EDCs);
(2) the persistent super-hydrophobic and other persistent organic pollutants (POPs), such as tetrachlorodibenzo-*p*-dioxin (TCDD), octachlorodibenzo-*p*-dioxin (OCDD), Mirex, and polychlorinated norbornanes (Toxaphene);
(3) tetrachlorobenzyltoluenes (TCBTs);
(4) polybrominated benzenes (PBBz) and polybrominated biphenyls (PBBs);
(5) polybrominated diphenylethers (PBDEs);
(6) polychlorinated diphenylethers (PCDEs);
(7) nitromusk compounds (NMCs);
(8) polycyclic musk fragrances (PMFs), and
(9) sunscreen agents (SSAs).

8.1 Bioconcentration of Natural Hormones, Synthetic Hormones, and Endocrine-Disrupting Chemicals (EDCs)

It has been known for many decades that some pesticides and other chemicals can act as weak hormones. These man-made environmental chemicals can alter in organisms the balance of natural endogenous hormones, such as estrogens, androgens, thyroxine etc., if their concentration exceeds certain threshold levels. In these cases they show physiological responses normally associated with high circulating concentrations of hormones and are capable of disrupting endocrine systems of aquatic and terrestrial animals, possibly including humans [101–106].

The following terms regarding hormones can be distinguished:

(i) **Natural hormones** are produced in

(a) animals, including humans: e.g. estrogens, androgens, progesterone, glucocorticoid, thyroxine etc.,
(b) plants (phytohormones): e.g. gossypol, an effective male contraceptive agent found in cotton seed, and especially phytoestrogens, e.g. genistein, coumestrol, equol, daidzein, etc., and
(c) fungi (mycohormones, especially mycoestrogens): zearalenone, zearalanone, α-zearalenol, β-zearalenol, α-zearanalol, β-zearanalol, etc.

(ii) **Synthetic hormones,** particularly *synthetic estrogens,* androgens and antiandrogens are synthesized or produced by man and are or were mainly used as medical pharmaceuticals or drugs for contraception and treatment of various diseases. Examples of synthetic estrogens are diethylstilbestrol (DES), hexestrol, dienestrol, 17α-ethinylestradiol, and mestranol. Synthetic androgens used in therapy are 17α-methyltestosterone, methandrostenolone, fluoxymesterone, methyltrienolone etc. Synthetic compounds with antiandrogenic activity are cyproterone acetate, flutamide and its metabolite 2-hydroxyflutamide.

(iii) **Endocrine-disrupting chemicals** (*contaminants, compounds*) **(EDCs)** are also named endocrine disrupters (EDs) or xenohormones. There are different definitions of EDCs or EDs:

(1) Definition of the U.S. Environmental Protection Agency [101]: An environmental endocrine disrupter is defined as an exogenous agent that interferes with the synthesis, secretion, transport, binding, action, or elimination of natural hormones in the body, that are responsible for the maintenance of homeostasis, reproduction, development, and/or behavior.
(2) Definition of another Expert Working Group [107] and extended by the authors: The endocrine-disrupting chemicals (EDCs) can be broadly defined as exogenous compounds or agents that can interfere with the action, binding, production, release, metabolism, and/or elimination of natural endogenous hormones of aquatic and terrestrial organisms, including humans. By these EDCs the maintenance of homeostasis, the regulation of reproduction, physiological, anatomical, sexual, and other developmental processes can be disrupted [107], if their concentration or the body burden exceed a threshold level.
(3) Definition by Experts of the European Workshop in Weybridge, UK: The workshop was organized by the European Commission, the European Environmental Agency, the WHO European Centre for Environment and Health, the OECD, national authorities and agencies of the UK, Germany, Sweden, and the Netherlands as well as CEFIC and ECETOC. It was agreed that an endocrine disrupter could be adequa-

tely defined only in terms of effects on intact animals, although identification of potential endocrine disrupters was possible *in vitro*. The following definitions were endorsed:

(a) An endocrine disrupter is an exogenous substance that causes adverse health effects in an intact organism, or its progeny, secondary to changes in endocrine function.
(b) A potential endocrine disrupter is a substance that possesses properties that might be expected to lead to endocrine disruption in an intact organism. Adverse hormonal effects may relate to disturbances in any of the major endocrine systems, including the reproductive, thyroid, and adrenal systems.

(iv) **Proendocrine-disrupting chemicals (PEDCs)** are compounds that are not bound to steroid receptors. Example are methoxychlor and some non-planar polychlorinated biphenyls, which are actually **proestrogens** which after metabolization to mono- and diphenol metabolites can be bound to the estrogen receptor and produce estrogenic effects. That means not the parent compound but in most cases their hydroxylated metabolites are responsible for endocrine e.g. estrogenic activity. This phenomenon has to be noted if the binding of a chemical to an estrogen receptor *in vitro* is evaluated.

All natural hormones, all synthetic hormones, and many endocrine-disrupting chemicals (EDCs) achieve their effects by binding to a receptor and/or hormone binding protein [108, 109]. However, **it should be noted that binding to the receptor is necessary, but not sufficient for activity. The activity of a hormone or EDC in an organism does not only depend on the binding behavior (strong or weak) of itself or a metabolite to the receptor but is affected by a variety of other factors** [110]:

(a) Absorption including metabolism relative to the route of exposure,
(b) partitioning between lipid or fat and aqueous compartments of the organism,
(c) plasma and tissue binding,
(d) effective concentration determined by how it is carried in circulation, and
(e) especially the concentration at the target tissue/receptor.

Evidence is accumulating that many chemicals released into the aquatic environment can disrupt normal endocrine function in different fish species and other aquatic organisms. Some of the effects observed in aquatic life that may be caused by chemicals with endocrine-disrupting properties are summarized [115a, d, e, f]:

(1) Decreased hatching success in fish
(2) Decreased fertility in fish and shellfish
(3) Abnormal thyroid function in fish
(4) Feminization and demasculinization of fish
(5) Defeminization and masculinization of fish and gastropods.

It is known that many environmental endocrine-disrupting chemicals have weak estrogenic or antiestrogenic activity [111], or they act as androgens (e.g. tributyltin) or possess antiandrogenic activities (e.g. linuron, 3,4-dichloroacetanilide etc.). Some chemicals can block the effects of male sex hormones, the androgens [112, 113, 115f]. These special chemicals are described in the next sections.

8.1.1
Chemicals with Estrogenic Activity (Xenoestrogens)

A major group of endocrine-disrupting chemicals in the aquatic environment mimic the effects of estrogens [121, 122]. Therefore, this section deals especially with environmental estrogens or the so-called xenoestrogens.

Estrogens are female sex hormones which have multiple sites of activity and biological actions on the reproductive cycle, reproductive function, mammary gland, and on the neuroendocrine system. The biological synthesis of the female steroid sex hormones, the estrogens, starts with cholesterol, which is metabolized to progestogens (e.g. pregnenolone, progesterone, 17α-hydroxyprogesterone etc.) and to the male steroid sex hormones, the androgens, such as 5α-dehydroepiandrosterone, 4-androstene-3,17-dione, testosterone, 5α-dihydrotestosterone, 11-ketotestosterone etc. Under normal conditions the androgens can be metabolized to the estrogens. The different biological syntheses pathways are catalyzed by special enzymes in special tissues or glands, and the hormones are secreted into the circulating blood. It is important to note that the biological pathways (s. Fig. 10) of sexual hormones are very complicated. Their release is regulated by feedback mechanisms from the hypothalamus and hypophysis.

The steroid sex hormones play an important role especially during the fetal, embryonic, and neonatal developmental stage and can elicit their physiologic effects at very low blood concentrations (ng ml^{-1} to pg ml^{-1}; i.e.; 10^{-9} to 10^{-12} g ml^{-1}). Therefore, during these developmental stages the embryo, fetus, and neonate are very sensitive to exogenous environmental hormones which can interfere or disrupt the endocrine function and act on the natural endogeneous hormones of the body. Natural hormones achieve their effects by binding to a special receptor lodged in the nuclei of cells. Nuclear receptors are ligand-activated transcription factors, which regulate the expression of target genes by binding to specific response elements.

Over the last decades, large amounts of different man-made chemicals which can act as weak estrogens have been released into the terrestrial and aquatic environment and are distributed world-wide. Classical environmental estrogens are pesticides, such as *o,p′*-DDT, and its metabolites *o,p′*-DDE and *o,p′*-DDD, methoxychlor and its metabolites, chlordecone (Kepone®), dieldrin, Toxaphene, and endosulfan [126, 135, 136]. It is also known that many chemicals with very weak or no measurable estrogenic activity can be metabolized in organisms especially to hydroxylated compounds which may have much more estrogenic potency than the parent compound. Examples are methoxychlor and its mono- and di-demethylated derivatives [126, 127] as well as the alkylphenol

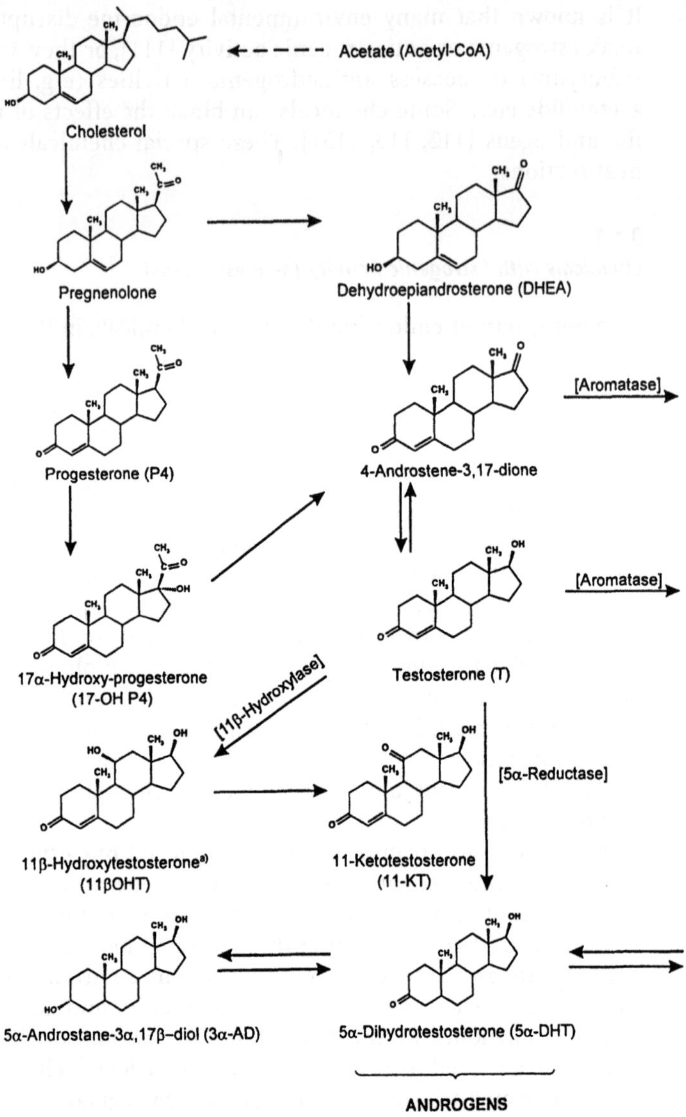

Fig. 10. Biosynthesis and main metabolic pathways of natural male and female steroidal sex hormones (androgens and estrogens) starting with cholesterol. The main enzymes, which catalyze these reactions, are given in angular brackets. Taken with modifications from Forth et al. [175], Schlumpf and Lichtensteiger [176], Turan [177], Bradlow et al. [178] and extended by the authors

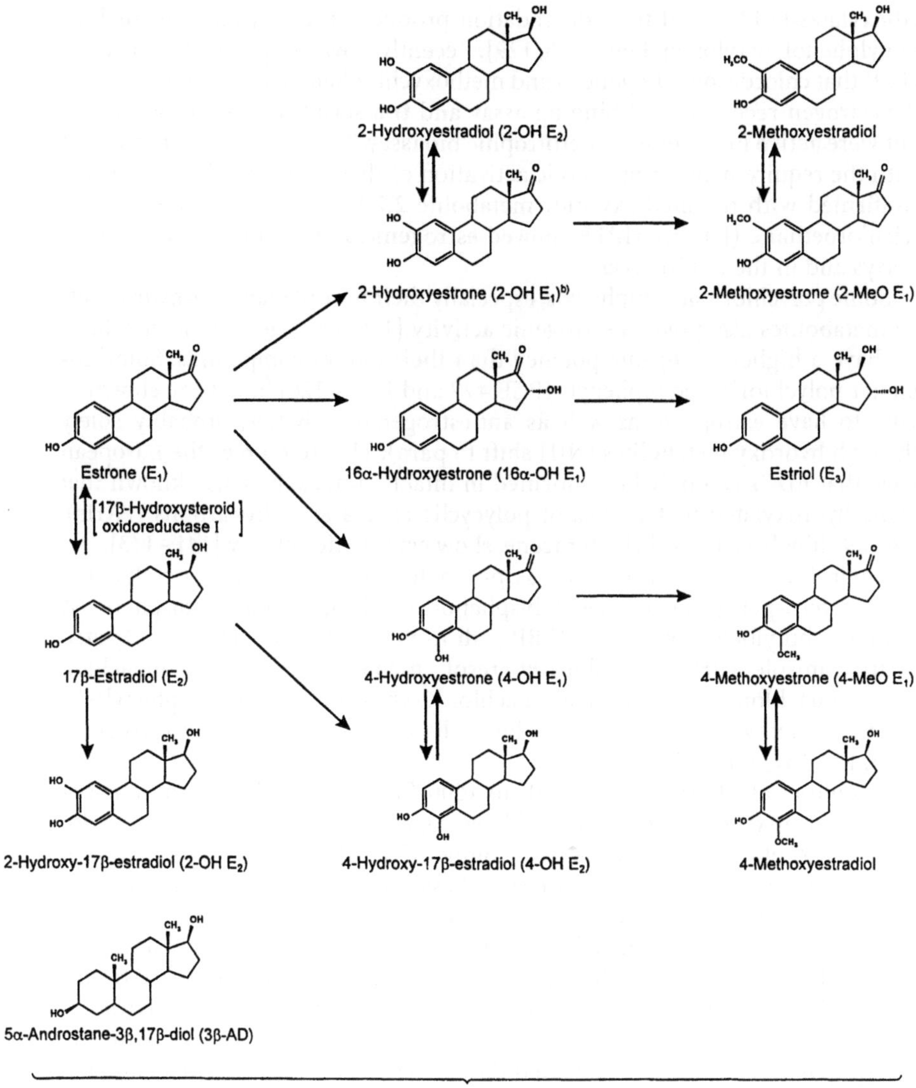

ESTROGENS + METABOLITES

ethoxylates (APEs) and their degradation products, the alkylphenols, such as nonylphenol, octylphenol etc. [128, 129]. Recently it was shown by Shelby et al. [149] that chlordecone (Kepone®) and methoxychlor had no estrogenic effects in the estrogen receptor (ER) binding assay and transcriptional activation assay, but were active in the mouse uterotrophic bioassay. These results are consistent with the requirement for metabolic activation of these two chemicals. This was confirmed with the methoxychlor metabolite 2,2-bis(*p*-hydroxyphenyl)-1,1,1-trichloroethane (HPTE). HPTE showed estrogenic activity in the two in vitro assays and in the in vivo assay.

Some polychlorinated biphenyls, especially their non-planar *para*-hydroxylated metabolites also possess estrogenic activity [126, 135, 140]. These metabolites have a higher estrogenic potency than their parent compounds. Some coplanar polychlorinated biphenyls (PCB #77 and PCB #126) have been shown in vivo to have estrogenic as well as antiestrogenic activities, probably solely through hydroxy metabolites (NIH shift to para). This reinforces the European view that EDCs can only be confirmed in intact animals. It is also known that some hydroxylated metabolites of polycyclic aromatic hydrocarbons (PAHs), e.g. 3,9-dihydroxybenzo[a]anthracene, show estrogenic activity [141–143].

Environmental chemicals such as *p*-nonylphenol (NP), 4-*tert.*-octylphenol (OP), 4-*tert.*-pentylphenol (TPP), bisphenol-A (BPA), tetrabromobisphenol-A (TBBA), butylbenzylphthalate (BBP), di-*n*-butylphthalate (DBP), butylated hydroxyanisole (BHA), *p*-chloro-*m*-cresol, *p*-chloro-*o*-cresol, *cis*-nonachlor, *trans*-nonachlor, and the herbicide alachlor [2-chloro-*N*-(2,6-diethylphenyl)-*N*-(methoxymethyl) acetamide] have been discovered to be weakly estrogenic [128, 129, 137, 138].

Arnold et al. [144] reported 150- to 1600-fold synergistic interactions between 1:1 mixtures of the very weakly estrogenic insecticides dieldrin, endosulfan, Toxaphene, and chlordane in competitive estrogen binding assays and in an estrogen-responsive assay in yeast. Less synergistic interactions between two weakly estrogenic hydroxy polychlorinated biphenyls (HO-PCBs) were also observed by Arnold et al. [144] in human endometrial cancer cells and in the yeast assay. However, Safe et al. [146, 147] could not confirm these results for 10 different estrogen-responsive assays. They found that the activities of combination of these weakly estrogenic pesticides are not synergistic but additive. Ashby et al. [148] evaluated the estrogenic effects of dieldrin and endosulfan using two standard assays. They found also no synergism. It is important to note that very recently McLachlan et al. [145a] have just formally withdrawn their report. In his laboratory the coworkers have conducted experiments duplicating the conditions of their earlier work, but were unable to replicate their original results.

The natural female steroid hormone with the greatest estrogenic activity is 17ß-estradiol. It is important to note that some synthetic estrogens, such as diethylstilbestrol (DES), moxestrol, and 11β-chloromethyl estradiol show 10 times more estrogenic activity than 17β-estradiol in the E-SCREEN assay [136]. Ethinylestradiol has the same estrogenic activity as 17β-estradiol, whereas the activity of the synthetic EDCs is by some orders of magnitude lower [136]. In this context it is important to note that most, if not all, effluents of sewage treatment plants [145b] and some rivers [145c] in the United Kingdom are estrogenic to fish. As a biomarker of estrogen exposure the induction of vitellogenin synthesis in caged male trout was used. Expression of the yolk protein vitellogenin (VTG) gene under normal conditions is not found in male fish. However, if in the water are estrogenic substances, male fish can produce VTG in quantities approaching those of mature females. This response is not confined to fish held in undiluted effluents of sewage-treatment plants but is evident in fish caged downstream of discharge, in some cases several kilometers from the input. It was suggested that industrial chemicals such as nonylphenol, one of the degradation compounds of a widely used surfactant, are likely endocrine disrupters. Recent evidence has indicated that this effect is due predominantly to the natural hormones 17β-estradiol and estrone and the synthetic hormone 17α-ethinylestradiol [145b]. The source of 17β-estradiol, estrone, and 17α-ethinylestradiol was believed to be anthropogenic, probably being excreted largely in women's urine. These hormones were present in a biologically active, unbound (free) form and not in the inactive, bound form in which the hormones would have been excreted. It was shown [404a, b] that inactive steroid metabolites can be re-activated in the sewage system and/or the sewage treatment plants. However, conclusions regarding the degree of sewage treatment and hormone concentrations in final domestic sewage effluents cannot be drawn due to the small number of sewage treatment plants evaluated. Although alkylphenols, such as nonylphenol and octylphenol, were also measured in effluents, their concentrations as estrogen equivalents (EQs) were between 140 to 500 times lower than the concentrations of natural and synthetic hormones (see Table 5). However, it has to be noted that this result is only a rough estimation because the concentrations as estrogen equivalents are based upon relative estrogen receptor binding affinity and not on estrogenic potencies of these compounds in whole animals (see also page 32).

John P. Giesy, Shane Snyder and coworkers from the Institute of Environmental Toxicology at Michigan State University studied effluents from several different types of municipal waste water treatment plants in central Michigan. They also came to the conclusion that human hormones (17β-estradiol) and synthetic hormones (ethinylestradiol), not industrial chemicals with estrogenic activity, in the effluents caused male fish to produce vitellogenin, a well-accepted indicator of endocrine disruption [145d].

In Table 6 the physico-chemical properties, chemical structures, and some other relevant data of some natural estrogens, synthetic estrogens, and of some environmental man-made EDCs are compiled, as is the estrogenic activity measured as the relative proliferative potency on human breast cancer MCF 7 cells in the E-SCREEN assay [136] and in the recombinant yeast cell estrogen screening assay (RCBA) [138b]. In the last column of Table 6 their occurrence in the aquatic environment and their bioconcentration factors in fish and mussels are presented also as far as these data were published.

Table 5. Concentrations of major estrogenic compounds in effluents of seven United Kingdom sewage treatment plants (STPs)

Estrogenic compound	N[a]	Concentration [ng l^{-1}] range	Concentration [ng l^{-1}] mean	Estrogen equivalent factor (EEF_i)[b]	Concentration as estrogen equivalents [ng l^{-1}] EQ[h]	Σ EQ
1. **Natural and synthetic hormones**						
17β-Estradiol (E_2)[c]	21	2.7–48	11.0	1.0	11.0	11.2–15.0
Estrone (E_1)[c]	21	1.4–76	17.3	1) 0.01 2) 0.1	1) 0.2 2) 1.7	
17α-Ethinylestradiol[c] (EE_2)	14	< 0.2–7	< 0.2[d]	1.0	< 0.2	
	3	0.2–0.8	0.5[e]		0.5	
	3	0.6–4.3	2.3[f]		2.3	
2. **Alkylphenols**						
Nonylphenol[g]	4	150–2,800	943	$3 \cdot 10^{-5}$	0.03	0.03–0.08
Octylphenol[g]	4	40–280	163	1) $3 \cdot 10^{-4}$ 2) $4 \cdot 10^{-6}$	1) 0.05 2) 0.0007	

Source: Adopted with modifications from Desbrow et al. [145b].

[a] N; number of effluent samples.
[b] The estrogen equivalent factors (EEFs) were established by the authors by using the RPP values from Table 6.
[c] The hormones were present in effluents in a biologically-active unbound (free) form. No other significant estrogenic activity was found.
[d] The concentrations of EE_2 in 14 effluent samples of 7 sewage treatment plants were below the detection limit of 0.2 ng l^{-1}.
[e] Mean of 3 effluent samples of 1 sewage treatment plant.
[f] Mean of 3 effluent samples of 1 sewage treatment plant.
[g] Dissolved alkylphenol in sewage treatment plant effluents.
[h] $EQ = \sum_{i=1}^{n} c_i \cdot EEF_i$. For more information on the estrogen equivalent (EQ) approach for estrogenic compounds see ref. [145e].

Table 6. Common name, abbreviation, IUPAC name, CAS No., chemical structure, molecular formula, molecular mass, water solubility, n-octanol/water partition coefficient (log K_{OW}), occurrence in the aquatic environment, and bioconcentration factor (BCF) of **natural** and **synthetic Estrogens** and of hydroxylated **Chemicals** with **estrogen-like activity** or other **endocrine-disrupting effects** as otherwise noted

Chemical (common name, abbreviation, and/or IUPAC name)	CAS no	Chemical structure	Molecular formula and molecular mass [g mol^{-1}]	RPP[a] (%)	Water Solubility (mg l^{-1})	log K_{OW}	Detected in water, sludge, sediment, algae, mussel, or fish. Bioconcentration factor (BCF)
1. **Natural steroidal estrogens**							
17*β*-Estradiol (E_2) 1,3,5(10)-Estratriene-3,17*β*-diol	50–28–2	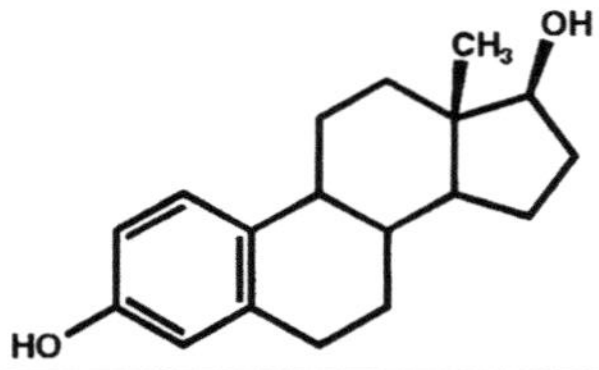	$C_{18}H_{24}O_2$ 272.39	100	1.7 4.7	4.0[b]	raw sewage water, effluents from municipal waste-water treatment plants
Estriol (E_3) 1,3,5(10)-Estratriene-3,16*α*,17*β*-triol	50–27–1	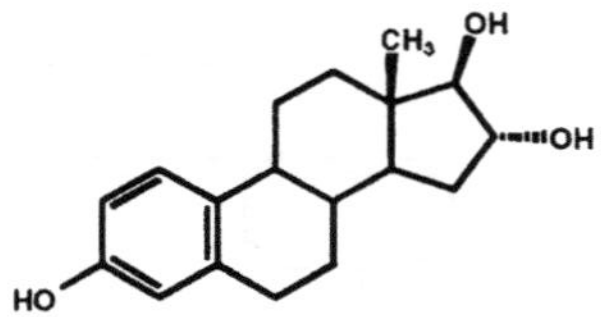	$C_{18}H_{24}O_3$ 288.39	10 0.63[1]	13.25	3.84	effluents from waste-water treatment plants
Estrone (E_1) 3-Hydroxy-1,3,5(10)-estratriene-17-one	53–16–7	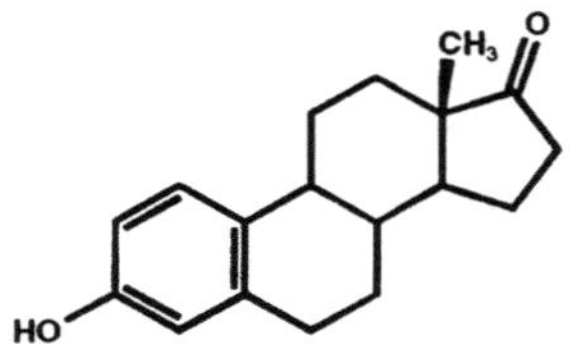	$C_{18}H_{22}O_2$ 270.37	1.0 9.6[1]	12.42	4.10	effluents from waste-water treatment plants

Table 6 (continued)

Chemical (common name, abbreviation, and/or IUPAC name)	CAS No.	Chemical structure	Molecular formula and molecular mass [g mol⁻¹]	RPP[a] (%)	Water Solubility (mg l⁻¹)	log K_{OW}	Detected in water, sludge, sediment, algae, mussel, or fish. Bioconcentration factor (BCF)
2. Synthetic steroidal estrogens							
17α-Ethinylestradiol (EE_2) 3-Hydroxy-19-nor-17α-pregna-1,3,5(10)-trien-20-in-17-ol	57–63–6	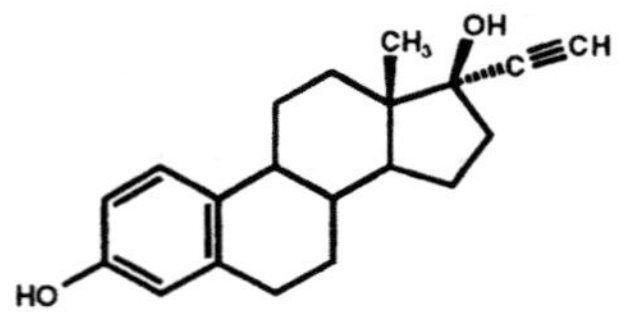	$C_{20}H_{24}O_2$ 296.41	100 88.8[1]	4.83 4.7	4.20[b] (pH 7)	river water, activated sludge, effluents from STPs[W] BCF in fish (experim.): fathead minnow BCF_W: 610[x] BCF_W: 660[y] BCF (predicted): BCF_W (5% L): 790[s] BCF_W (10% L): 1590[s] BCF_W (20% L): 3170[s] BCF_L: 15,850[s]
Mestranol 3-Methoxy-17α-ethinylestradiol (MEE_2) 3-Methoxy-19-nor-17α-pregna-1,3,5(10)-trien-20-in-17-ol	72–33–3	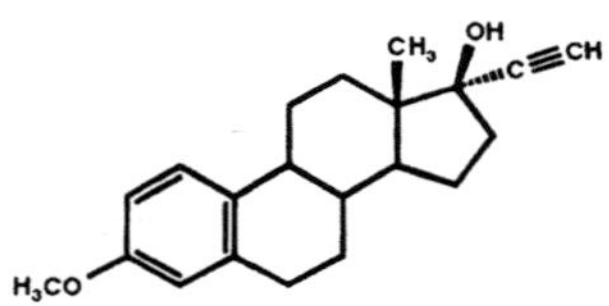	$C_{21}H_{26}O_2$ 310.42	7.3[1]	0.32 0.31	4.80 (calc.)	raw sewage water, effluents from waste-water treatment plants, river water BCF in fish BCF_W (5% L): 3160[s] BCF_W (10% L): 6310[s] BCF_L: 63,100[s]

3. Synthetic non-steroidal estrogens							
Diethylstilbestrol (DES) *α, β*-Diethyl-4.4′-dihydroxystilbene (*E*)-4,4′-(1,2-Diethyl-1,2-ethenediyl)bis-phenol [z]	56–53–1	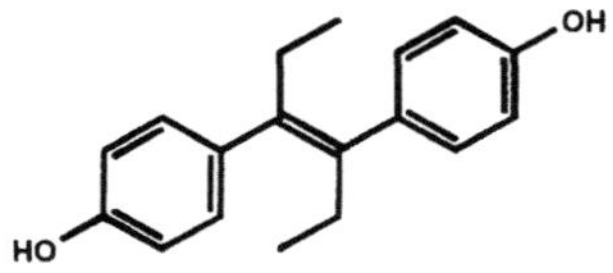	$C_{18}H_{20}O_2$ 268.36	1000 74.3 [l]	(12.5) [o]	5.07	was found in effluents in USA, when it was used as a growth – promoting agent in large amounts. ^{14}C-DES was bioconcentrated in algae, snail, and fish [n]
Hexestrol (HE) 4,4′-(1,2-Diethyl-ethylene)diphenol	84–16–2	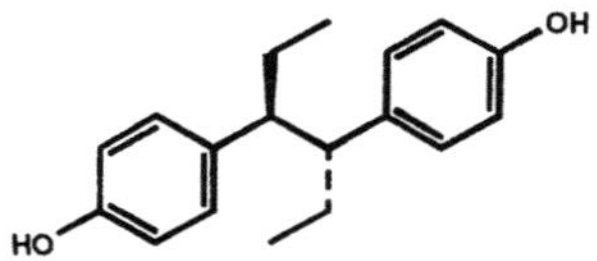	$C_{18}H_{22}O_2$ 270.37	30.6 [l]		5.03 (calc.)	no data available
Chlorotrianisene (CTA) Chloro-tris (4-methoxy-phenyl)-ethylene	569–57–3	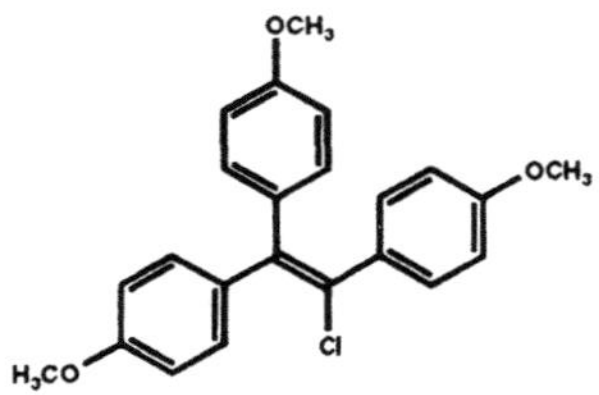	$C_{23}H_{21}ClO_3$ 380.87			5.65 (calc.)	no data available
4. *p*-Alkylphenols and its polyethoxylate derivatives							
Nonylphenol mono-ethoxylate (NP1EO)	27986–36–3	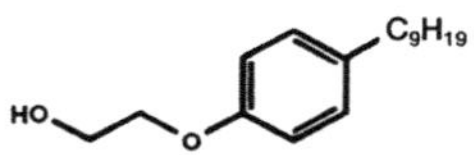	$C_{17}H_{28}O_2$ 264.41		3.0	4.17 [c]	sediment, sludge, algae, mussel, fish BCF in mussels: BCF_w: 170 [u]

Table 6 (continued)

Chemical (common name, abbreviation, and/or IUPAC name)	CAS no	Chemical structure	Molecular formula and molecular mass [g mol^{-1}]	RPP[a] (%)	Water Solubility (mg l^{-1})	log K_{OW}	Detected in water, sludge, sediment, algae, mussel, or fish. Bioconcentration factor (BCF)
Nonylphenol di-ethoxylate (NP2EO)	27176–93–8	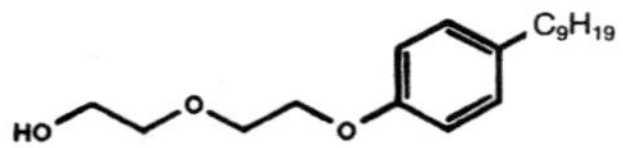	$C_{19}H_{32}O_3$ 308.46	0.0006[p]	3.38	4.21[c]	sediment, sludge. algae, mussels, fish BCF in mussels BCF_W: 100[u] BCF_L: 10,000
4-Nonylphenol (NP) *p*-Nonylphenol (straight chain)	25154–52–3	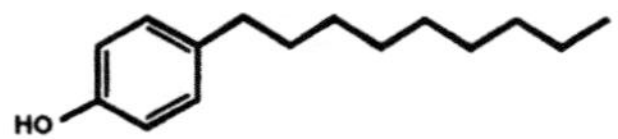	$C_{15}H_{24}O$ 220.36	0.0022[l] 0.026[k]	5.4	4.48[c]	sediment, sludge, effluents, algae, mussels, fish BCF in mussels: BCF_W: 340[u] BCF_L: 34,000[u] BCF_W: 3,430[d,q] BCF_L: 193,000[d,q]
4-Nonylphenol (technical grade)		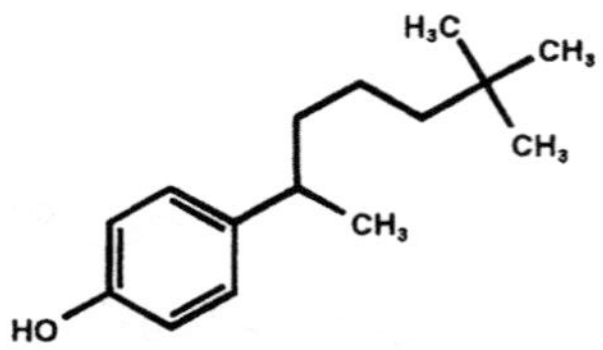 and other branched isomers		0.003 0.005[l] 0.0009[p]			BCF in fish: BCF_W: 800[v] BCF_W: 1,250[d,q] BCF_W: 1,890[e] BCF_L: 17,090[v] BCF_L: 17,250[d,q] BCF_L: 23,200[e]

4-Octylphenol (OP) *p-tert*-Octylphenol	140–66–9	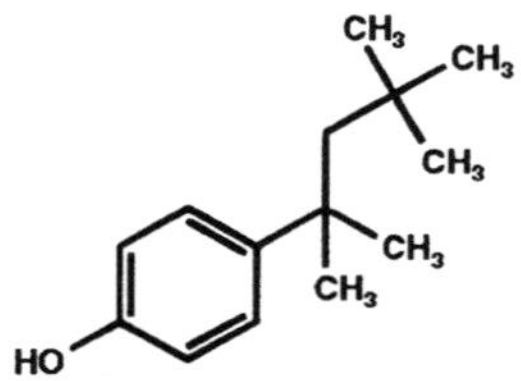	$C_{14}H_{22}O$ 206.33	0.03 0.072[k] 0.003[l] 0.0004[l] 0.0037[p]	12.6	4.12[c]	sediment, sludge
p-tert-Butylphenol (4-BP) 4-*tert*-Butylphenol	98–54–4	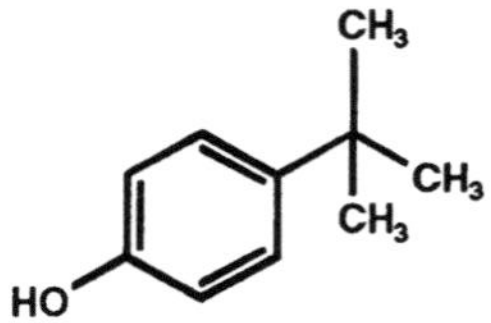	$C_{10}H_{14}O$ 150.22	0.016[p]		3.41[g] 3.31	surface water (river Rhein and Main), wastewater and sediment BCF in algae (*Chlorella fusca*) BCF_W: 34 BCF_D: 170 BCF in fish (golden ide; uptake 3 days) BCF_W: 120[q] BCF_L: 1970[q] BCF in zebra fish (kinetic approach) BCF_W: 74[e] BCF_L: 1850[e]
5. **Miscellaneous chemicals**							
Bisphenol – A (BPA) 2,2-Bis-(4-hydroxy-phenyl)propane	80–05–7	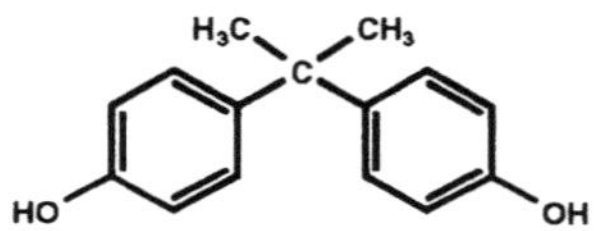	$C_{15}H_{16}O_2$ 228.28	0.003 0.005[j] 0.006[k] 0.005	0.12 (pH: 7, t = 20–25 °C	3.32 3.40	waste-water, river water BCF_{max} (15 µg/l; uptake 6 weeks) in fish (carp): BCF_W: 68 BCF_L: 1,700

Table 6 (continued)

Chemical (common name, abbreviation, and/or IUPAC name)	CAS no	Chemical structure	Molecular formula and molecular mass [g mol^{-1}]	RPP[a] (%)	Water Solubility (mg l^{-1})	log K_{OW}	Detected in water, sludge, sediment, algae, mussel, or fish. Bioconcentration factor (BCF)
Tetrabromobisphenol-A (TBBA) 2,2-Bis-(3,5-dibromo-4-hydroxyphenyl) propane	79–94–7	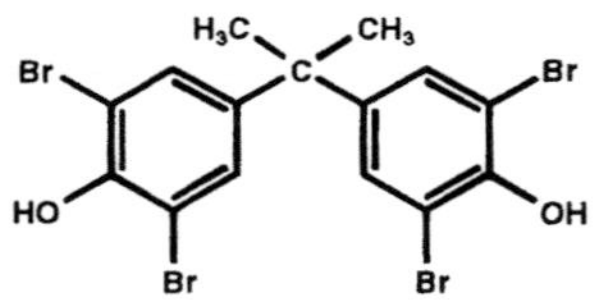	$C_{15}H_{12}Br_4O_2$ 543.87	0.002[1]		5.21[g] 4.54	sediment, mussel and fish BCF in fish: zebrafish (kinetic approach): BCF_W: 960[f] BCF_L: 28,300[f] fathead minnows (24 days uptake) BCF_W: 1,200[q] BCF_L: 24,000[q] oyster (*Crassostrea virginica*), 14 days uptake BCF_W: 720[q] BCF_L: 60,000[q]

Pentachlorophenol (PCP)	87–86–5	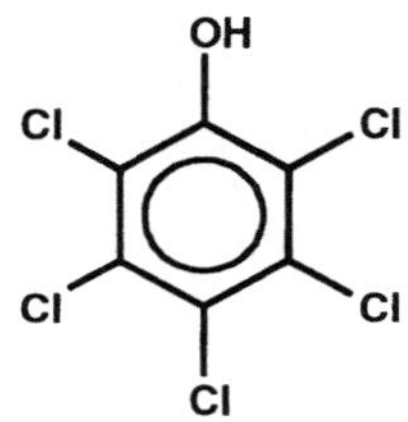	C_6HCl_5O 266.34	no effect[t]	14 20 (pH: 7)	3.81 3.69 (pH: 7)	fresh water, sea water, sediment, mussel, and fish BCF in algae (*Chlorella fusca*): BCF_W: 1,250 BCF_D: 7,250 BCF in mussels (*Mytilus edulis*): BCF_W: 170 BCF_L: 20,000 (*Anodonta anatina*) BCF_W: 80 BCF_L: 7,340 (*Pseudanodonta complanata*) BCF_W: 61 BCF_L: 5,690 BCF in fish Golden orfe: BCF_W: 219 BCF_W: 334 BCF_L: 5,000 BCF_L: 5,510 Fathead minnow: BCF_W: 770 BCF_L: 7,330 BCF in mussels and fish: BCF_W: 50–780[h] BCF_L: 7,300[i] (mean)

Table 6 (continued)

Chemical (common name, abbreviation, and/or IUPAC name)	CAS no	Chemical structure	Molecular formula and molecular mass [g mol^{-1}]	RPP[a] (%)	Water Solubility (mg l^{-1})	log K_{OW}	Detected in water, sludge, sediment, algae, mussel, or fish. Bioconcentration factor (BCF)
Chlordecone (Kepone®)[m] 1,1a,3,3a,4,5,5,5a,5b,6-decachlorooctahydro-1,3,4-metheno-2*H*-cyclobuta[*cd*] pentalen-2-one	143–50–0	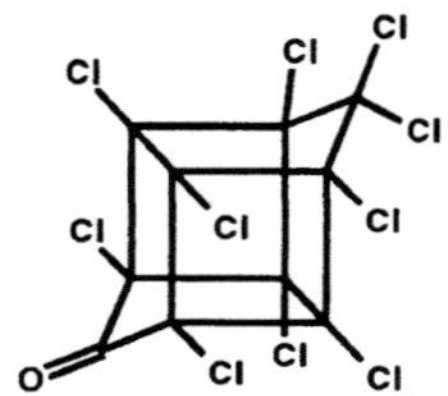	$C_{10}Cl_{10}O$ 490.64	0.0001	$3.0 \cdot 10^{-5}$ (pH: 7) $3.5 \cdot 10^{-5}$ (pH: 8)	5.50	BCF in fish[r] (fathead minnows and bluegills) BCF_w (range): 10,440–16,590 BCF_W: 13,000 (mean) BCF_L (range): 130,200–348,100 BCF_L: 196,000 (mean)

Source: Rippen [156] and The Merck Index [152], as otherwise noted.

[a] RPP: Relative proliferative potency is the ratio between 17*β*-estradiol and the xenobiotic doses needed to produce maximal cell yields ×100 (E-SCREEN assay of Soto et al.) [136]. All data from Ref. [136] as otherwise noted.

[b] Schweinfurth et al. (Shake-flask method) [168].

[c] Data from Ahel and Giger [169].

[d] Bioconcentration of ^{14}C-NP determined by Ekelund et al. [170].

[e] Bioconcentration in zebrafish determined by the kinetic method by Butte et al. [171].

[f] Out-door experiment, pH: 7.5, concentration of TBBA in filtered water: 30.3 µg l^{-1} (Butte et al.) [172].

[g] Determined by the HPLC method by Butte et al. [172].

[h] Range of BCF_W values of PCP in mussels and different fish species compiled from the literature by Geyer et al. [173a].

[i] Mean BCF_L value calculated by Geyer et al. [173a].

[j] Tetrabromobisphenol-A showed no estrogenic effects in an eucaryotic test system (K. Rehmann personal communication 1996). However, Körner et al. [138a] found estrogenic potency of this chemical in the proliferation assay with the MCF-cell line (purity of TBBA: 97%).

[k] Relative binding affinity (RBA) assay in serum-free medium determined by Nagel et al. [110].

[l] Relative estrogenic potency compared to 17β-estradiol (100) by molar mass determined with the recombinant yeast cell bioassay (RCBA) [138b].
[m] Trade name.
[n] Laboratory model **static** ecosystem used by Metcalf [174]. It is not possible to calculate "real" steady-state BCF values.
[o] The water solubility value seems too high in comparison to the high log K_{OW} value.
[p] Relative potency compared to 17β-estradiol (100%). Data from Jobling and Sumpter (1993): Aquat. Toxicol 27: 361.
[q] The bioaccumulation included all the metabolites of the test substance because ^{14}C-labeled chemical was used. It should be noted that the synthesized ^{14}C-NP yields ca. 50% NP, along with several other compounds including dinonyl phenol [400]. Therefore, the BCF in mussels may be higher than the factor determined in the field study.
[r] For single BCF_W and BCF_L values of chlordecone in fish see Table 10.
[s] Worst-case BCF values predicted by means of equation (26) for fish. It is important to note that the inactivation of these synthetic steroids in liver and other tissues is relatively slow.
[t] PCP shows no estrogen-like activity. However, pure PCP decreased thyroxine (T_4), triiodothyronine (T_3), and thyrothropine (TSH) levels in serum of rats [173b] and is known as an endocrine-disrupting chemical (EDC).
[u] Field study, uptake for 7 weeks in caged mussels (*Mytilus edulis*) [399].
[v] Bioconcentration factor in fathead minnows determined by Brooke [395].
[w] STPs; sewage treatment plants.
[x] Bioconcentration factor on a wet weight basis of ^{14}C-ethinylestradiol (^{14}C–EE_2 mean measured concentration in water 12 ng l^{-1}) in fathead minnow after 239 days post hatching determined by R. Länge, T.H. Hutchinson, C.P. Croudace, G.H. Panter, J.P. Sumpter (1999) Environ Toxicol Chem (in preparation).
[y] Bioconcentration factor of ^{14}C–EE_2 after 153 days post hatching determined by R. Länge et al. (1999) Environ Toxical Chem (in preparation).
[z] Chemical Abstracts name.

8.1.2
Chemicals with Antiestrogenic Activity (Xenoantiestrogens)

In contrast to substances which exert estrogen-like activity, a variety of non-steroidal chemicals possess antiestrogenic activity. These chemicals are also named antiestrogens or xenoantiestrogens. The term antiestrogen can be applied to several classes of chemicals that modify, modulate, inhibit, or antagonize the actions and effects of natural estrogens. These include

(a) competitive antagonists that can bind to the estrogen receptor (ER) without activating it, and simultaneously prevent binding of endogenous estrogens,
(b) antagonists that act through binding to the aryl hydrocarbon (Ah or dioxin) receptor,
(c) inhibitors of estrogen synthesis (e.g. gonadotropin-releasing hormone, GnRH; aromatase inhibitors),
(d) chemicals that influence estrogen-dependent processes by altering estrogen metabolism and availability, or
(e) chemicals that exert opposing physiological actions (e.g. androgens and progestins).

In this section we will refer only to antiestrogenic compounds of groups (a) and (b). Competitive estrogen antagonists are the most specific antiestrogens. Such compounds are used in the treatment of infertility, breast cancer and osteoporosis. Examples are *trans*-clomiphene and its metabolite *trans*–4-hydroxyclomiphene, tamoxifen and raloxifene. Clomiphene is used as a fertility agent. This drug can bind to the estrogen receptor, thereby blocking activation by endogenous estrogens. The breast cancer adjuvant non-steroidal pharmaceutical agent tamoxifen and its metabolite, 4-hydroxy tamoxifen, exhibit both antiestrogenic and estrogenic activities [149, 294]. Raloxifene is a nonsteroidal estrogen receptor mixed agonist/antagonist depending on the tissue. This drug is useful in preventing further bone loss when the onset of osteoporosis has been detected in woman after menopause. Furthermore, raloxifene may prevent women older than 60 from getting breast cancer. Pure anti-estrogens are ICI 164384 and ICI 182780 (Fulvestrant, Faslodex®).

Beside antiestrogens that may elicit their activity through the ER, a growing number of environmental chemicals are being shown to possibly cause antiestrogenic effects indirectly through the aryl hydrocarbon receptor (AhR). There is currently no known endogenous ligand for the AhR. Polychlorinated dibenzo-*p*-dioxins (PCDDs), such as 2,3,7,8-tetrachlorodibenzo-*p*-dioxin(TCDD) [116– 118], polychlorinated dibenzofurans (PCDFs) [116–118], and the coplanar polychlorinated biphenyls (PCBs), such as 3,3′,4,4′-tetrachlorobiphenyl, 3,4,4′,5-tetrachlorobiphenyl, 3,3′,4,4′,5-pentachlorobiphenyl, and 3,3′,4,4′,5,5′-hexachlorobiphenyl [119, 120a] are examples of antiestrogenic chemicals which may alter the estrogenic response through binding to the AhR (see Table 8). It is important to note that these compounds belong to a larger group of so-called persistent organic pollutants (POPs) which possess a very high bioaccumulation potential in aquatic and terrestrial organisms including humans (see Sect. 8.2 and Table 8).

The polycyclic aromatic hydrocarbons (PAHs) encompass a further group of environmental chemicals with antiestrogenic activity. Examples are benzo[a] pyrene, benz[a]anthracene, 3-methyl-cholanthrene, and 7,12-dimethylbenz[a] anthracene [120b].

Indole-3-carbinol (I3 C) and its derivatives are examples of an important group of natural antiestrogens. The antiestrogenic activity of all these Ah receptor ligands is directly correlated to their binding affinity to the Ah receptor and the associated CYP 1 A and CYP 2B1 inducing potency. The evidence suggests that the structure of the Ah receptor is heterogeneous among species. However, it is not known if these structural differences influence species susceptibility to antiestrogenic compounds, such as PCDDs, PCDFs, and PCBs. The Ah receptor is necessary but not sufficient for eliciting some of the toxic and biological responses caused by these compounds. Therefore, other factors are involved in these processes.

8.1.3
Chemicals with Androgenic Activity (Xenoandrogens)

At this time one known, non-steroidal environmental chemical with androgenic activity is tributyltin (TBT). This compound is supposed to be responsible for negative effects on reproduction of marine neogastropods. The imposition of male sex organs, including a penis and vas deferens, on female mud snails was linked to TBT. The phenomenon was termed "imposex" or "pseudohermaphrodism" [65]. Recent studies by Oehlmann et al. [114] indicated that TBT increased the testosterone titers in female gastropods. Simultaneous exposure to TBT and to the antiandrogen cyproterone acetate suppressed imposex development completely in *Nucella lapillus* and greatly reduced imposex in *Hinia reticulata*. These results proved that the imposex inducing-effects of TBT are mediated by an increasing androgen level and are not caused directly by TBT itself. Furthermore, imposex development by TBT was suppressed in both snails by adding estrogens to the water. It was also shown by Oehlmann et al. [114] that the specific aromatization inhibitor 1-methyl-1,4-androstadien-3,17-dione was able to induce imposex in marine snails. These results suggested that TBT causes an inhibition of the cytochrome P-450 dependent aromatase system which catalyses the aromatization of androgens (e.g. testosterone and 4-androstene-3,17-dione) to estrogens (see Fig. 10) with a subsequent shift of the androgen/estrogen balance in favor of androgens [114]. Studies by Ronis and Mason [115b] of the tributyltin effects on testosterone metabolism have indicated that this chemical is enhancing the conversion of testosterone to other androgenic steroid hormones.

The n-octanol/water partition coefficients (log K_{OW}) of tinorganic compounds are dependent on the pH of the water and were compiled by Fent [65]. The bioconcentration factors (BCF_W) of some tinorganic compounds in fish, mussels, and other organic organisms were also compiled by the same author [65]. The BCFs on a wet weight basis are in the range between 200 and ca. 10,000. The present knowledge leads to the conclusion that biomagnification of TBT in the aquatic environment does not seem to occur, or only to a minor extent.

The abnormal occurrence of pseudohermaphrodism is not restricted to gastropods but has been reported among populations of crustaceans, including copepods, isopods, amphipods, and penaeid shrimps. A very high (93%) incidence of intersex was reported among copepods inhabiting an area receiving sewage discharge. It was speculated that chemicals in the sewage were responsible for the pseudohermaphrodism. Laboratory experiments by LeBlanc et al. [115c, d] have shown that exposure of the crustacean *Daphnia magna* to a variety of chemicals, such as fungicides (pentachlorophenol), detergents (4-nonylphenol), and agricultural effluents significantly inhibited the metabolic clearance of exogenously administered ^{14}C-testosterone and enhanced the production of androgenic metabolites. This phenomenon of androgenization is identical to that observed with tributyltin and gastropods. It is suggested that a variety of chemicals have the potential to disrupt the hormonal balance of sensitive organisms [115d, e].

8.1.4
Chemicals with Antiandrogenic Activity (Xenoantiandrogens)

The action of androgens are mediated via the androgen receptor (AR). This is essential for normal development of the male reproductive system. Testosterone and 5α-dihydrotestosterone (5α-DHT) are the primary androgens that activate the AR under normal physiological conditions. In teleost fishes 11-ketotestosterone (11-KT) shows a greater androgenic potency than testosterone and is considered to be the main androgen in teleost. Reduction of the 11-keto group to 11β-hydroxytestosterone (11β-OHT) is a first step toward deactivation of 11-KT. It was shown that chemicals can influence the androgen levels. Those chemicals having antagonistic properties with the androgen receptor (AR) are of particular concern. These antiandrogens can bind to the AR without activating it, and simultaneously prevent binding of natural androgens, such as 5α-dihydrotestosterone, testosterone, and/or 11-ketotestosterone.

Examples of chemicals of antiandrogenic activity are the nonsteroidal pharmaceutical flutamide and its metabolite 2-hydroxyflutamide. The agricultural fungicides vinclozolin and procymidone with some of their metabolites, some phenylurea herbicides (PUHs), such as linuron and diuron and their metabolites 3,4-dichloroaniline and 3,4-dichloroacetanilide bind to the androgen receptor and prevent binding of natural androgens (Table 7). However, their binding affinity to the androgen receptor is relatively low compared to 5α-DHT. Furthermore, the measured and/or predicted bioconcentration factors in fish are low or moderate (Table 7). However, the toxicological studies by Allner with stickleback have shown that 3,4-dichloroaniline and the metabolite 3,4-dichloroacetanilide are bioconcentrated in considerable amounts in the fish brain [125e, f]. It is predicted that other phenylurea herbicides (PUHs), such as monolinuron, monuron, neburon, chlorotoluron, fluometuron, isoproturon, metobromuron, and diflubenzuron as well as their metabolites, especially the acetanilides, have weak antiandrogenic activity.

Table 7. Chemical structure, molecular formula, molecular mass, relative binding affinity (RBA) to the androgen receptor (AR), n-octanol/water partition coefficient (log K_{OW}), and bioconcentration factors on a wet weight basis (BCF_W) and on a lipid basis (BCF_L) of **Steroidal Androgens** and **Nonsteroidal Chemicals** with **Antiandrogenic Activity**

Chemical (abbreviation) [use, metabolites etc.]	Chemical structure	Molecular formula and molecular mass [g mol^{-1}]	RBA[a]	log K_{OW}	Bioconcentration factor (BCF) in fish	
					BCF_W	BCF_L
1. Natural steroidal Androgens						
5α-Dihydrotestosterone (5α-DHT) [Natural androgen]	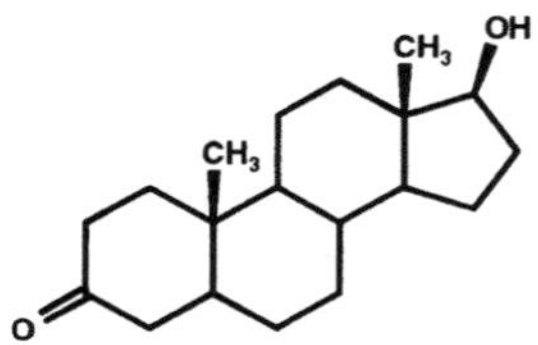	$C_{19}H_{30}O_2$ 290.43	1.00	3.40	126[b]	2510[k]
Testosterone (T) [Androgen]	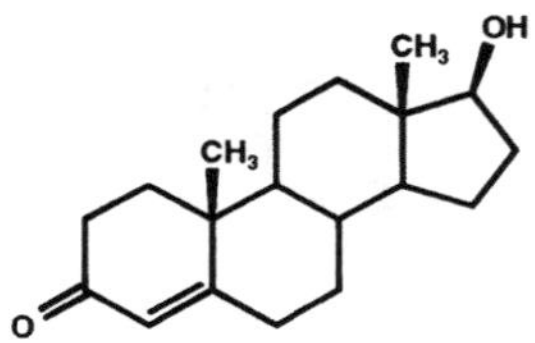	$C_{19}H_{28}O_2$ 288.43	0.333 0.74	3.32	104[b]	2090[k]

Table 7 (continued)

Chemical (abbreviation) [use, metabolites etc.]	Chemical structure	Molecular formula and molecular mass [g mol^{-1}]	RBA[a]	log K_{OW}	Bioconcentration factor (BCF) in fish	
					BCF_W	BCF_L
2. Synthetic steroidal antiandrogens						
Cyproterone [Antiandrogenic drug]	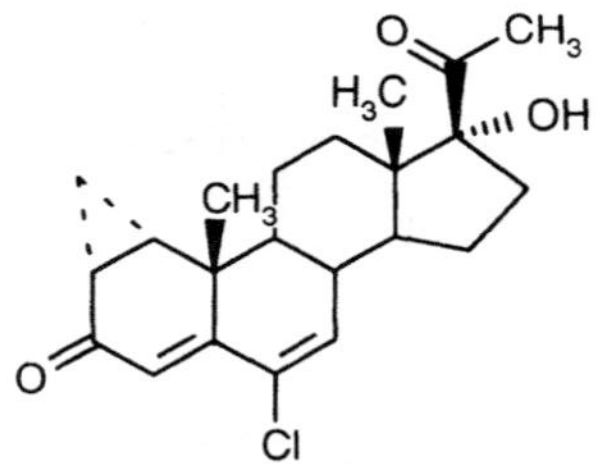	$C_{22}H_{27}ClO_3$ 374.91		3.64[g]	220[b]	4370[k]
3. Synthetic nonsteroidal antiandrogens						
Flutamide 2-Methyl-*N*-[4-nitro-3-(trifluoromethyl) phenyl]propanamide [Antiandrogenic drug]	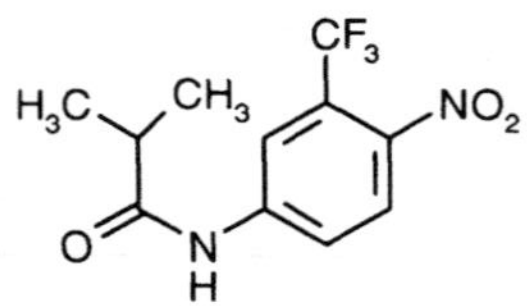	$C_{11}H_{11}F_3N_2O_3$ 276.22	0.00008 0.00009 (negative $c < 10^{-4}$ M)	3.35[h]	112[b]	2240[k]
2-Hydroxyflutamide [Metabolite of flutamide]	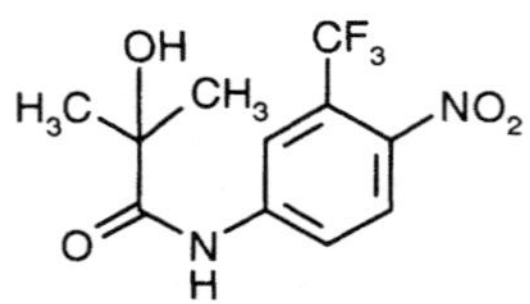	$C_{11}H_{11}F_3N_2O_4$ 292.22	0.0056 0.0027 0.0004	2.70[h]	25[b]	500[k]

Nilutamide [Antiandrogenic drug]	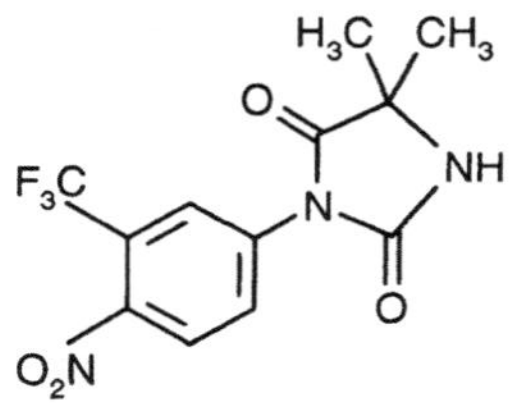	$C_{12}H_{10}F_3N_3O_4$ 317.22		1.92[g]	4[b]	83[k]
4. Agrochemicals and/or their metabolites with antiandrogenic activity						
Linuron 2-(3,4-Dichlorophenyl)-1-methoxy-1-methylurea [Herbicide]	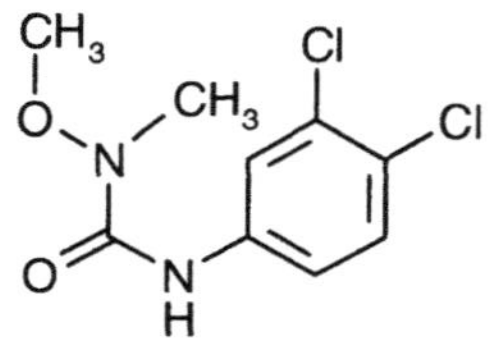	$C_9H_{10}Cl_2N_2O_2$ 249.10	0.00005 0.00002 0.00013	3.20	80[b]	1600[k]
Hydroxylinuron 3-(3,4-Dichlorphenyl)-1-hydroxy-1-methylurea [Metabolite of linuron]	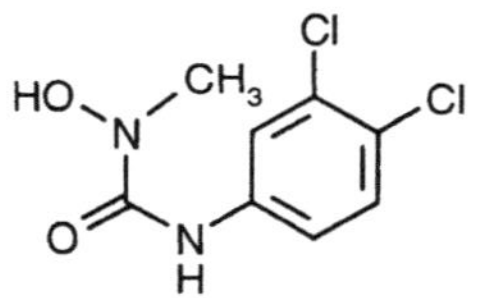	$C_8H_8Cl_2N_2O_2$ 235.07	0.00046	2.90[g]	40[b]	790[k]
3-(3,4-Dichlorophenyl)-1-methoxyurea [Metabolite of linuron]	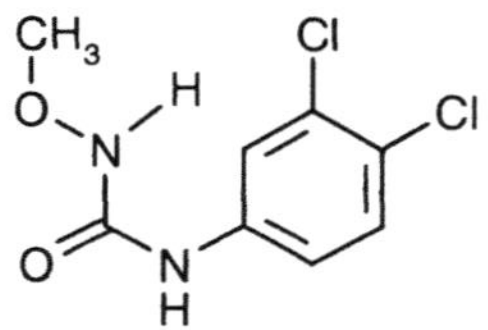	$C_8H_8Cl_2N_2O_2$ 235.07	0.00001	2.99[g]	50[b]	980[k]

Table 7 (continued)

Chemical (abbreviation) [use, metabolites etc.]	Chemical structure	Molecular formula and molecular mass [g mol^{-1}]	RBA[a]	log K_{OW}	Bioconcentration factor (BCF) in fish	
					BCF_W	BCF_L
3-(3,4-Dichlorophenyl)-1-methylurea [Metabolite of linuron]	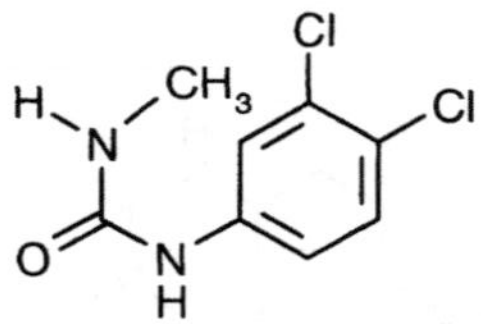	$C_8H_8Cl_2N_2O$ 219.07	0.000005	2.54	17[b]	350[k]
3,4-Dichloroaniline (3,4-DCA) [Metabolite of linuron and intermediate for synthesis of pesticides etc.]	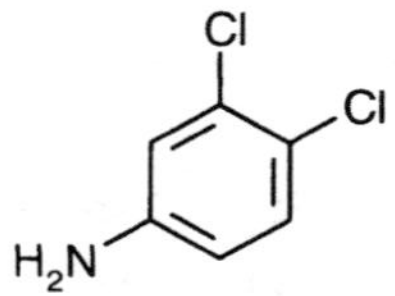	$C_6H_5Cl_2N$ 162.02	0.00001 0.000066	2.85	30[c] 28[j]	810 820
3,4-Dichloroacetanilide (3,4-DCAc) [Metabolite of 3,4-DCA]	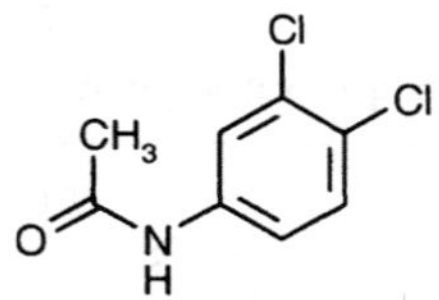	$C_8H_7Cl_2NO$ 204.04	0.000135	2.54	17[b]	350[k]
3-Chloro-4-methyl-aniline[f] (3-CMA) 3-Chloro-*p*-toluidine[f] [Pesticide, avicide]	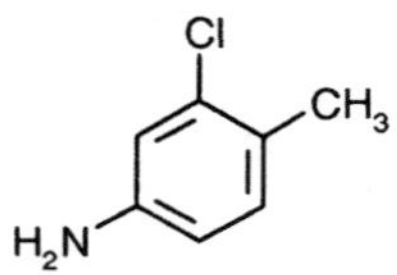	C_7H_8ClN 141.60		2.75[g]	90[e]	2700[e]

3-Chloro-4-methyl-acetanilide[f] [Metabolite of 3-CMA]	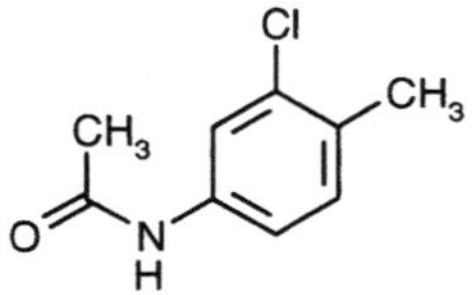	$C_9H_{10}ClNO$ 183.64		2.69[l]	24[b]	490[k]
Diuron 3-(3,4-Dichlorophenyl)-1,1-dimethylurea [Herbicide]	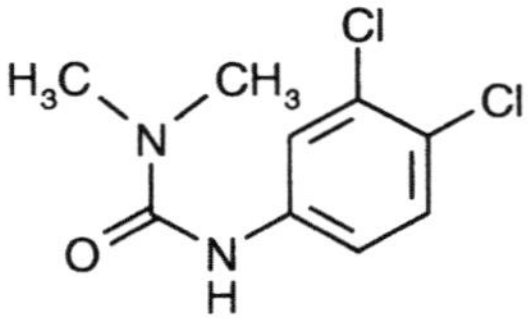	$C_9H_{10}Cl_2N_2O$ 233.10	0.000034	2.89	157[d]	4910[d]
Vinclozolin (RS)-3-(3,5-Dichloro-phenyl)-5-ethenyl-5-methyl-2,4-oxazolidinedione [Agricultural fungicide]	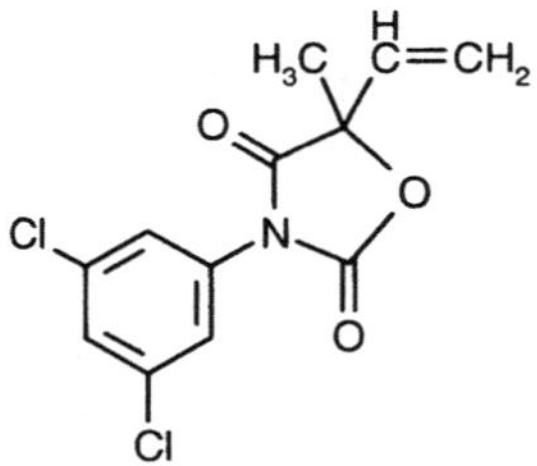	$C_{12}H_9Cl_2NO_3$ 286.11	0.00001	3.10 (3.03; pH 6.5)	60[b]	1260[k]
3′,5′-Dichloro-2-hydroxy-2-methylbut-3-en anilide *N*-(3,5-Dichlorophenyl)-2-hydroxy-2-methyl-3-butenamide [Metabolite of vinclozolin]	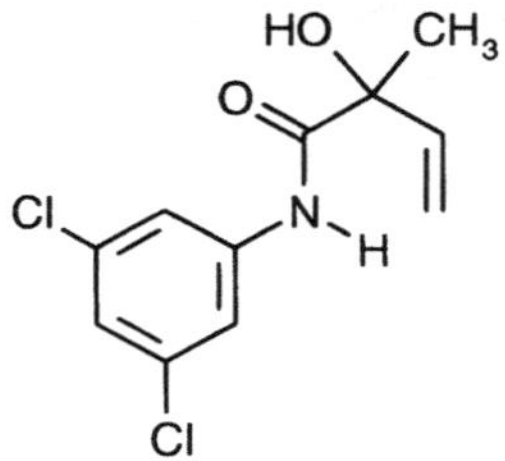	$C_{11}H_{11}Cl_2NO_2$ 260.12	0.003	2.15[g]	70[b]	140[k]

Table 7 (continued)

Chemical (abbreviation) [use, metabolites etc.]	Chemical structure	Molecular formula and molecular mass [g mol^{-1}]	RBA[a]	log K_{OW}	Bioconcentration factor (BCF) in fish	
					BCF_W	BCF_L
2-{ [(3,5-Dichloro-phenyl)-carbamoyl] oxy}-2-methyl-3-butenoic acid [Metabolite of vinclozolin]	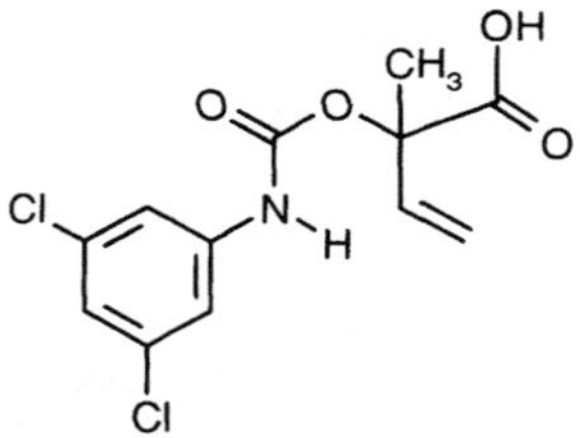	$C_{12}H_{11}Cl_2NO_4$ 304.11	0.00012	3.52[g]	170[b]	3300[k]
Procymidone 3-(3,5-Dichlorophenyl)-1,5-dimethyl-3-aza-bicyclo[3.1.0]hexane-2,4-dione [Agricultural fungicide]	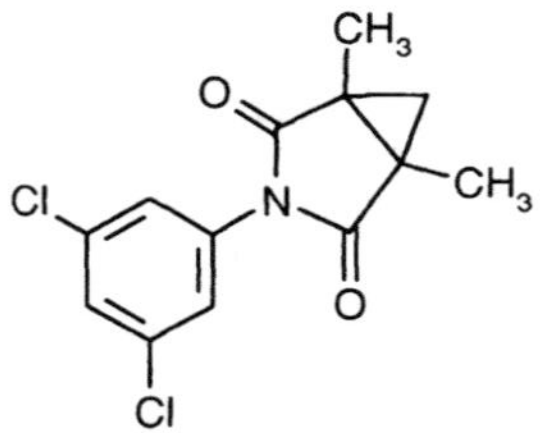	$C_{13}H_{11}Cl_2NO_2$ 284.14	0.00001	3.14	70[b]	1380[k]
3,5-Dichlorobenz-anilide-2-cyclopropane-carboxylic acid [Metabolite of procymidone]	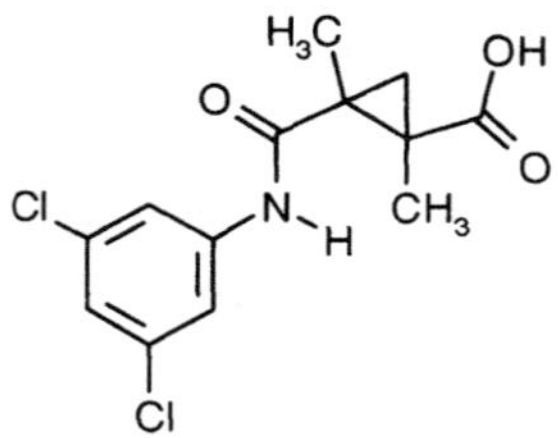	$C_{13}H_{13}Cl_2NO_3$ 302.14	0.0001	3.27[g]	90[b]	1860[k]

Iprodione[f] 3-(3,5-Dichlorophenyl)-*N*-isopropyl-2,4-dioxoimidazolidine-1-carboxamide [Agricultural fungicide]	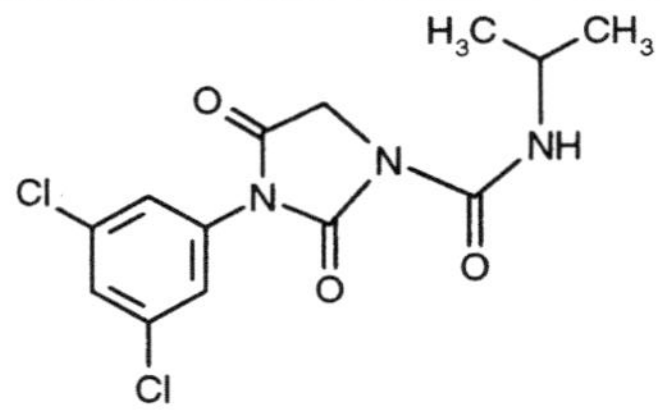	$C_{13}H_{13}Cl_2N_3O_3$ 330.17		3.10	63[b]	1260[k]
1,1-Dichloro-2,2-bis(p-chlorophenyl)ethylene (*p,p*′-DDE) [Metabolite of *p,p*′-DDT]	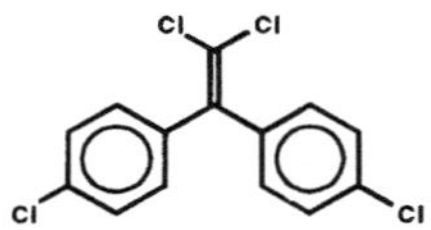	$C_{14}H_8Cl_4$ 318.04	4×10^{-7}	6.96	8.1×10^4	1.1×10^6

[a] RBA; Relative binding affinity to the androgen receptor in comparison to 5α-dihydrotestosterone (DHT). The RBA values were calculated from Ref. [125b, c, d, j].
[b] Worst-case BCF_W value predicted for fish with 5% lipid.
[c] BCF value of ^{14}C-3,4-DCA in zebrafish from Ref. [125 g].
[d] BCF value of ^{14}C-Diuron in fathead minnows (3.2% lipid) from Ref. [125 h].
[e] BCF value of ^{14}C-3-CMA in bluegill sunfish (1.2 g body weight; pH 6.9–7.1) from Ref. [125i].
[f] Supposed to have an antiandrogenic activity.
[g] Calculated according to Ref. [232 a, b].
[h] Measured by Morris et al. [232 c].
[i] Measured by Nakagawa et al. [232 d].
[j] BCF value of ^{14}C-3,4-DCA in three-spined stickleback [232 e].
[k] Worst-case BCF_L value for fish predicted from the log K_{ow} value if no metabolism occurs or is negligible.

Recently, it was found by Kelce et al. [125a] that the persistent *p,p′*-DDT metabolite *p,p′*-DDE is a potent androgen receptor antagonist. This compound is highly hydrophobic and has a very high bioconcentration potential. Furthermore, it is important to note that also estrogens, such as estradiol, diethylstilbestrol, Kepone, *o,p′*-DDT, and methoxychlor can bind also to the androgen receptor [125b].

8.1.5 *Chemicals Which Interact with Different Hormonal Receptors and/or Hormone-Binding Proteins*

It is known that natural and synthetic steroidal hormones can interact with more than one steroid receptor and exert different physiological actions in organisms. For example, mifepristone (RU-486), an abortifacient for use in early pregnancy, is a progesterone antagonist, glucocorticoid antagonist, and posseses weak antiandrogenic activity. Other synthetic steroids, such as tibolone, have weak androgenic, estrogenic and progestogenic activity. There are also indications that organic chemicals can interact with the binding of natural hormones to two or more receptors.

Some environmental endocrine-disrupting chemicals classified as estrogens can bind to more than one steroid receptor. For example, chlordecone (Kepone) and *o,p′*-DDT can bind to the estrogen (ER) and progesterone receptors (PR) with each chemical having IC_{50} values that are nearly identical for the two receptors [123, 124a,b]. Nonylphenol and the metabolite of methoxychlor, 2,2′-bis(hydroxyphenyl)-1,1,1-trichloroethane, are capable of inhibiting the binding to the estrogen, androgen, and progesterone receptor with similar affinities [124b.] Other environmental chemicals, such as *p,p′*-DDT, *p,p′*-DDD, and *p,p′*-DDE can bind to the androgen receptor (AR) 14, 11, and 200 times more effectively, respectively, than to the estrogen receptor [124a, 125a].

The experiments by Danzo [109] have demonstrated that environmental chemicals interact in a specific and differential manner not only with the estrogen receptor (ER) but also with the androgen receptor (AR), androgen-binding protein (ABP), and sex hormone-binding globulin (SHBG). Several chemicals, such as γ-hexachlorocyclohexane (γ-HCH, lindane), δ-hexachlorocyclohexane (δ-HCH), *p,p′*-DDT, *p,p′*-DDE, *o,p′*-DDT, dieldrin, pentachlorophenol (PCP), and atrazine, were capable of inhibiting [^{3}H]5α-dihydrotestosterone (5α-DHT) binding to the androgen receptor. Methoxychlor, *o,p′*-DDT, pentachlorophenol, and nonylphenol significantly reduced [^{3}H]17β-estradiol binding to the estrogen receptor (ER) by 10, 20, 60, and 75%. Methoxychlor, nonylphenol, *p,p′*-DDT, and atrazine reduced [^{3}H]5α-DHT binding to the androgen-binding protein (ABP) by ca. 40%. Pentachlorophenol and *o,p′*-DDT resulted in a significant 20% inhibition of [^{3}H]5α-DHT binding to human sex hormone-binding globulin (hSHBG). These findings by Danzo [109] indicate that some environmental chemicals can interfere with the binding of natural hormones to two or more binding moieties, thus may be capable of disrupting physiological processes regulated by these pathways (see also Ref. 405).

8.1.6 *Conclusions*

It is important to note that many hydrophobic environmental synthetic organic chemicals with endocrine activity are relatively resistant to metabolic degradation, especially those which are highly chlorinated and/or those with many nitro groups. A negative characteristic of these EDCs is that many of them possess a long biological half-life ($t_{1/2}$) and can persist for a long time (some months to some years) in organisms. The half-life of a chemical in an organism is dependent on its resistance to metabolic degradation, on its lipophilicity (K_{OW} value), and especially on the total lipid content of the organism. The half-life of chemicals is increasing with their K_{OW} values, and with the organisms' lipid content [29, 40]. Those EDCs which persist in the environment are bioaccumulated in aquatic organisms [62, 130] and terrestrial vertebrates [131] including humans [132, 133] (see Table 8). They can accumulate to high concentrations in lipids of the organisms [62, 130–133] from which they can slowly be released to provide a low EDC level in blood. Such long-term continuous concentration of EDCs may be effective in stimulating certain estrogenic, antiestrogenic, androgenic, antiandrogenic or other hormonal responses.

8.2 Bioconcentration of Super-Hydrophobic Chemicals and Other Persistent Organic Pollutants (POPs)

Persistent organic pollutants (POPs) have become the focus of growing national and international concern (United Nations, Greenpeace, Environmental Protection Agencies of the USA, Germany and many other countries) [159, 160]. POPs are organic substances that

(1) have a long-range atmospheric transport,
(2) are volatile enough to evaporate and condense in air, water, and soil at environmental temperatures,
(3) have a high persistence in soil, water, and biota,
(4) have a very high lipophilicity ($\log K_{OW} > 5$),
(5) have a high bioaccumulation potential in aquatic and terrestrial organisms including human, and
(6) can have toxic or adverse effects on reproduction, development and/or immunological function of aquatic and terrestrial animals.
(7) It is also important to note that many of these POPs in relatively high concentrations have shown endocrine-disrupting effects *in vitro* and/or *in vivo* (Table 8). Furthermore, some POPs are carcinogenic in experimental animals.

The acronym, POP, is gaining world-wide acceptance, although some national agencies still use other terms, e.g. persistent environmental pollutants (PEPs), for these chemicals. The chemical industry, for instance, terms them **"persistent, bioaccumulative, toxic substances" (PBTs).** The US Environmental Protection Agency (EPA) prefers "bioaccumulative chemicals of concern" (BCCs). Much of

Table 8. Selected characteristics of **Persistent Organic Pollutants (POPs)**. All BCF data of aquatic organisms from laboratory experiments unless otherwise stated

No	Chemical or chemical class (abbreviation) [CAS No.]	Chemical structure	log K_{OW}	Molecular formula and molecular mass [g mol^{-1}]
1	Aldrin (HHDN) [309-00-2]	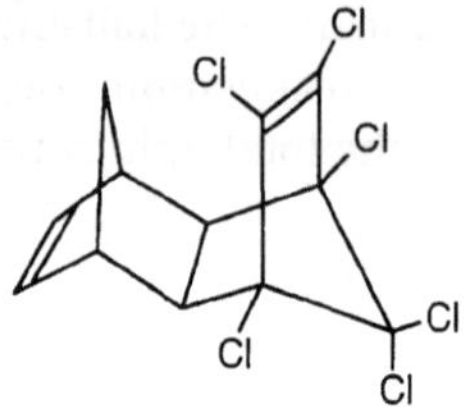	6.496[d]	$C_{12}H_8Cl_6$ 364.91
2	Dieldrin (HEOD) [60-57-1]	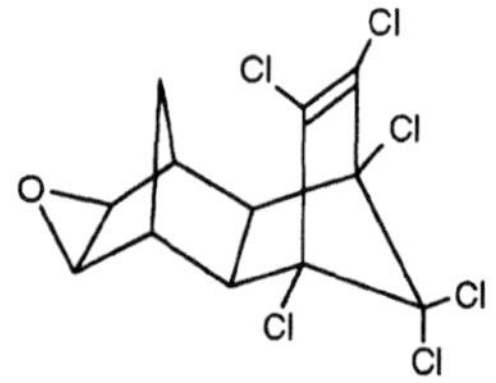	5.40[d]	$C_{12}H_8Cl_6O$ 380.91
3	Endrin [70-20-8]	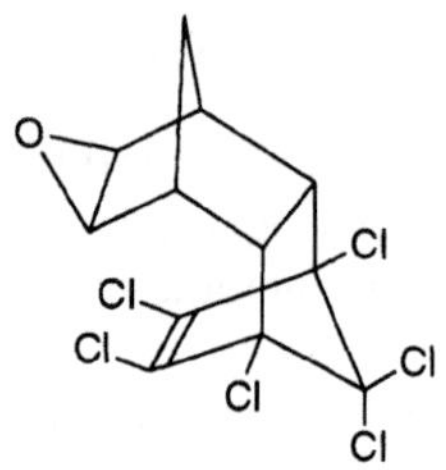	5.195[d]	$C_{12}H_8Cl_6O$ 380.91

Half-life or persistence in soil (*S*), sediment (SE) or environment (*E*) ($t_{1/2}$ in years)	Endocrine – disrupting effects[a] and effects on enzymes (TOEI)[j]		Bioconcentration Factor (BCF)[b]		
	in vitro	*in vivo*	Organisms	BCF_W	BCF_L
5–9 (*S*)		not estrogenic in rats;	algae (*Chlorella*)	12,300	(dry wt.) 61,000
			Daphnia	35,300	3,530,000
		TOEI: PB-type	mussel (*Mytilus edulis*)	4,570	457,000
5–7 (S)	estrogenic in the E-SCREEN assay and antiandrogenic (reduction of 5α-dihydrotestosterone binding to specific prostatic nuclear and cytoplasmic receptors)	not estrogenic in rats;	algae (*Chlorella*)	2,300	(dry wt.) 11,500
			mussel (*Mytilus edulis*)	3,100	310,000
			mussel (*Mytilus edulis*)	3,750	375,000
		TOEI: PB-type	*Daphnia*	3,490	349,000
			oyster (*Crassostrea virginica*)	2,880	240,000
			oyster (*Crassostrea virginica*)	2,070	172,500
			oyster (*Crassostrea virginica*)	5,000	417,000
			fish		
			guppy (f)	12,700	180,000
			carp	26,000	260,000
			human (fat)	49	71
			(range)	38–77	55–155
up to 12 (*S*)		not estrogenic in rats;	mussel (*Mytilus edulis*)	1,920	192,000
		TOEI: PB-type	oyster (*Crassostrea virginica*)	1,670	139,000
			oyster (*Crassostrea virginica*)	2,780	232,000
			clam	2,625	
			fish		
			fathead minnow (uptake 300 d)	4,570	152,000

Table 8 (continued)

No	Chemical or chemical class (abbreviation) [CAS No.]	Chemical structure	log K_{OW}	Molecular formula and molecular mass [g mol^{-1}]
4	Chlordane; 1,2,4,5,6,7,8,8-Octachloro-2,3,3a,4,7,7a-hexahydro-4,7-methano-1*H*-indene [57–74–9]	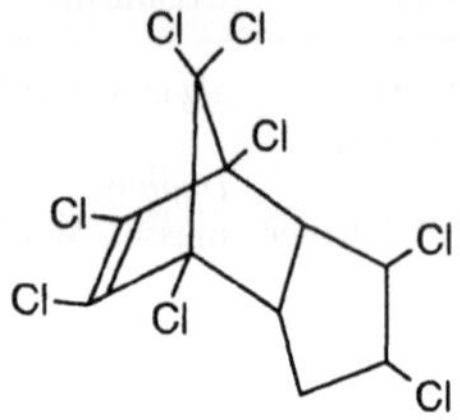	6.16[d]	$C_{10}H_6Cl_8$ 409.78
4.1	*cis*-Chlordane; α-Chlordane [5103–71–9]	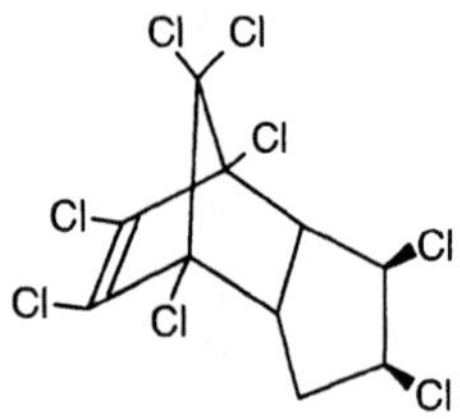	6.10[d]	$C_{10}H_6Cl_8$ 409.78
4.2	*trans*-Chlordane; γ-Chlordane [5103–74–2]	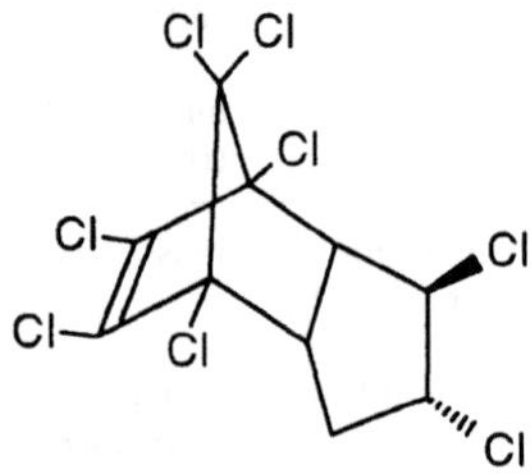	6.22[d]	$C_{10}H_6Cl_8$ 409.78
4.3	*cis*-Nonachlor [5103–73–1]	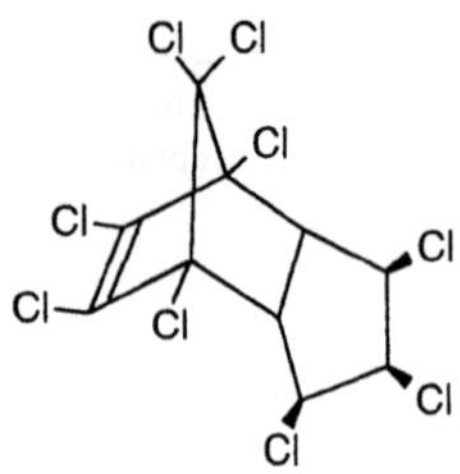	6.08[d]	$C_{10}H_5Cl_9$ 444.23

Half-life or persistence in soil (*S*), sediment (SE) or environment (*E*) ($t_{1/2}$ in years)	Endocrine – disrupting effects[a] and effects on enzymes (TOEI)[j]		Bioconcentration Factor (BCF)[b]		
	in vitro	*in vivo*	Organisms	BCF_W	BCF_L
1–20 (*S*) 5–15 (*E*)	not estrogenic in the E-SCREEN assay	decrease of plasma testosterone, estrone, and 17β-estradiol levels by increasing steroid hydroxylase in rats and mice. Decrease of thyroxine; TOEI: PB-type	fish fathead minnow (uptake 32 d, no steady-state)	> 37,800	> 360,000 (no steady-state)
			human (fat) range	540[f] 414–656[f]	780[f] 600–950[f]
extremely persistent	not estrogenic in the E-SCREEN assay		*Daphnia*	24,000	1,600,000
			Fish rainbow trout (kinetic)	28,000	384,000
			chum salmon (*Oncorhynchus keta*) (9.1% lipid) marine environment	184,000	2,020,000
very persistent but this compound is more unstable than *cis*-chlordane			algae (*Ankistrodesmus ammalloides*)	1,100	(dry wt.) 5,500
			Daphnia	20,130	2,013,000
			fish rainbow trout (kinetic approach)	16,200	222,000
			chum salmon (*Oncorhynchus keta*) (9.1% lipid) marine environment	111,000	1,320,000
persistent			fish (5% lipid)	60,000[c]	1,200,000[c]

Table 8 (continued)

No		Chemical or chemical class (abbreviation) [CAS No.]	Chemical structure	log K_{OW}	Molecular formula and molecular mass [g mol^{-1}]
	4.4	*trans*-Nonachlor [39765-80-5]	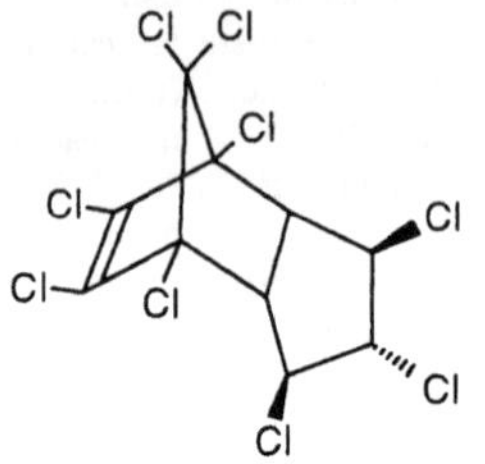	6.35[d]	$C_{10}H_5Cl_9$ 444.23
5	5.1	Heptachlor [76-44-8]	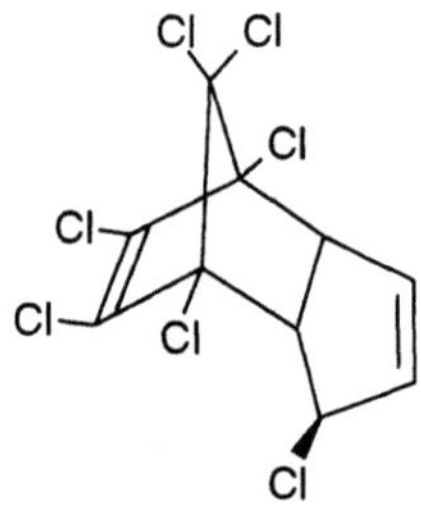	6.10[d]	$C_{10}H_5Cl_7$ 373.32
	5.2	Heptachlor epoxide [1024-57-3]	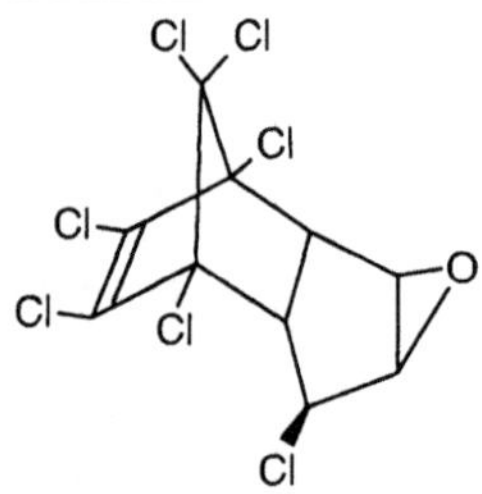	5.40[d]	$C_{10}H_5Cl_7O$ 389.32
6		DDT (technical)			
	6.1	1,1,1-Trichloro-2,2-bis (*p*-chlorophenyl)ethane; (*p,p*-DDT) 1,1′-(2,2,2-Trichloroethylidene)bis(4-chlorobenzene) [50-29-3]	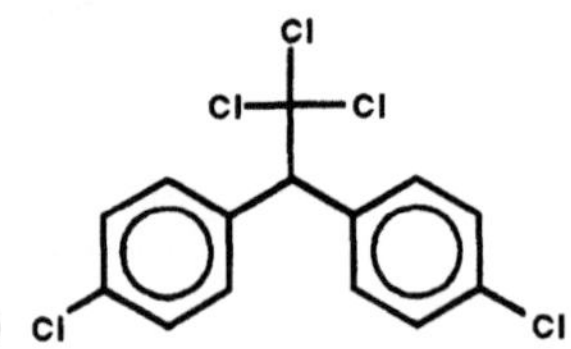	6.91[d]	$C_{14}H_9Cl_5$ 354.49

Half-life or persistence in soil (*S*), sediment (SE) or environment (*E*) ($t_{1/2}$ in years)	Endocrine – disrupting effects[a] and effects on enzymes (TOEI)[j]		Bioconcentration Factor (BCF)[b]		
	in vitro	*in vivo*	Organisms	BCF_W	BCF_L
very persistent			zooplankton (1.9 % lipid) environment	190,000	10,000,000
			fish chum salmon (*Oncorhynchus keta*) (9.1 % lipid) environment	1,100,000	12,000,000
7–14 (*S*)	not estrogenic in the E-SCREEN assay, but the metabolite 1-hydroxy-chlordane is estrogenic	TOEI: PB-type	oyster (*Crassostrea virginica*) (uptake 6 months)	17,000	1,400,000
			fish fathead minnow (uptake 276 d)	20,000	710,000
ca. 3 (*S*)		estrogenic in rats; TOEI: PB-type	fish fathead minnow (uptake 32 d, no steady-state)	> 14,400	> 137,000
3–35[h] (*S*); >60 (*E*)	estrogenic	estrogenic		very high	very high
very persistent 3–35[h] (*S*) 0.4–3.9[h] of ^{14}C-*p,p*′-DDT (tropical S)	antiandrogenic and weak estrogenic	estrogenic in rodents; decrease of thyroxine;	algae (*Chlorella*)	9,350	(dry wt.) 64,800
			Daphnia	28,500	2,850,000
			oyster (flow-through 6 months)	127,000	10,600,000
			(flow-through 6 months)	152,000	12,600,000
			(mean)	139,500	11,600,000
			fish rainbow trout (kinetic approach)	93,000[h]	4,700,000[h]
			human (fat)	870[h]	1,280[h]
			range	447–1,280[h]	670–1,920[h]

Table 8 (continued)

No	Chemical or chemical class (abbreviation) [CAS No.]	Chemical structure	log K_{OW}	Molecular formula and molecular mass [g mol^{-1}]
6.2	1,1-Dichloro-2,2- bis (*p*-chlorophenyl)ethylene; 1,1′-(Dichloroethenylidene) bis (4-chlorobenzene) (*p,p*′-DDE) [72–55–9]	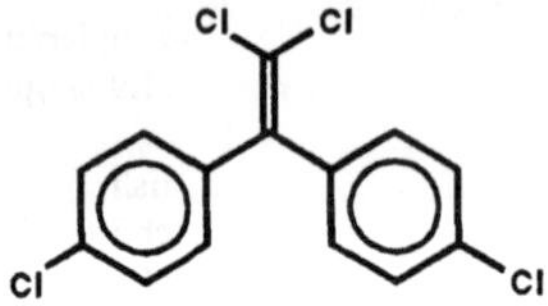	6.96[d]	$C_{14}H_8Cl_4$ 318.04
6.3	1,1-Dichloro-2,2-bis (*p*-chlorophenyl)ethane; 1,1′-(2,2-Dichloroethyl-idene)bis(4-chlorobenzene) (*p,p*′-DDD; *p,p*′-TDE); Rothane [72–54–8]	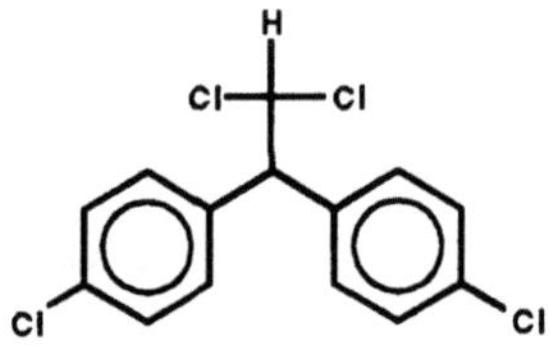	6.22[d]	$C_{14}H_{10}Cl_4$ 320.04
6.4	1,1,1-Trichloro-2-(*o*-chloro-phenyl)-2-(*p*-chloro-phenyl)ethane; (*o,p*′-DDT) [789–02–6]	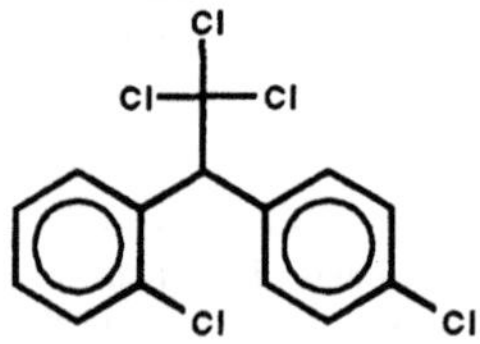	6.76[e]	$C_{14}H_9Cl_5$ 354.49
6.5	1,1-Dichloro-2-(o-chloro-phenyl)-2-(p-chlorophenyl) ethylene; (*o,p*′-DDE) [3424–82–6]	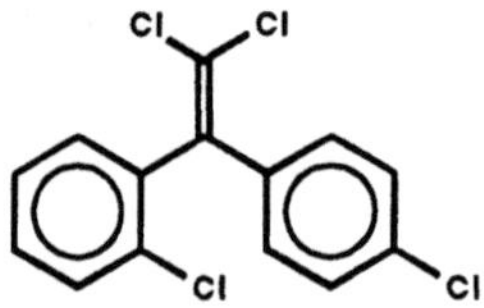	6.94[e]	$C_{14}H_8Cl_4$ 318.04
7	Hexachlorobenzene (HCB) [118–74–1]	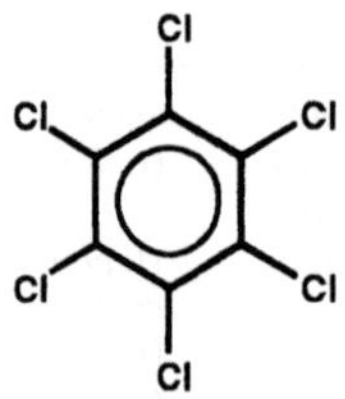	5.73[d]	C_6Cl_6 287.78

Half-life or persistence in soil (*S*), sediment (SE) or environment (*E*) ($t_{1/2}$ in years)	Endocrine – disrupting effects[a] and effects on enzymes (TOEI)[j]		Bioconcentration Factor (BCF)[b]		
	in vitro	*in vivo*	Organisms	BCF_W	BCF_L
very persistent > 20 (*S*) 0.4–1.7 of ^{14}C-*p,p′*-DDE (tropical S)	antiandrogenic and weak estrogenic	estrogenic and anti-androgenic; TOEI: PB-type	fish rainbow trout (Kinetic K_1/K_2)	89,950	1,110,000
			fathead minnow 32 d uptake (no steady-state)	> 60,000	> 570,000
very persistent	antiandrogenic and weak estrogenic	estrogenic in rats	mussel (*Mytilus edulis*)	12,500 14,420	1,250,000 1,620,000
			oyster (*Crassostrea virginica*) (56 weeks)	47,900	1,600,000
	estrogenic (ER binding 0.1%)	estrogenic (uterotropic) in rodents; decrease of thyroxine; TOEI: PB-type	fish fathead minnow (10.5% lipid) (28 d; no steady-state)	> 37,200	> 354,000
	estrogenic	estrogenic in rats			
3–6 (*S*)	not estrogenic in the E-SCREEN assay, Ah receptor binding, TOEI: mixed type	decrease in plasma thyroxine level, porphyrogenic; TOEI: mixed type	algae (*Chlorella*)	24,000	(dry wt.) 120,000
			Daphnia	9,600	960,000
			mussel (*Mytilus edulis*) (21-d, no steady-state)	> 3,430	> 343,000
			fish golden orfe (*Leuciscus idus melanotus*) lipid 0.95%)	4,850	510,000
			human (fat) range	448 259–742	674 373–1,146

Table 8 (continued)

No		Chemical or chemical class (abbreviation) [CAS No.]	Chemical structure	log K_{OW}	Molecular formula and molecular mass [g mol^{-1}]
8		Mirex [2385-85-5]	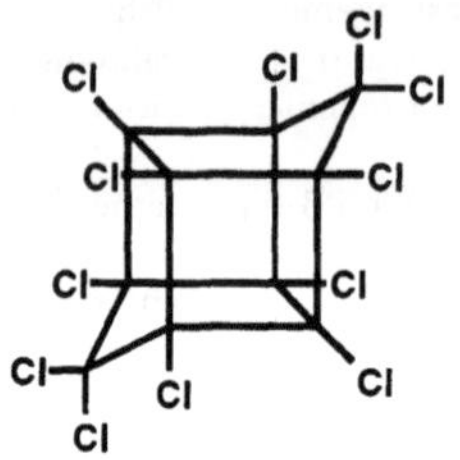	7.50	$C_{10}Cl_{12}$ 545.54
9		Toxaphene [8001-35-2] Mixture of	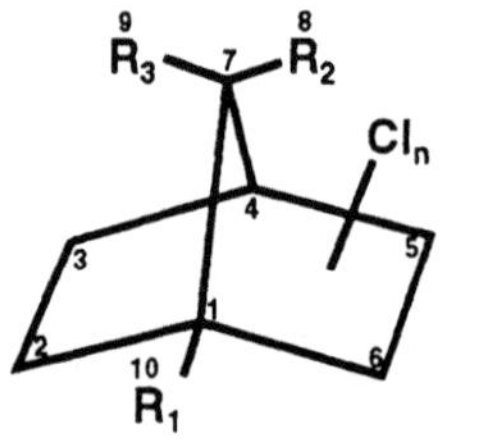 R_1,R_2,R_3 = CH_3, CH_2Cl, $CHCl_2$	6.50 (mean) 5.2-7.8 (range)	$C_{10}H_{10}Cl_8$ (average) $C_{10}H_{18-n}Cl_n$
	9.1	Polychlorobornanes; (chemical structure is presented)			n = 6-10 414 (average)
	9.2	Polychlorobornenes;			$C_{10}H_{16-n}Cl_n$ n = 6-10
	9.3	Polychlorinated camphenes and other chlorinated compounds			
10		**Polychlorinated Biphenyls** (PCB IUPAC/ Ballschmiter No.) [1336-36-3]	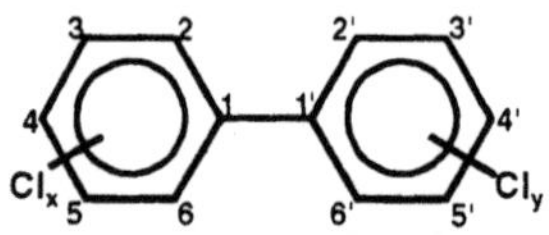 x = 1-5 y = 0-5	4.50-8.30	$C_{12}H_{10-n}Cl_n$ n = 1-10
	10.1	**Technical Mixtures**			
	10.1.1	Aroclor 1221 (21% Cl) [11104-28-2]	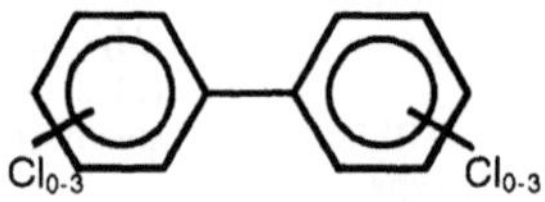	4.40 (4.10-4.70)	average 192
	10.1.2	Aroclor 1242 (42% Cl) [53469-21-9]	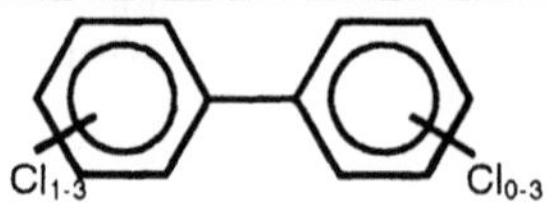	5.58	average 261

Half-life or persistence in soil (*S*), sediment (SE) or environment (*E*) ($t_{1/2}$ in years)	Endocrine – disrupting effects[a] and effects on enzymes (TOEI)[j]		Bioconcentration Factor (BCF)[b]		
	in vitro	*in vivo*	Organisms	BCF_W	BCF_L
extremely persistent 8.2 (*S*)	not estrogenic in the E-SCREEN assay	not estrogenic in rats; increased testosterone metabolic clearance; TOEI: PB-type	fish guppy (6.5% lipid)	940,000	14,500,000
			human (fat) (predicted)	6,200 – 18,000	9000 – 25,000 (predicted) no steady-state reached
10 – 18 (*E*) 0.8 – 14 (*S*) < 0.03[i]	estrogenic in MCF-7 cells, increased estrogen and progesterone levels	not estrogenic, induction of enzymes (e.g. androgen hydroxylase) which are involved in the metabolism of steroid hormones, increased hepathic metabolism; TOEI: weak PB-type	oyster (*Crassostrea virginica*) (flow-through 6 months)	32,900	2,740,000
			(flow-through 6 months)	37,500	3,130,000
			fish fathead minnow 98 d (flow-through)	69,200	1,150,000
			150 d (flow-through) 13 ng/l, no steady-state	63,000	630,000
			channel catfish (adult) 100 d (flow-through) 49 ng/l, no steady-state	> 54,000	> 690,000
			human (fat)	1,100	1,600
3 – 17 (*S*)			aquatic and terrestrial organisms		high/extremely high bioconcentration/bioaccumulation potential
	estrogenic, not antiestrogenic in MCF-7 cells	estrogenic, TOEI: weak PB-type			
	not antiestrogenic in MCF-7 cells	estrogenic; decreased T_3 (hypothyroidism) in rats. TOEI: mixed-type	fish (flow-through) (assuming 5% lipid)	49,000	980,000

Table 8 (continued)

No	Chemical or chemical class (abbreviation) [CAS No.]	Chemical structure	log K_{OW}	Molecular formula and molecular mass [g mol^{-1}]
10.1.3	Aroclor 1248 (48% Cl) [12672-29-6]	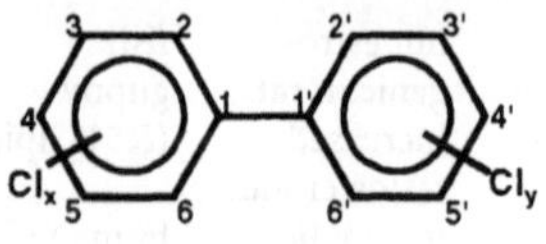 x + y = 2 – 6	6.11	average 288
10.1.4	Aroclor 1254 (54% Cl) [11097-69-1]	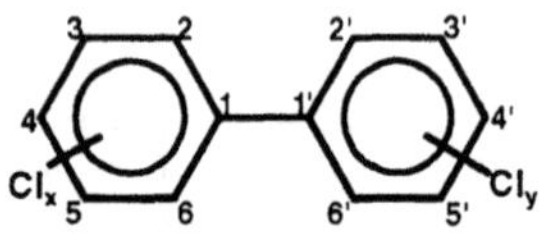 x + y = 3 – 8	6.47	average 327
10.1.5	Aroclor 1260 (60% Cl) [11096-82-5] Chlophen A 60 (60% Cl)	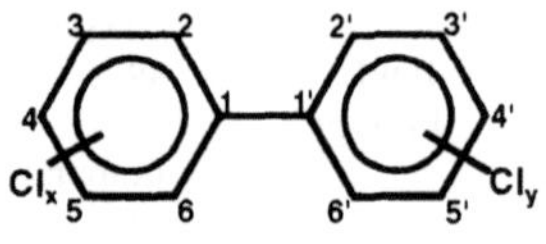 x + y = 4 – 9	6.91 (6.3–7.5)	average 372
10.2	**Group I. Estrogenic PCBs** (low chlorinated PCB congeners with non-*para*-, one-*para*-, or di-*para*-substituted positions and two adjacent nonsubstituted lateral C atoms)	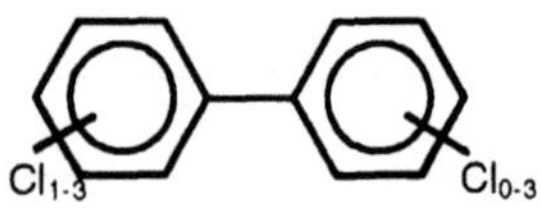	4.5–5.9	$C_{12}H_{5-9}Cl_{1-5}$
10.2.1	4-Chlorobiphenyl (PCB # 3) [2051-62-9]	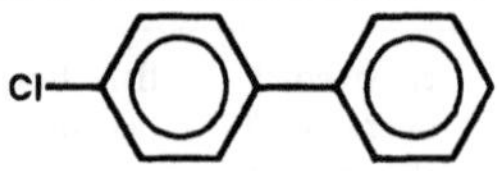	4.64	$C_{12}H_9Cl$ 188.65
10.2.2	2,3-Dichlorobiphenyl (PCB # 5) [16605-91-7]	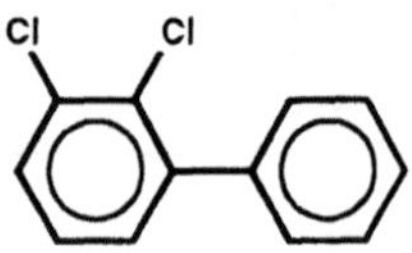	5.17	$C_{12}H_8Cl_2$ 223.1

Half-life or persistence in soil (*S*), sediment (SE) or environment (*E*) ($t_{1/2}$ in years)	Endocrine – disrupting effects[a] and effects on enzymes (TOEI)[j]		Bioconcentration Factor (BCF)[b]		
	in vitro	*in vivo*	Organisms	BCF_W	BCF_L
	not antiestrogenic in MCF-7 cells	estrogenic; decreased serum progesterone and thyroxine in rats. TOEI: mixed-type	fish (uptake 250 d)		
			fathead minnow (male)	63,000	1,190,000
			fathead minnow (female)	120,000	1,200,000
	not antiestrogenic in MCF-7 cells	not estrogenic in female rats, decreased serum T_3 and/or T_4 and multiple steroid hormone abnormalities in rats. TOEI: strong mixed-type	oyster (*Crassostrea virginica*) (flow-through 56 weeks)	89,000	4,500,000
			fish		
			fathead minnow (32 d uptake, no steady-state)	> 100,000	> 952,000
	not antiestrogenic in MCF-7 cells	not estrogenic in female rats, increased length of estrus in rats. TOEI: mixed-type	oyster	63,000	5,300,000
			fish (uptake 250 d)		
			fathead minnow male	167,000	3,150,000
			fathead minnow female	270,000	3,380,000
			human (fat)	175	251
			range	128–277	192–317
	estrogenic	estrogenic in rodents	aquatic and terrestrial organisms		high/very high bioconcentration/ bioaccumulation potential
not persistent		estrogenic in rats	fish (assuming 5% lipid)	590[c]	11,800[c]
	estrogenic	estrogenic (predicted)	oyster (predicted)	1,200	214,000
			rainbow trout	13,000	159,000

Table 8 (continued)

No	Chemical or chemical class (abbreviation) [CAS No.]	Chemical structure	log K_{OW}	Molecular formula and molecular mass [g mol^{-1}]
10.2.3	2,4′-Dichlorobiphenyl (PCB # 8) [34883-43-7]	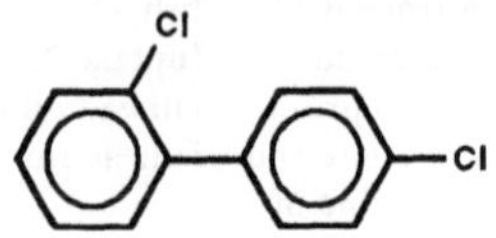	5.24	$C_{12}H_8Cl_2$ 223.1
10.2.4	2,5-Dichlorobiphenyl (PCB # 9) [34883-39-1]	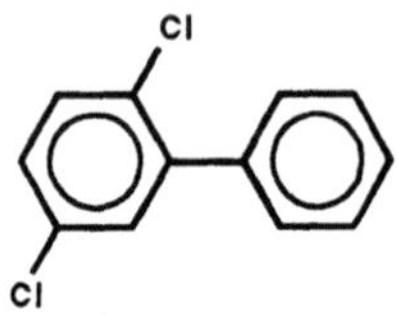	5.18	$C_{12}H_8Cl_2$ 223.1
10.2.5	4,4′-Dichlorobiphenyl (PCB # 15) [2050-68-2]	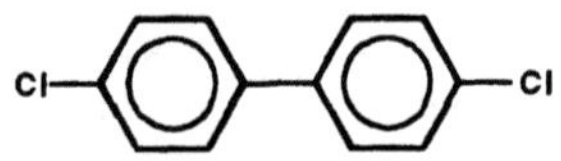	5.36	$C_{12}H_8Cl_2$ 223.1
10.2.6	2,2′,5-Trichlorobiphenyl (PCB # 18) [37680-65-2]	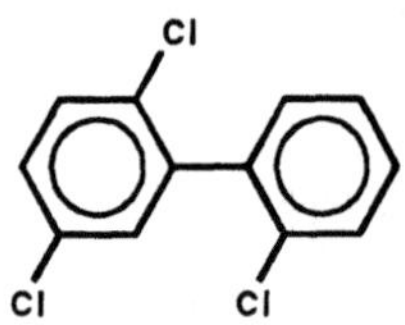	5.64	$C_{12}H_7Cl_3$ 257.54
10.2.7	2,3,4-Trichlorobiphenyl (PCB # 21) [55702-46-0]	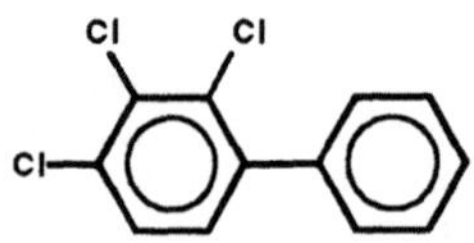	5.86[d]	$C_{12}H_7Cl_3$ 257.54
10.2.8	2,4,4′-Trichlorobiphenyl (PCB # 28) [7012-37-5]	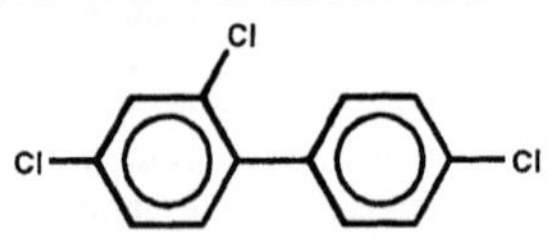	5.67	$C_{12}H_7Cl_3$ 257.54
10.2.9	2,4,5-Trichlorobiphenyl (PCB # 29) [15826-07-4]	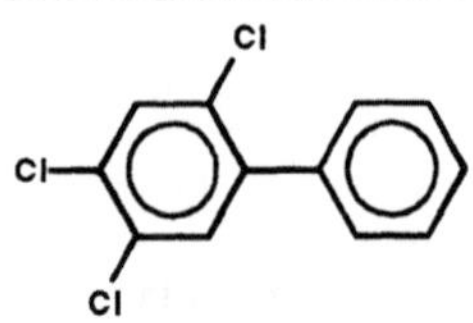	5.90[d]	$C_{12}H_7Cl_3$ 257.54
10.2.10	2,4′,5-Trichlorobiphenyl (PCB # 31) [15862-07-4]	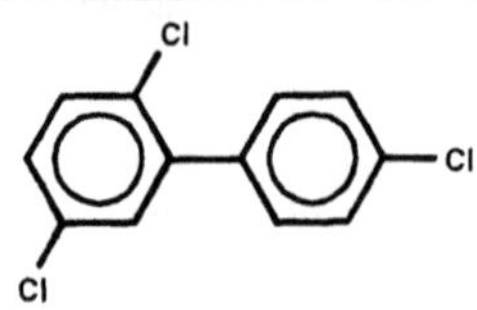	6.00	$C_{12}H_7Cl_3$ 257.54

Half-life or persistence in soil (*S*), sediment (SE) or environment (*E*) ($t_{1/2}$ in years)	Endocrine – disrupting effects[a] and effects on enzymes (TOEI)[j]		Bioconcentration Factor (BCF)[b]		
	in vitro	*in vivo*	Organisms	BCF_W	BCF_L
		estrogenic in rats	algae (*Chlorella*)	6,760	33,800 (dry wt.)
			Daphnia	3,720	372,000
		estrogenic (predicted)	fish		
			rainbow trout (flow-through 96d)	10,000	122,000
			zebrafish (kinetic approach)	7,710	264,000
		estrogenic in rats	fish (5% lipid)	11,500[c]	230,000[c]
not persistent	metabolite estrogenic (binding to the estrogen receptor < 0.004%)	estrogenic (uterotropic) in rodents	fish		
			rainbow trout (flow-through 96d)	17,000	210,000
			goldfish	19,800	400,000
			zebrafish (kinetic approach)	12,900	441,000
	estrogenic	estrogenic (predicted)			
persistent 0.8–2.5 (SE)		estrogenic (predicted)	fish (5% lipid)	20,800[c]	417,000[c]
		TOEI: PB-type	mussel (*Mytilus edulis*)	5,500	458,000
	estrogenic (predicted)	estrogenic (predicted)	guppies (f) (kinetic)	18,000	340,000
	estrogenic	estrogenic	algae (*Chlorella*)	8,950	(dry wt.) 44,800
			Daphnia	17,100	171,000
			mussel (*Mytilus edulis*)	12,000[k]	1,100,000[k]
			goldfish	42,200	848,000
			zebrafish (kinetic approach)	45,600	1,560,000

Table 8 (continued)

No	Chemical or chemical class (abbreviation) [CAS No.]	Chemical structure	log K_{OW}	Molecular formula and molecular mass [g mol^{-1}]
10.2.11	2′,3,4-Trichlorobiphenyl (PCB # 33) [38444-86-9]	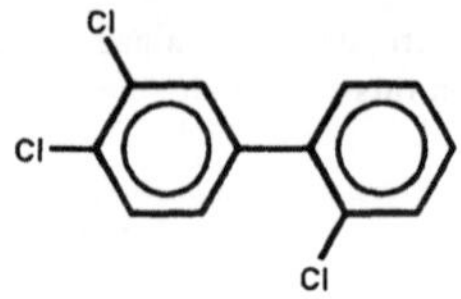	5.87[d]	$C_{12}H_7Cl_3$ 257.54
10.2.12	2,2′,3,5′-Tetrachlorobiphenyl (PCB # 44) [41464-39-5]	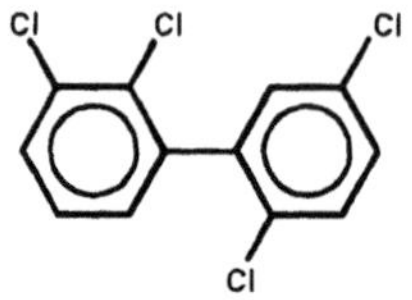	6.35	$C_{12}H_6Cl_4$ 291.99
10.2.13	2,2′,4,4′-Tetrachlorobiphenyl (PCB # 47) [2437-79-8]	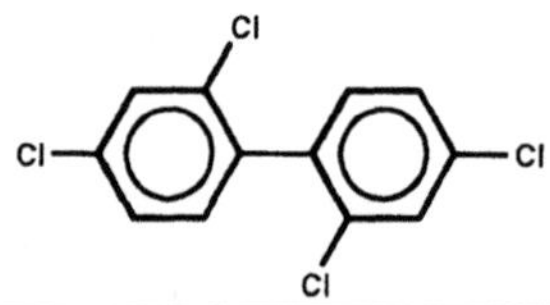	5.94	$C_{12}H_6Cl_4$ 291.99
10.2.14	2,2′,4,5-Tetrachlorobiphenyl (PCB # 48) [70362-47-9]	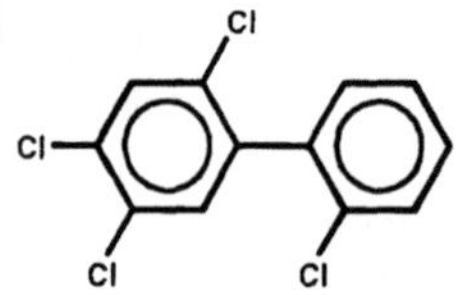	5.71	$C_{12}H_6Cl_4$ 291.99
10.2.15	2,2′,4,5′-Tetrachlorobiphenyl (PCB # 49) [41464-40-8]	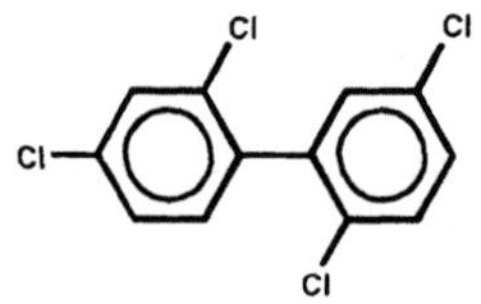	6.36[d]	$C_{12}H_6Cl_4$ 291.99
10.2.16	2,2′,5,5′-Tetrachlorobiphenyl (PCB # 52) [35693-99-3]	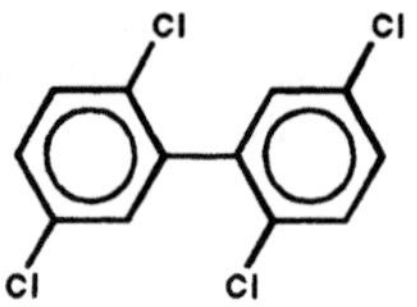	6.36	$C_{12}H_6Cl_4$ 291.99
10.2.17	2,3′,4′,5-Tetrachlorobiphenyl (PCB # 70) [32598-11-1]	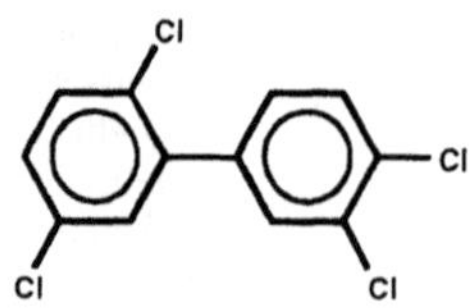	6.62	$C_{12}H_6Cl_4$ 291.99

Half-life or persistence in soil (*S*), sediment (SE) or environment (*E*) ($t_{1/2}$ in years)	Endocrine – disrupting effects[a] and effects on enzymes (TOEI)[j]		Bioconcentration Factor (BCF)[b]		
	in vitro	*in vivo*	Organisms	BCF_W	BCF_L
		estrogenic (predicted)	oyster	6,200	1,100,000
			fish (5% lipid)	37,000[c]	740,000[c]
	estrogenic	estrogenic	oyster (*Crassostrea virginica*)	11,000	1,960,000
			zebrafish (kinetic approach)	69,400	2,377,000
persistent	estrogenic (ER-binding 0.003%)	estrogenic (uterotropic) in rodents; TOEI: PB-type	rainbow trout (8–10 g) (a) muscle (3% lipid)	9,950	332,000
			(b) whole fish (10–15 g) (kinetic)	28,700	360,000
	estrogenic in MCF-7 cells		fish (5% lipid)	26,000[c]	510,000[c]
	estrogenic	estrogenic	zebrafish (kinetic approach)	69,870	2,393,000
1.4–10 (SE)	estrogenic (ER-binding < 0.001%)	estrogenic (uterotropic) in rodents	*Daphnia* (21 day renewal)	4,000	400,000
			oyster	7,400	1,320,000
			mussel (*Mytilus edulis*)	19,000[k]	1,710,000[k]
			mussel	26,300	2,190,000
			guppy	43,000	860,000
			goldfish	49,300	986,000
			zebrafish (kinetic approach)	83,500	2,860,000
	estrogenic	estrogenic TOEI: PB-type	zebrafish (kinetic approach)	119,300	4,085,000

Table 8 (continued)

No	Chemical or chemical class (abbreviation) [CAS No.]	Chemical structure	log K_{OW}	Molecular formula and molecular mass [g mol^{-1}]
10.2.18	2,2′,4,5,5′-Pentachlorobiphenyl (PCB # 101) [37680-73-2]	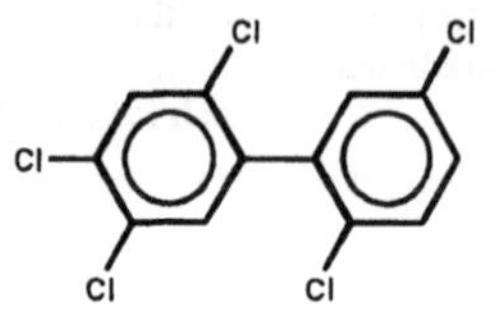	6.86	$C_{12}H_5Cl_5$ 326.43
10.2.19	2,3,3′,4′,6-Pentachlorobiphenyl (PCB # 110) [38380-03-9]	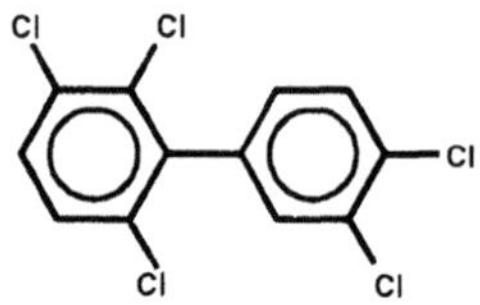	6.48	$C_{12}H_5Cl_5$ 326.43
10.2.20	2,2′,3,3′,6,6′-Hexachlorobiphenyl (PCB # 136) [38411-22-2]	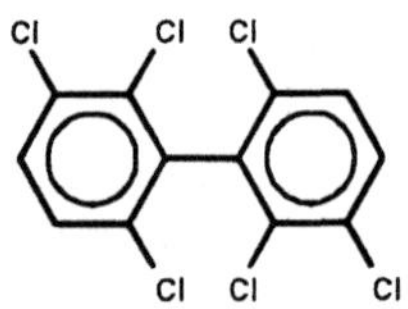	7.12[d]	$C_{12}H_4Cl_6$ 360.88
10.3	**Group II A of PCBs** Non-ortho- and di-para-substituted coplanar (dioxin-like anti-estrogenic) PCBs (Ballschmiter/IUPAC No.) [CAS No.]	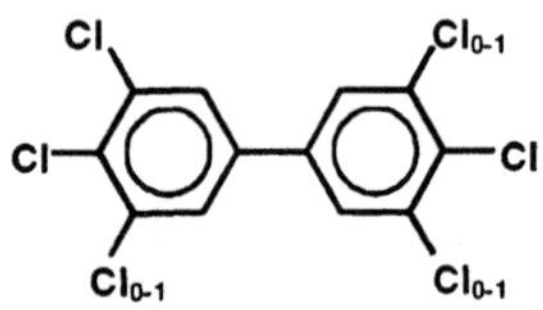	5.83–7.41	$C_{12}H_{4-7}Cl_{3-6}$
10.3.1	3,4,4′-Trichlorobiphenyl (PCB # 37) [38444-90-5]	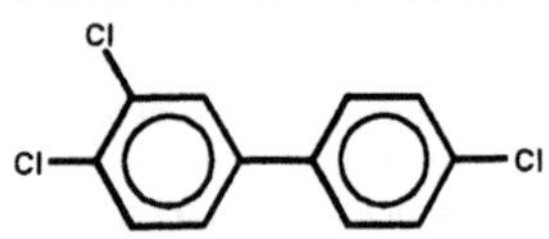	5.90	$C_{12}H_7Cl_3$ 257.54
10.3.2	3,4,4′,5-Tetrachlorobiphenyl (PCB # 81) [70362-50-4]	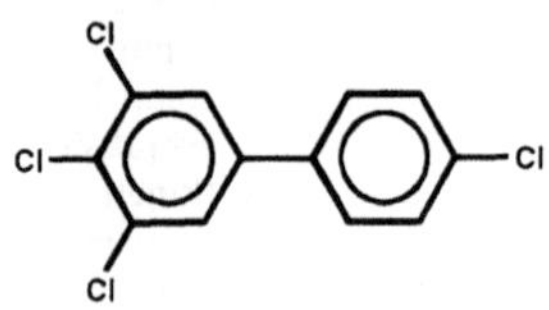	6.40	$C_{12}H_6Cl_4$ 291.99

Half-life or persistence in soil (*S*), sediment (SE) or environment (*E*) ($t_{1/2}$ in years)	Endocrine – disrupting effects[a] and effects on enzymes (TOEI)[j]		Bioconcentration Factor (BCF)[b]		
	in vitro	*in vivo*	Organisms	BCF_W	BCF_L
	estrogenic (ER–binding < 0.001%)	estrogenic (predicted) TOEI: weak PB-type	*Daphnia* (21 day renewal) zebrafish (kinetic approach) mussel (*Mytilus edulis*) (from the environment)	11,400 295,000 126,000[c]	1,140,000 10,000,000 10,500,000
	estrogenic (ER–binding < 0.002%)	estrogenic (uterotropic) in rodents and a modest depleter of thyroxine (T4) TOEI: strong PB-type	fish (5% lipid)	151,000[c]	3,000,000[c]
	estrogenic in the E-SCREEN assay		zebrafish (kinetic approach)	267,000	9,150,000
1–6 (*S*)	antiestrogenic in MCF-7 cells	antiestrogenic; metabolites may be estrogenic; decreased thyroid hormones TOEI: MC-type	aquatic and terrestrial organisms		very high bioconcentration/ bioaccumulation potential
	antiestrogenic (predicted)		fish (5% lipid)	38,000[c]	790,000[c]
	antiestrogenic (predicted)		fish (5% lipid)	126,000[c]	2,500,000[c]

Table 8 (continued)

No	Chemical or chemical class (abbreviation) [CAS No.]	Chemical structure	log K_{OW}	Molecular formula and molecular mass [g mol^{-1}]
10.3.3	3,3′,4,4′-Tetrachlorobiphenyl (PCB # 77) [32598-13-3]	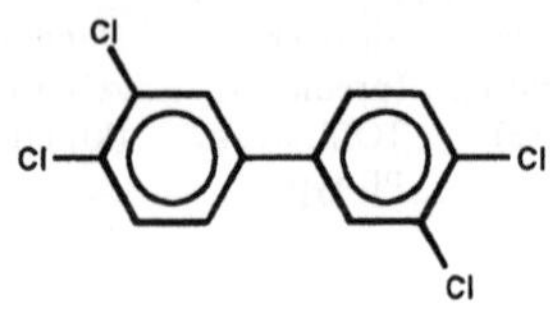	6.63[d]	$C_{12}H_6Cl_4$ 291.99
10.3.4	3,3′,4,4′,5-Pentachlorobiphenyl (PCB # 126) [57465-28-8]	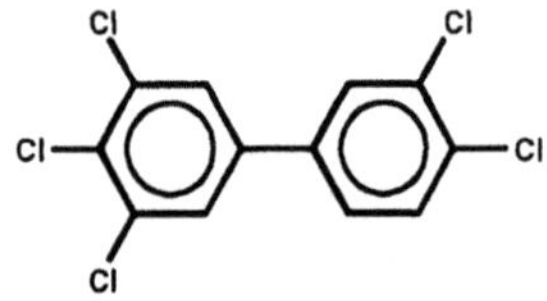	7.20	$C_{12}H_5Cl_5$ 326.43
10.3.5	3,3′,4,4′,5,5′-Hexachlorobiphenyl (PCB # 169) [32774-16-6]	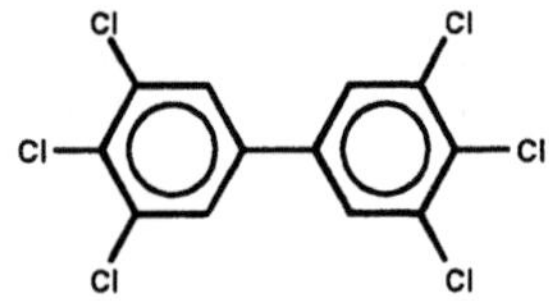	7.41[d] 7.68	$C_{12}H_4Cl_6$ 360.88
10.4	**Group II B of PCBs** Mono-*ortho* and di-*para*-substituted coplanar dioxin-like PCBs	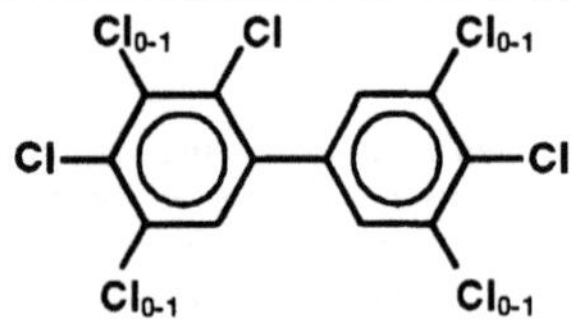	6.65–7.71	$C_{12}H_{3-5}Cl_{5-7}$
10.4.1	2,3,3′,4,4′-Pentachlorobiphenyl (PCB # 105) [32598-14-4]	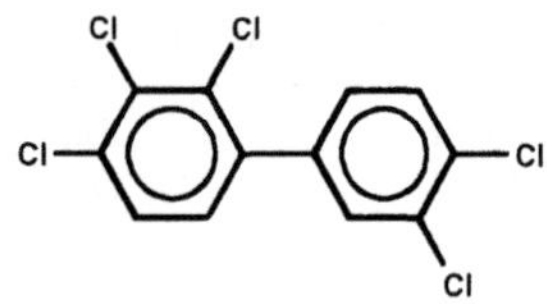	6.65	$C_{12}H_5Cl_5$ 326.43
10.4.2	2,3,4,4′,5-Pentachlorobiphenyl (PCB # 114) [74472-37-0]	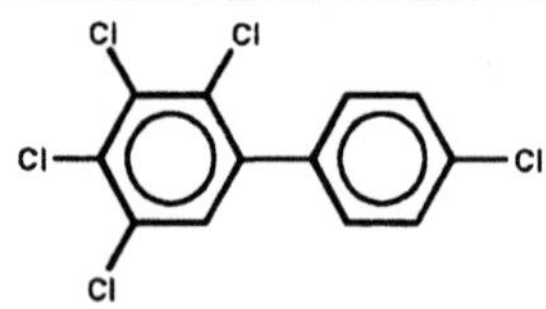	6.65	$C_{12}H_5Cl_5$ 326.43
10.4.3	2,3,3′,4,4′,5-Hexachlorobiphenyl (PCB # 156) [38380-08-4]	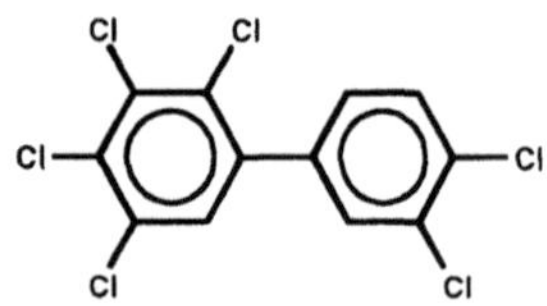	7.18	$C_{12}H_4Cl_6$ 360.88

Half-life or persistence in soil (*S*), sediment (SE) or environment (*E*) ($t_{1/2}$ in years)	Endocrine – disrupting effects[a] and effects on enzymes (TOEI)[j]		Bioconcentration Factor (BCF)[b]		
	in vitro	*in vivo*	Organisms	BCF_W	BCF_L
not very persistent	porphyrogenic; antiestrogenic and estrogenic	antiestrogenic and estrogenic (the major metabolite 3,3′,4′,5-tetrachloro-4-biphenylol is estrogenic)	zebrafish (kinetic approach)	230,400	7,890,000
			mussel (*Mytilus edulis*) (kinetic approach)	26,000[k]	2,360,000[k]
			human (fat)	420	590
	antiestrogenic, (ER-binding < 0.001%)	antiestrogenic and uterotropic in rodents	zebrafish (kinetic approach)	652,000	22,340,000
	antiestrogenic; porphyrogenic in hepatocytes	antiestrogenic in rodents; decrease of serum testosterone in male rats	zebrafish (kinetic approach)	940,000	32,200,000
	antiestrogenic TOEI: strong mixed-type inducers	antiestrogenic; decrease of thyroxine TOEI: strong mixed-type inducers	aquatic and terrestrial organisms		very high bioconcentration/ bioaccumulation potential
7 (SE)	antiestrogenic		fish (5% lipid)	220,000[c]	4,460,000[c]
	antiestrogenic		fish (5% lipid)	220,000[c]	4,460,000[c]
	antiestrogenic		fish (5% lipid)	760,000[c]	15,000,000[c]

Table 8 (continued)

No	Chemical or chemical class (abbreviation) [CAS No.]	Chemical structure	log K_{OW}	Molecular formula and molecular mass [g mol^{-1}]
10.4.4	2,3′,4,4′,5,5′-Hexachlorobiphenyl (PCB # 167) [52663-72-6]	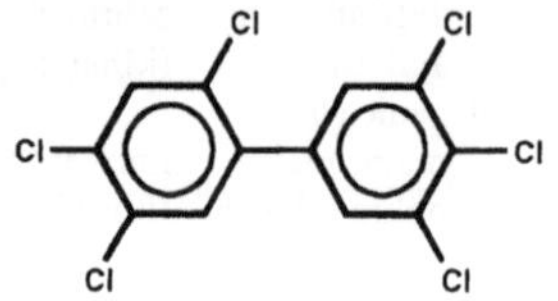	7.27	$C_{12}H_4Cl_6$ 360.88
10.4.5	2,3,3′,4,4′,5,5′-Heptachlorobiphenyl (PCB # 189) [39635-31-9]	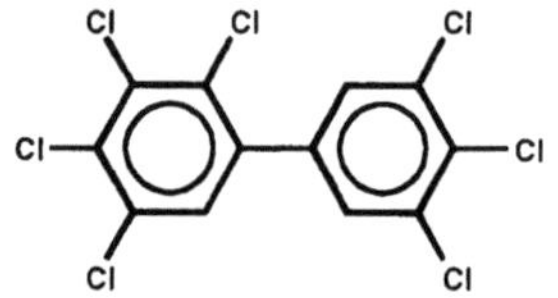	7.71	$C_{12}H_3Cl_7$ 395.32
10.5	**Group III of PCBs** Highly chlorinated (>5 Cl) non- coplanar mono- or di-*para*-substituted biologically persistent PCB congeners	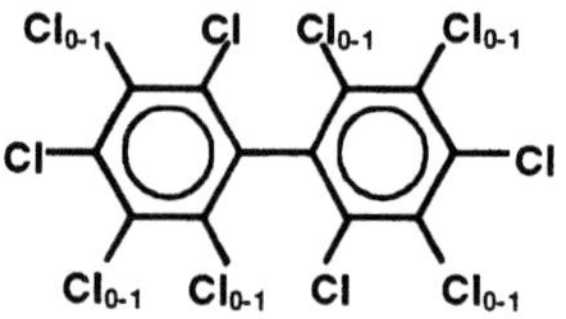	6.8–8.3	$C_{12}H_{0-4}Cl_{6-10}$
10.5.1	2,2′,3,3′,4,4′-Hexachlorobiphenyl (PCB # 128) [38380-07-3]	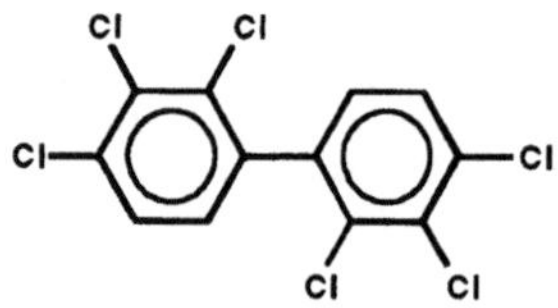	7.32[d]	$C_{12}H_4Cl_6$ 360.88
10.5.2	2,2′,3,4,4′,5-Hexachlorobiphenyl (PCB # 138) [35065-28-2]	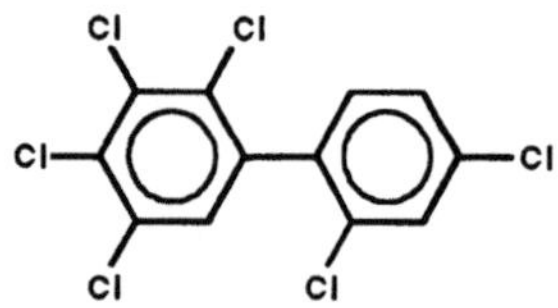	7.44	$C_{12}H_4Cl_6$ 360.88
10.5.3	2,2′,4,4′,5,5′-Hexachlorobiphenyl (PCB # 153) [35065-27-1]	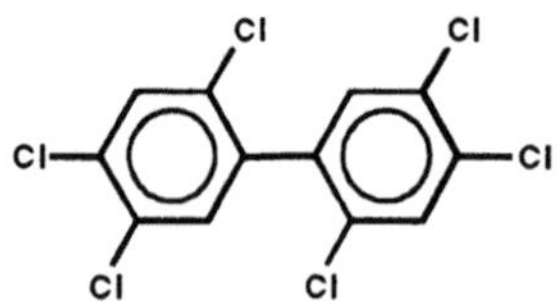	7.23	$C_{12}H_4Cl_6$ 360.88

Half-life or persistence in soil (*S*), sediment (SE) or environment (*E*) ($t_{1/2}$ in years)	Endocrine – disrupting effects[a] and effects on enzymes (TOEI)[j]		Bioconcentration Factor (BCF)[b]		
	in vitro	*in vivo*	Organisms	BCF_W	BCF_L
			fish (5% lipid)	930,000[c]	18,600,000[c]
			fish (5% lipid)	2,600,000[c]	51,000,000[c]
biologically very or extremely persistent	TOEI: strong PB-type inducers; may be weak MC-type inducer	decrease of thyroxine; TOEI: strong PB-type inducer; may be weak MC-type	aquatic and terrestrial organisms		extremely high bioconcentration/ bioaccumulation potential
			zebrafish (kinetic approach) mussel (*Mytilus edulis*) (data from the environment)	589,600 263,000	20,200,000 29,900,000
19–25 (SE)			zebrafish (kinetic approach) mussel (*Mytilus edulis*) (data from the environment)	764,400 282,000	26,180,000 23,500,000
very persistent; 19–25 (SE)	ER-binding 0.004%	estrogenic (uterotropic) in rodents	oyster mussel (data from the environment) guppy (kinetic approach) zebrafish (kinetic approach)	48,000 302,000 450,000 448,000	8,600,000 25,200,000 9,800,000 15,350,000

Table 8 (continued)

No	Chemical or chemical class (abbreviation) [CAS No.]	Chemical structure	log K_{OW}	Molecular formula and molecular mass [g mol^{-1}]
10.5.4	2,2′,4,4′,6,6′-Hexachloro-biphenyl (PCB # 155) [33979-03-2]	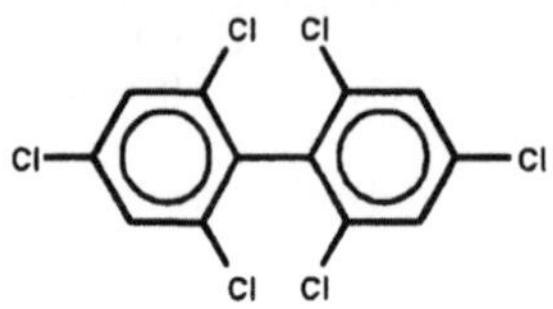	7.29 [d]	$C_{12}H_4Cl_6$ 360.88
10.5.5	2,2′,3,4,4′,5,5′-Heptachloro-biphenyl (PCB # 180) [35065-29-3]	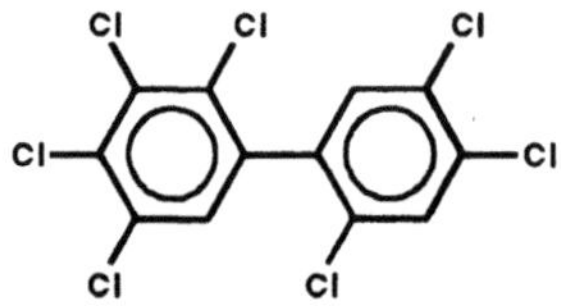	7.36	$C_{12}H_3Cl_7$ 395.32
10.5.6	2,2′,3,4,4′,5′,6-Heptachloro-biphenyl (PCB # 183) [52663-69-1]	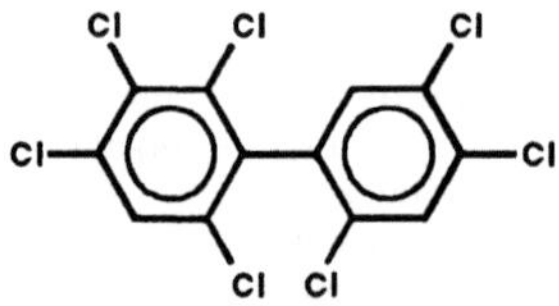	7.47	$C_{12}H_3Cl_7$ 395.32
10.5.7	2,2′,3,4,5,5′,6′-Heptachloro-biphenyl (PCB # 185) [52712-05-7]	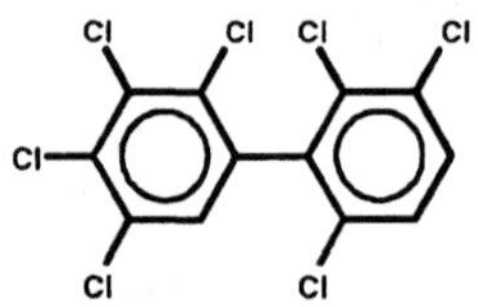	7.43	$C_{12}H_3Cl_7$ 395.32
10.5.8	2,2′,3,3′,4,4′,5,5′-Octachlorobiphenyl (PCB # 194) [35694-08-7]	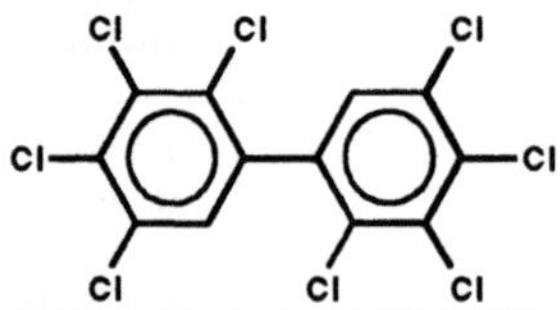	7.62	$C_{12}H_2Cl_8$ 429.77
10.5.9	2,2′,3,3′,5,5′,6,6′-Octachlorobiphenyl (PCB # 202) [2136-99-4]	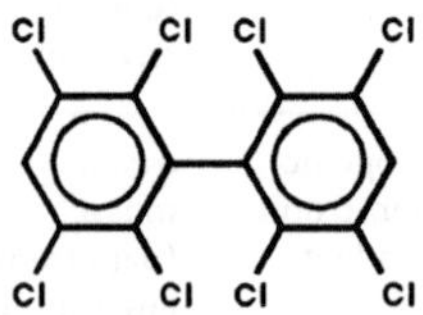	7.73[d]	$C_{12}H_2Cl_8$ 429.77
10.5.10	2,2′,3,3′,4,4′,5,5′,6,6′-Decachlorobiphenyl (PCB # 209) [2051-24-3]	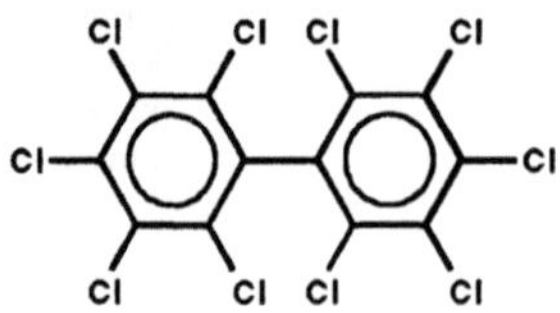	8.27[d]	$C_{12}Cl_{10}$ 498.66

Half-life or persistence in soil (*S*), sediment (SE) or environment (*E*) ($t_{1/2}$ in years)	Endocrine – disrupting effects[a] and effects on enzymes (TOEI)[j]		Bioconcentration Factor (BCF)[b]		
	in vitro	*in vivo*	Organisms	BCF_W	BCF_L
very persistent	antiestrogenic, ER-binding 0.28%	antiestrogenic, in rats	fish (5% lipid)	975,000[c]	19,500,000[c]
very persistent; 25 (SE)			*Daphnia* magna (21 days of static renewal exposure)	28,000	2,800,000
			fish (5% lipid)	1,150,000[c]	22,900,000[c]
very persistent			zebrafish (kinetic approach)	685,000	23,460,000
			zebrafish (kinetic approach)	858,000	29,400,000
very persistent			zebrafish (kinetic approach)	652,000	22,330,000
			zebrafish (kinetic approach)	658,400	22,600,000
extremely persistent			zebrafish (kinetic approach)	> 276,000[g]	> 9,440,000[g]
			guppy	> 340,000	> 9,800,000
			fish (5% lipid) (predicted)	9,000,000[c]	180,000,000[c]
			human (fat) (steady-state not reached during whole life)	2,100	9,400

Table 8 (continued)

No		Chemical or chemical class (abbreviation) [CAS No.]	Chemical structure	log K_{OW}	Molecular formula and molecular mass [g mol^{-1}]
11		Polychlorinated Dibenzo-*p*-dioxins (PCDDs)	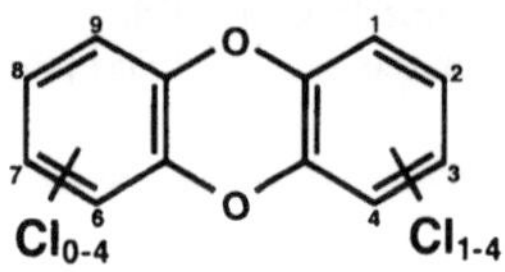	5.10–8.60	$C_{12}H_{8-n}Cl_nO_2$ n = 1–8
	11.1	2,3,7,8-Tetrachloro-dibenzo-*p*-dioxin (2,3,7,8-TCDD) [1746–01–6]	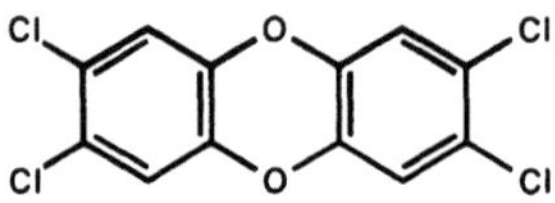	6.64	$C_{12}H_4Cl_4O_2$ 321.97
	11.2	1,2,3,4,6,7,8,9-Octachloro-dibenzo-*p*-dioxin (OCDD) [3268–87–9]	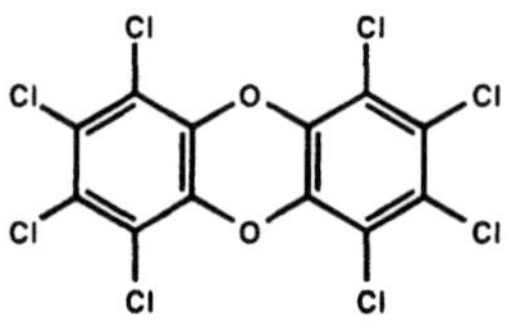	8.60	$C_{12}Cl_8O_2$ 459.75
12		Polychlorinated Dibenzofurans (PCDFs)	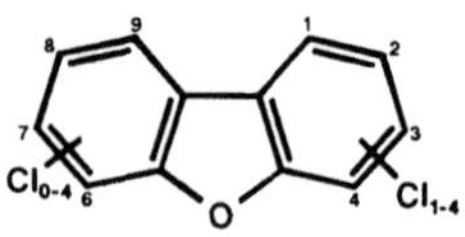	4.90–8.78	$C_{12}H_{8-n}Cl_nO$ n = 1–8
	12.1	2,3,7,8-Tetrachlorodi-benzofuran (2,3,7,8-TCDF) [51207–31–9]	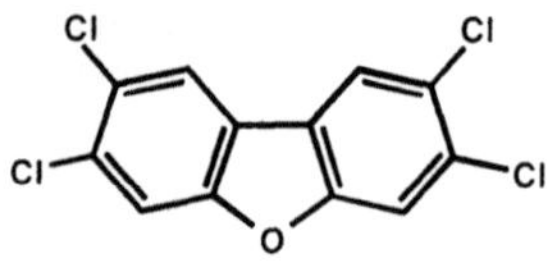	6.53[d]	$C_{12}H_4Cl_4O$ 305.89
	12.2	2,3,4,7,8-Pentachlorodi-benzofuran (2,3,4,7,8-PeCDF) [51207–31–4]	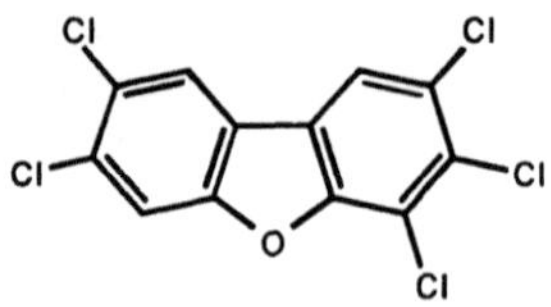	6.92[d]	$C_{12}H_3Cl_5O$ 340.34

Half-life or persistence in soil (*S*), sediment (SE) or environment (*E*) ($t_{1/2}$ in years)	Endocrine – disrupting effects[a] and effects on enzymes (TOEI)[j]		Bioconcentration Factor (BCF)[b]		
	in vitro	*in vivo*	Organisms	BCF_W	BCF_L
the 2,3,7,8-chlorinated PCDD congeners are highly persistent	antiestrogenic. TOEI: MC-type	antiestrogenic in rodents; decrease of thyroxine; TOEI: MC-type	aquatic and terrestrial organisms		very high bioconcentration/ bioaccumulation potential, especially the 2,3,7,8-chlorinated PCDDs
10 (*S*) 9.9–98 (SE)	antiestrogenic. TOEI: strong MC-type	antiestrogenic in rodents; decrease of plasma testosterone in male rats and of serum thyroxine. Porphyrogenic in mice and rats. TOEI: strong MC-type	adult human (fat) range fish medeka (10% lipid)	390 104–670 510,000	430 115–740 5,100,000
0.02–143 (SE) > 10 (*S*)		TOEI: weak MC-type	human (fat) fish (5% lipid)	2,930 14,000,000[c]	4,100 after 80 years; no steady-state reached 280,000,000[c]
the 2,3,7,8-chlorinated PCDF congeners are very persistent	antiestrogenic. TOEI: strong MC-type	antiestrogenic in rodents; decrease of thyroxine. TOEI: strong MC-type	aquatic and terrestrial organisms		very high bioconcentration/ bioaccumulation potential, especially the 2,3,7,8-chlorinated PCDFs
very persistent 61 (SE)	antiestrogenic. TOEI: strong MC type	TOEI: strong MC type	fish (5% lipid)	170,000[c]	3,400,000[c]
very persistent 60 (SE)	antiestrogenic. TOEI: strong MC type	antiestrogenic in rodents	fish (10% lipid) (5% lipid)	 830,000[c] 415,000[c]	 8,300,000[c]

Table 8 (continued)

No	Chemical or chemical class (abbreviation) [CAS No.]	Chemical structure	log K_{OW}	Molecular formula and molecular mass [g mol^{-1}]
12.3	1,2,3,4,6,7,8-Heptachlorodibenzofuran (1,2,3,4,6,7,8-HepCDF) [67462-39-4]	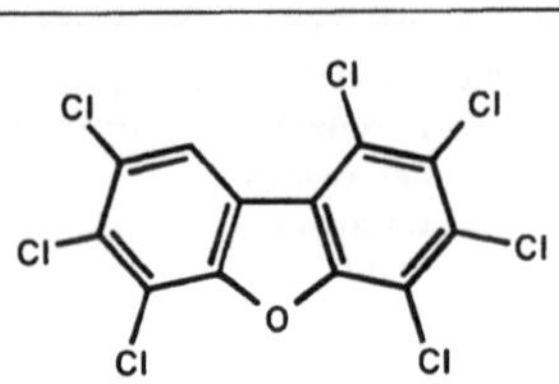	7.92[d]	$C_{12}HCl_7O$ 409.30
12.4	1,2,3,4,6,7,8,9-Octachlorodibenzofuran (OCDF) [39001-02-0]	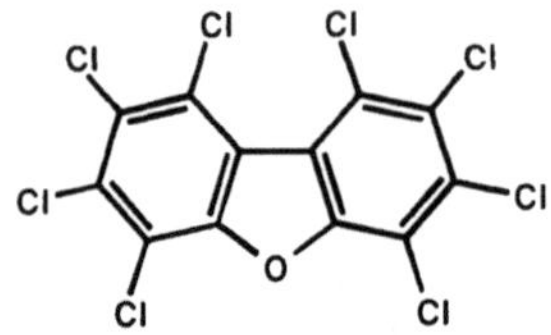	8.78	$C_{12}Cl_8O$ 443.76

Source: Selected data from Ref. [152–156] unless otherwise noted.

[a] Ref. [125–129, 136, 157, 158a, 158b, 158c].

[b] Bioconcentration factors (BCFs) (see Ref. [153–157]).
BCF values in algae (*Chlorella fusca*) taken from Ref. [74].
BCF values in *Daphnia* taken from Ref. [75].
BCF values in mussels (*Mytilus edulis*) taken from Ref. [76, 161, 162, 401].
BCF values in oysters (*Crassostrea virginica*) taken from Ref. [163] and the original literature.
BAF values in human fat taken from Ref. [150, 151].

$$\text{Bioaccumulation factor} = \frac{\text{concentration in human (fat)} \ [\text{ng} \times \text{kg}^{-1}]}{\text{concentration in total diet} \ [\text{ng} \times \text{kg}^{-1}]}$$

[c] BCF values in fish predicted from the log K_{OW} value.

[d] "Slow-stirring" method.

[e] Calculated K_{OW} value.

concern involves 12 chemicals and chemical classes, the so-called "dirty dozen": Aldrin, dieldrin, endrin, chlordane, heptachlor, DDT, hexachlorobenzene (HCB), Mirex, Toxaphene, polychlorinated biphenyls (PCBs), polychlorinated dibenzo-*p*-dioxins (PCDDs), and polychlorinated dibenzofurans (PCDFs).

Although the use of the persistent pesticides is restricted or banned in many developed countries, they are still manufactured for export. However, they remain in wide and relatively unregulated use in developing countries. Among developed countries, there exist largely consensus for restrictions on production and use of these persistent organic pollutants.

In Table 8, the CAS numbers, the chemical structures, n-octanol/water partition coefficients, molecular formula, molecular mass, half-lives or persistence, endocrine effects, and selected bioconcentration factors (BCFs) of these POPs are compiled. It is important to note that the half-life ($t_{1/2}$) or persistence of a chemical in soil, sediment and/or sludge depends not only on the properties of the chemical, but also on the surrounding environment. Main factors which af-

Half-life or persistence in soil (*S*), sediment (SE) or environment (*E*) ($t_{1/2}$ in years)	Endocrine – disrupting effects [a] and effects on enzymes (TOEI) [j]		Bioconcentration Factor (BCF) [b]		
	in vitro	*in vivo*	Organisms	BCF_W	BCF_L
very persistent 36 (SE)			fish (10% lipid) (5% lipid)	 8,300,000 [c] 4,150,000 [c]	 83,000,000 [c]
very persistent 29 (SE)			fish (10% lipid) (5% lipid)	 60,000,000 [c] 30,000,000 [c]	 600,000,000 [c]

[f] Sum of *cis*-chlordane, *trans*-chlordane, *cis*-nonachlor, *trans*-nonachlor, including 7 persistent compounds of technical chlordane, and the metabolites heptachlor epoxide and oxychlordane.
[g] This chemical was tested by the kinetic approach. However, the BCF value is underestimated because the concentration in water was above its water solubility.
[h] Σ DDT + DDE + DDD.
[i] Anaerobic degradation in sewage sludge.
[j] TOEI: type of enzyme induction; (1) PB-type: Phenobarbital inducer; induction of cytochrome P450 1A (CYP1A). (2) MC-type: 3-Methylcholanthrene type inducer; induction of cytochrome P450 2B (CYP2B). (3) Mixed-type: some of both PB- and MC-type inducers; mixed CYP1A/CYP2B induction.
[k] Kinetic approach Ref. [401] and personal communication from M. Gilek to H. Geyer.

fect the half-life or persistence of an organic compound are the temperature, sunlight intensity, nature of the microbial community, and oxygen content of the environment. It is known that the biodegradation of some organic chemicals in soil, sediment and/or sludge under anaerobic conditions is much faster than under aerobic conditions [402]. Therefore, it is misleading to document a single reliable half-life of a chemical. Mackay et al. [153–155] recommend to suggest a semi-quantitative classification of half-lives into groups, assuming average environmental conditions to apply. Obviously, a different class will generally apply in air, water, and sediment. The BCF data of algae (*Chlorella* sp.), water fleas (*Daphnia* sp.), mussels (*Mytilus edulis*), oysters (*Crassostrea virginica*), and different fish species are from controlled laboratory experiments. In some cases, BCF data from outdoor (marine environment) investigations are presented. It is obvious that the POPs have a high or very high bioconcentration potential in these aquatic organisms. It is also known that these POPs are bioaccumulated in the human body, especially in adipose fat. For comparison and risk assessment the **bioaccumulation factors (BAFs) in humans** of some POPs

Table 9. Physico-chemical properties, bioconcentration factors on a wet weight basis (BCF_W) and on a lipid basis (BCF_L), and estrogenic or anti-estrogenic effects of **2,3,7,8-Tetrachlorodibenzo-*p*-dioxin** (TCDD) and **Octachlorodibenzo-*p*-dioxin** (OCDD)

Chemical	2,3,7,8-Tetrachlorodibenzo-*p*-dioxin (TCDD)	Octachlorodibenzo-*p*-dioxin (OCDD)
Chemical structure	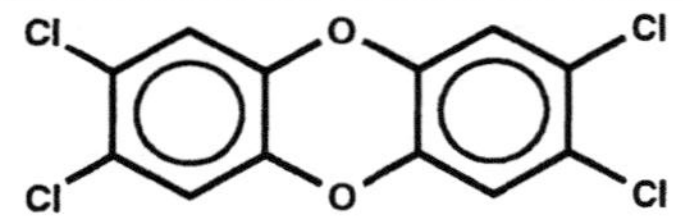	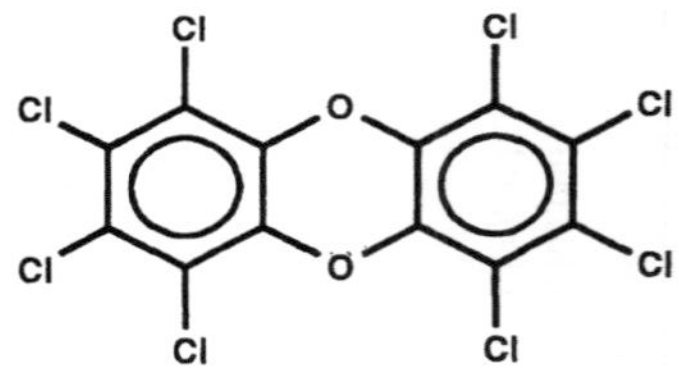
CAS No.	1746–01–6	3268–87–9
Chemical name	2,3,7,8-Tetrachlorodibenzo-*p*-dioxin	Octachlorodibenzo-*p*-dioxin
Trade name	–	–
Molecular formula	$C_{12}H_4Cl_4O_2$	$C_{12}Cl_8O_2$
Molecular mass [g mol^{-1}]	321.97	459.75
Melting point [MP: °C]	305.0	332.0
Water solubility [ng l^{-1}]	7.9 ± 2.7 (20–22 °C) 8.9 ± 1.9 (25 °C) 12.5 – 13.3 (22 °C) } mean: 9.9 (20–25 °C)	0.074 (25 °C)
Sorption coefficient on organic carbon (log K_{OC})	6.28[a]	7.90[g] 8.24[a]
n-Octanol/water partition coefficient (log K_{OW})	6.64	8.60
Bioaccumulation factor in human fat $\left(BAF_L = \frac{C_{fat}}{C_{diet}}\right)$	115–740 mean: 430	no steady-state reached during the whole life 4100 (after 80 years) 83,000–165,000[e]

Bioconcentration factor (BCF) in aquatic organism: on a wet weight basis (BCF_W)	510,000 medaka (10% lipid)[m]	4,300,000 (5% lipid) 9,000,000 (10% lipid) (extrapolated) 14,000,000 (5% lipid) (predicted from log K_{OW})
BCF in fish on a lipid basis (BCF_L)	5,100,000[f] medaka	85,000,000 250,000,000[b] mean : 280,000,000[b] range: 158,000,000 – 398,000,000[b]
Endocrine disrupting effects	anti-estrogenic effects[c,d]	suspected very weak anti-estrogenic effects

Source: Geyer et al. [36], Geyer and Muir [37], Geyer et al. [38], Geyer et al. [191], Geyer et al. [192], Rippen [156], The Merck Index [152], and selected data from Mackay et al [154, 155], as otherwise cited.

[a] Estimated log K_{OC} value according to the equation of Karickhoff [194] : $\log K_{OC} = 0.989 \log K_{OW} - 0.346$.

[b] Estimated BCF_L value in fish from the n-octanol/water partition coefficient.

[c] Gallo et al. [197].

[d] Safe et al. [183].

[e] Predicted BAF_L value according to the equation of Geyer et al [191, 192] : $\log BAF_L = 0.745 \log K_{OW} - 1.19$.

[f] Schmieder et al. [193].

[g] Measured log K_{OC} value of Broman et al. [198].

are also presented in Table 8. These BAF data were calculated by dividing the concentration of the chemicals in human fat (µg/kg) by the concentration in total diet (µg/kg) [150, 151]. It is important to note that these BAF data are comparable if the concentration in human fat is divided by the chemical amount (µg) which is taken up per day by an adult human because an average adult human in industrial countries eats between 0.6 and 1 kg food per day.

In the following sections, the physico-chemical properties and the bioconcentration of selected super-hydrophobic persistent organic pollutants (POPs) such as 2,3,7,8-tetrachlorodibenzo-*p*-dioxin (TCDD), octachlorodibenzo-*p*-dioxin (OCDD), Mirex, and Toxaphene in fish and other animals will be discussed.

8.2.1
Bioconcentration of 2,3,7,8-tetrachlorodibenzo-p-dioxin (TCDD)

2,3,7,8-Tetrachlorodibenzo-*p*-dioxin (TCDD), also known as "dioxin" or "Seveso poison", is not produced for commercial purposes and has no reported use other than as a test chemical in research. However, it is formed during the thermolysis of 2,4,5-trichlorophenol and 2,4,5-trichlorophenoxy acetic acid (2,4,5-T) and has been found in fly ash and flue gases [179, 180]. TCDD can also be formed by the combustion of chlorinated organic compounds, and municipal and industrial wastes. This compound has been detected also as an unwanted trace contaminant in 2,4,5-trichlorophenol (2,4,5-TCP) and other products made from 2,4,5-TCP such as 2,4,5-T and related herbicides (silvex) as well as in the germicide hexachlorophene. Another potential source of TCDD and other polychlorinated dibenzodioxins is the occurrence of fires involving electrical transformers containing a mixture of chlorobenzenes and PCBs as insulating fluid. TCDD shows little potential for metabolic alteration to less toxic forms in mammals and has a potential to promote carcinogenicity and genotoxicity, as well as to cause teratogenic effects. Based on acute toxicity studies in several species of animals, TCDD is the most toxic man-made chemical known. The acute toxicity (LD_{50}) ranges from 0.6 µg TCDD kg body wt^{-1} for young male guinea pigs, to 5051 µg TCDD kg body wt^{-1} for Golden Syrian hamsters, and to >8000 µg TCDD kg body wt^{-1} for **adult** Wistar rats (for review see [181, 182]). This marked species difference in TCDD toxicity has been an unresolved problem for more than a decade.

In spite of extensive investigations in recent years, the cause of liver injury and lethality, the mode of action, and the mechanism of toxicity of TCDD are not fully known. It is assumed that most, if not all toxic effects of TCDD are mediated through binding to the aryl hydrocarbon (Ah) receptor [183]. However, the binding affinity of TCDD to this Ah receptor alone can not explain the species differences to the toxicity of TCDD. Recently Geyer et al. [181, 182] found a significant positive relationship between the acute toxicity (30d-LD_{50}) of TCDD in different mammals and their total body fat content (%). That means, the higher the fat content of an organism, the more resistant is this organism to toxic effects of TCDD and other lipophilic persistent chemicals. It is concluded that the storage of TCDD and related chemicals in lipids of aquatic and terrestrial organisms is, in a sense, a detoxification mechanism by which the chemicals are removed from receptors, target organs such as liver and nerves, or other sites of action [181, 182, 184].

TCDD and related compounds are also very toxic to aquatic organisms, especially to newly fertilized eggs, newly hatched, and young fish [185, 186]. This sensitivity to toxicity of TCDD could also be explained by the lower lipid content of the younger fish as compared to the older ones. However, it is predicted that adult eels which become also very fat (up to 30%) should be also very resistant to toxic effects of TCDD although this fish species can bioconcentrate this compound and other lipophilic chemicals to a very high amount. Although no systematic intense TCDD toxicity studies with fish of different age, body weight, lipid content etc. have been carried out, this can be concluded from our investigations of toxicity of lindane (γ-HCH) to different fish species [187, 188]. Geyer et al. [187, 188] found a significant positive linear relationship between the lipid content (%) of 16 fish species and their susceptibility to the acute toxic effects of γ-HCH. These authors found also a significant positive correlation between the bioconcentration factor of lindane and the lipid content of different fish species [40].

Recently, Lassiter and Hallam [189] proposed the "survival of fattest model", which means that organisms with higher body fat/lipid content will survive longer, since they are more resistant to toxic effects of lipophilic chemicals than organisms with lower lipid content. Our results confirm and corroborate this hypothesis of the "fattest model" proposed by Lassiter and Hallam [189].

The physico-chemical properties of TCDD are compiled in Table 9. TCDD has a very low water solubility (between ca. 8 and 19.3 ng l^{-1}) and a very high lipophilicity (n-octanol/water partition coefficient log $K_{OW} = 6.64$). TCDD belongs to the group of so-called super-hydrophobic compounds. Due to these physico-chemical properties and its high stability against biotic and abiotic degradation, TCDD can be bioaccumulated in terrestrial organisms, such as rats, beef cattle, monkeys, and human [190–192]. TCDD is also bioconcentrated in aquatic organisms such as algae, *Daphnia*, mussels, and fish [193].

The bioconcentration factors of TCDD in various fish species were compiled by Schmieder et al. [193]. The BCF values on a wet weight basis range from 9,270 to 510,000 and the BCF values on a lipid basis are between 81,300 and 5,100,000. Although the BCF values differ by some orders of magnitude they show clearly that this persistent super-hydrophobic compound is bioconcentrated in fish to a very high extent. We came to the conclusion that the steady-state bioconcentration factor on a lipid basis (BCF_L) of 5,100,000 for TCDD in fish measured by Schmieder et al. [193] is the best one because these authors used the flow-through system, the kinetic method, and a very long depuration time of 175 days. It was also important for this study that the TCDD concentration in the exposure aquarium (101 ± 26 pg l^{-1}) was lower than the maximal published water solubility. Furthermore, the generator column method without solvent carrier was used and correction for growth dilution was applied. No toxic effects were observed and the BCF_L value is in excellent agreement with the n-octanol/water partition coefficient of TCDD (log $BCF_L = 6.70$, log $K_{OW} = 6.64$).

8.2.2
Bioconcentration of Octachlorodibenzo-p-dioxin (OCDD)

Octachlorodibenzo-*p*-dioxin (OCDD) at this time is not produced for commercial purposes and has no reported use although this chemical and other octahalogenated dibenzo-*p*-dioxins were proposed by a Canadian company as chemical intermediates, biocides, and flame-retardants [199]. However, it is not known, if these compounds had been intentionally produced at any time. Usually, OCDD is the most prevalent polychlorinated dibenzo-*p*-dioxin congener found in pentachlorophenol (PCP), fly ash, sediments, fish, and other biotic samples (for review see [200]). This chlorinated compound is highly persistent and resistant to biotic and abiotic degradation, except for photolysis. OCDD belongs to the group of super-hydrophobic or super-lipophilic compounds with an octanol/water partitioning coefficient (log K_{ow}) of 8.6 and water solubility of 74 pg l^{-1} (see Table 9).

Different research groups have determined the bioconcentration of OCDD in various fish species. Wet weight bioconcentration factors (BCF_W) of OCDD in various fish species were compiled by Geyer et al. [201, 202] from recent papers. Only steady-state BCF data obtained in flow-through systems were considered. For comparison, BCF_W values were transformed in BCF_L values. Table 10 contains body weights, lipid contents, BCF_L values, and corresponding ambient OCDD concentrations. In order to assess the most likely BCF_L (ambient OCDD concentrations < water solubility), experimental BCF_L data of OCDD were plot-

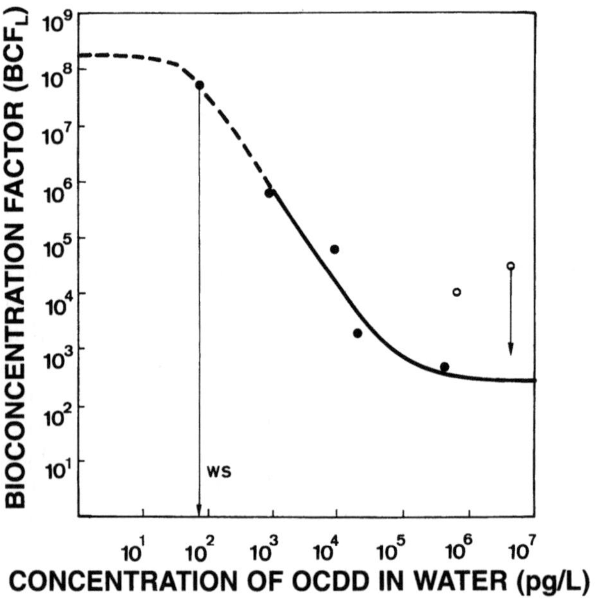

Fig. 11. Relationship between bioconcentration factor on lipid basis (BCF_L) of octachlorodibenzo-*p*-dioxin (OCDD) in fish and the OCDD concentration in ambient water (WS: water solubility of OCDD=74 pg/L). (With modifications from H. Geyer et al. [201, 202])

Table 10. Bioconcentration factors on a wet weight basis (BCF_W) and on a lipid basis (BCF_L) of octachlorodibenzo-*p*-dioxin (OCDD) in different fish species depending on OCDD concentrations in ambient water (C_W) (Kinetic approach, as otherwise cited)

Fish species	Mean body weight (g)	Lipid content (%)	Ambient OCDD conc. C_W (pg l^{-1})	Bioconcentration Factor		References
				BCF_W	BCF_L	
Guppy (male)	0.1	3.5	$4.0 \cdot 10^6$	<1050	$<3 \cdot 10^4$	[203]
Rainbow trout	0.3	6.9	$4.15 \cdot 10^5$	34	$4.9 \cdot 10^2$	[205]
Guppy (female)	0.079	7.5	$6.4 \cdot 10^5$	703	$9.4 \cdot 10^3$	[206]
Rainbow trout	0.3	6.9	$2.0 \cdot 10^4$	136	$2.0 \cdot 10^3$	[205]
Fathead minnow	1.7	3.5	$9.0 \cdot 10^3$	2226	$6.4 \cdot 10^4$	[205]
Guppy (female)	0.91	9.7	$8.0 \cdot 10^2$	1308[a]	$1.4 \cdot 10^{4}$[a]	[207]
Fathead minnow	1.7	3.5	$10 \cdot 10^3$	22,300	$6.4 \cdot 10^5$	[204]
Fish	0.73	5.0[b]	$7.4 \cdot 10$[c]	$4.3 \cdot 10^6$	$8.5 \cdot 10^{7}$[d]	[201]

Source: Taken with modifications from Geyer et al. [201, 202].
[a] Exposure time: 21 days (no steady-state reached).
[b] Assumed average lipid content (% on a wet weight basis) of fish.
[c] Water solubility of OCDD.
[d] Predicted by extrapolation from Eq. (27).

Table 11. Physico-chemical properties, bioconcentration factors on a wet weight basis (BCF_W) and on a lipid basis (BCF_L), and estrogenic or anti-estrogenic effects of **Mirex** and **Chlordecone (Kepone)**

Chemical	Mirex	Chlordecone
Chemical structure	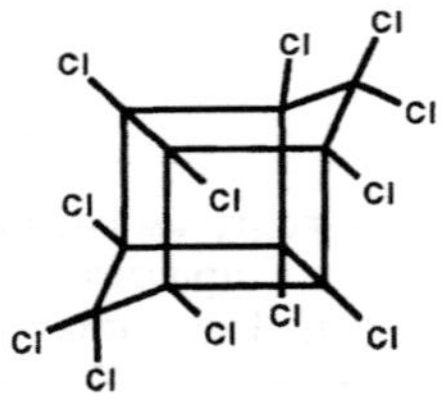	
CAS No.	2385–85–5	143–50–0
Chemical name	1,1a,2,2,3,3a,4,5,5,5a,5b,6-Dodecachlorooctahydro-3,4-metheno-1*H*-cyclobuta[*cd*] pentalene	1,1a,3,3a,4,5,5,5a,5b,6-decachlorooctahydro-1,3,4-metheno-2*H*-cyclobuta[*cd*] pentalen-2-one
Trade name	Dechlorane, Ferriamicide	Kepone
Molecular formula	$C_{10}Cl_{12}$	$C_{10}Cl_{10}O$
Molecular mass [g mol^{-1}]	545.54	490.64
Melting point [MP: °C]	485 (decomposition)	350 (decomposition)
Water solubility [ng l^{-1}]	1.3[a]	30 (pH: 7.0) 35 (pH: 8.0)
Sorption coefficient on organic carbon (log K_{OC})	7.38[j]	5.14[e]
n-Octanol/water partition coefficient (log K_{OW})	7.50	5.50
Bioaccumulation factor in human fat $\left(BAF_L = \frac{C_{fat}}{C_{diet}}\right)$	no steady-state reached during the whole life 9,000–25,000[h]	808[h]

Bioconcentration factor (BCF) in aquatic organism: on a wet weight basis (BCF_W)	940,000 guppy (6.5% lipid)[c] 15,000,000 rainbow trout (8% lipid)[b]	12,500 fathead minnows (9.5% lipid) 16,590 fathead minnows (9.5% lipid) 12,370 fathead minnows (9.5% lipid) 10,440 bluegills (3% lipid)
BCF in fish on a lipid basis (BCF_L)	14,500,000[c] 32,000,000[f] 188,000,000[d]	131,580 fathead minnows 174,580 fathead minnows 130,190 fathead minnows 348,100 bluegills
Endocrine disrupting effects	no estrogenic effects in rats and in the E-SCREEN assay[g,i]	estrogenic effects in rats and mice and in the E-SCREEN assay[g,i]

Source: Geyer et al. [36], Geyer and Muir [37], Geyer et al. [38], Geyer et al. [191], Geyer et al. [192], Rippen [156], The Merck Index [152], and selected data from Mackay et al [154, 155], as otherwise cited.

[a] Calculated from the equation of Yalkowsky and Banerjee [196]: $\log WS\ (mol\ l^{-1}) = 0.323 - 0.944 \log K_{OW} - 0.01\ MP\ (°C)$.

[b] Field BCF_W in rainbow trout from Lake Ontario calculated by Oliver and Niimi [14].

[c] Bioavailability-corrected bioconcentration factor experimentally determined in the guppy by Gobas et al. [91].

[d] Field BCF_L calculated from the lipid content (8%) and concentration data of Mirex in rainbow trout measured by Oliver and Niimi [14].

[e] Estimated $\log K_{OC}$ value according to the equation of Karickhoff [194] : $\log K_{OC} = 0.989 \log K_{OW} - 0.346$.

[f] Estimated BCF_L value in fish from the n-octanol/water partition coefficient.

[g] Soto et al. [136].

[h] Predicted BAF_L value according to the equation of Geyer et al [191, 192] : $\log BAF_L = 0.745 \log K_{OW} - 1.19$.

[i] Gellert et al. [195].

[j] Smith et al. [197].

ted against respective external OCDD on a log/log basis (s. Fig. 11). The most likely BCF_L value of OCDD in fish was obtained from an extrapolation of the linear relationship down to the OCDD water solubility of 74 pg l^{-1}. Although only few BCF data of OCDD are available, it is obvious that the experimentally obtained BCF_L values in different fish species depend on ambient OCDD concentrations (Table 10, Fig. 11). This means that BCF_L values are increasing with decreasing external concentration in the experiments. Using only the three highest BCF_L values (center of Fig. 11), experimentally determined under nearly identical conditions by Muir et al. [204, 205], a linear regression was found (Eq. 27):

$$\log BCF_L = 11.18 - 1.74 \cdot \log C_W \qquad (27)$$

where C_W is the OCDD concentration [pg l^{-1}] in ambient water. At a water solubility of 74 pg l^{-1}, this regression gives a BCF_L of 85,000,000. This BCF_L value – which is obviously the real or best one – exceeds even the published maximum value by two orders of magnitude and is in satisfactory agreement with the BCF value of OCDD predicted from its K_{ow} value.

8.2.3
Bioconcentration of Mirex

Mirex is an organochlorine insecticide that was used for imported fire ant control in large areas of the southeastern United States. Formulated as a bait, Mirex was intended to control the imported fire ant (*Solenopsis richteri Forel* and *Solenopsis invicta Buren*). In 1977, the US Environmental Protection Agency canceled the registrations of pesticides containing this chemical. In Europe, to the best of our knowledge, Mirex was never registered or used as an insecticide. This highly chlorinated compound was also used as flame retardant for rubber, paint, paper, electrical goods, polymers and plastics. However, when it was found that there is sufficient evidence for the carcinogenicity in experimental animals such as mice and rats of both sexes, the use of Mirex (Dechlorane) was discontinued. Mirex at doses up to 100 mg kg body $wt.^{-1}$ had no uterotrophic activity in weanling rats while Kepone produced significant uterine growth in animals 24 h after injection of 10 mg kg^{-1} body wt. [208]. Mirex failed to induce persistent vaginal estrus (PVE) syndrome following neonatal treatment [208]. From these experimental results Gellert concluded that Mirex does not possess estrogenic activity.

Mirex also belongs to the group of super-hydrophobic chemicals with log K_{ow} value of 7.50. This chemical is highly persistent and resistant to biotic and abiotic degradation and/or metabolism [209a, b, c]. ^{14}C Mirex administered for 15 months in the diet of rats was retained at high levels in the body, especially in adipose tissue, was not metabolized to any detectable extent, and no steady-state was reached [210a]. As a Lake Ontario contaminant, Mirex was first in 1974 identified by Kaiser with its observation in fish [210b]. A number of publications have reported this chemical in sediments, water, mussels, several fish species including eels, aquatic birds, and in beluga whales (*Delphinapterus leucas*) [209c, 210c].

The bioconcentration of Mirex in fish was investigated by several scientists in the laboratory. We have compiled and reexamined these BCF values of Mirex, and try to give some explanations for the apparent dependence of the BCF values on concentrations of these super-hydrophobic chemical, as well as to present the "true" bioconcentration factor (Table 12, Fig. 12). Huckins et al. [209b], Veith et al. [211a] and Skaar et al. [211b] published bioconcentration factors on a wet weight basis (BCF_W) for Mirex in fathead minnows (*Pimephales promelas*) and bluegill sunfish (*Lepomis macrochirus*). Huckins et al. [209b] exposed the fish in a flow-through system to different concentrations for 56 days. The BCF_W values were transformed into BCF values on a lipid basis. In this context it must be mentioned that Dr. Gregory Lien and Dr. Jimmie D. Petty provided us with lipid data of the fathead minnows and the bluegills. The bioconcentration factors, lipid data, and mean concentration data of Mirex in water and fish are presented in Table 12. The relationship between BCF_L values and concentration of Mirex in water is shown in Fig. 12. It is obvious that the bioconcentration factors (BCF_L) are increasing with decreasing concentration of Mirex in water. If the correlation between bioconcentration factors and ambient water concentrations is extrapolated to the maximal water solubility of Mirex, a BCF_W value of 1,380,000 and a bioconcentration factor on a lipid basis (BCF_L) of 13,800,000 is obtained. As expected, this BCF_L value at ambient concentrations not exceeding maximum water solubility lies clearly above the maximum values published thus far. However, this is no steady-state bioconcentration factor, because the uptake of Mirex in fathead minnows was determined after 56 days and after such a short time no steady-state can be reached for such a super-hydrophobic compound.

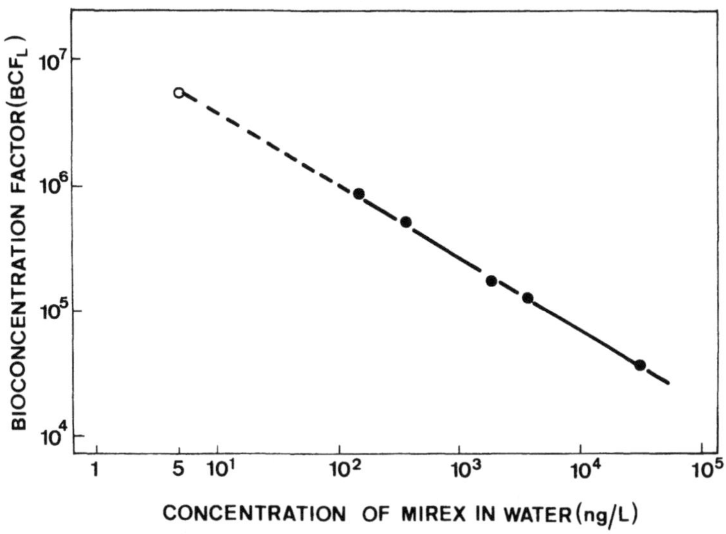

Fig. 12. Relationship between bioconcentration factors on a lipid basis (BCF_L) after 56 days of Mirex exposure in fish and the Mirex concentration in ambient water

Table 12. Bioconcentration factors on a wet weight (BCF_W) and on a lipid basis (BCF_L) of Mirex in fish in dependence on the concentrations in the ambient water (Flow- through system)

No.	Fish species	Body weight (g)	Lipid content (%)	Mean concentration of Mirex in water (ng l^{-1})	Mean concentration of Mirex in fish (ng kg^{-1} wet wt.)	Time of uptake (days)	Bioconcentration factor (BCF)[a] BCF_W	Bioconcentration factor (BCF)[a] BCF_L
1	Fathead minnow	5–6	9.5[b]	33,000	19,000,000	56	3,700	38,950
2	Fathead minnow	5–6	9.5[b]	3,800	47,600,000	56	12,530	131,900
3	Fathead minnow	0.5–0.65	10.5[c]	1,200	21,720,000	32	18,100	172,380
4	Fathead minnow	5–6	9.5[b]	370	19,000,000	56	51,350	540,540
5	Bluegill	0.5–1.0	3.0[b]	161	3,690,000[d]	56	22,900	764,000
6	Fish	5–6	10.0	1	1,380,000	56	1,380,000	13,800,000[e]

Source: References [210, 215, 216].

[a] Bioconcentration factor after 56 or 32 days: $BCF = \dfrac{\text{Concentration of Mirex in fish} \quad [ng \cdot kg^{-1}]}{\text{Concentration of Mirex in water} \quad [ng \cdot l^{-1}]}$.

BCF_W: Bioconcentration factor on a wet wt. basis. BCF_L: Bioconcentration factor on a lipid basis.

[b] Personal communication from Dr. Gregory Lien of the U.S. EPA's Duluth Laboratory, MN, to Dr. Jimmie D. Petty of the U.S. Department of Interior National Biological Service, Columbia, MO, USA.

[c] Personal communication from Dr. Gilman D. Veith, U.S. EPA's Duluth Laboratory, MN, USA.

[d] This residue of Mirex (only uptake from water) was calculated from the experiments of Skaar et al. [216] who determined the uptake of Mirex from food (*Daphnia*) or food plus water.

[e] Predicted by the authors of this work by equation: $\log BCF_L = 7.14–0.572 \log C_W$ (ng l^{-1}) (see Fig. 12) C_W = Water solubility of Mirex.

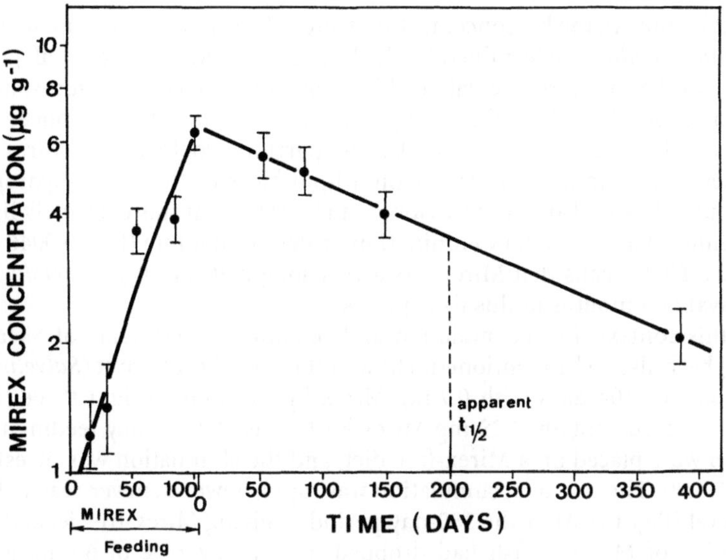

Fig. 13. Concentration of Mirex (μg g^{-1}) in brook trout (*Salvelinus namaycush*) which received Mirex contaminated feed for 100 days. Each point represents the mean ± S.D. (Adopted with modifications from Skea et al. [218])

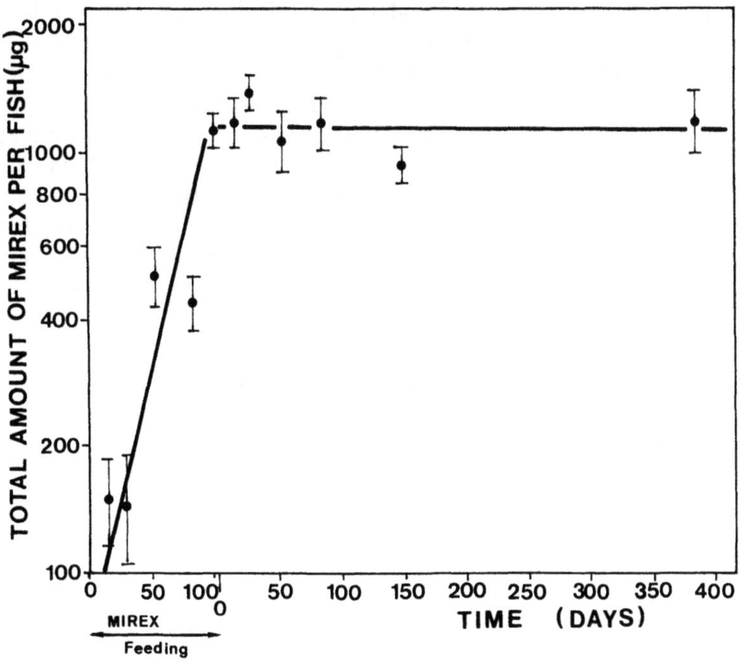

Fig. 14. Total amount of Mirex (μg) per fish (*Salvelinus namaycush*) which received Mirex contaminated feed for 100 days. Each point represents the mean ± S.D. (Adopted with modifications from Skea et al. [218])

In this context, the bioconcentration factor (BCF_L) of Mirex in male guppies (*Poecilia reticulata*) determined by the kinetic method of Gobas et al. [211c] is important. The authors calculated a bioavailability-corrected steady-state BCF_L value of 15,000,000. This value is in good agreement with the bioconcentration factor predicted from the n-octanol/water partition coefficient of Mirex and is the "best" bioconcentration factor on a lipid basis of this super-hydrophobic compound. It was also found by Gobas et al. [211c] that this extremely lipophilic chemical has a very slow elimination or depuration rate ($k_2 = 0.0046\ d^{-1}$) in guppies. That means that Mirex has a very long half-life ($t_{1/2} = 0.693/0.0046 = 150$ days) of 5 months in this fish species.

In this context the accumulation and retention experiments of Skea et al. [211d] have also to be mentioned. These authors fed brook trout (*Salvelinus namaycush*) for 104 days with 0.7 mg Mirex kg^{-1} of body weight three times a week. The feed contained 29 mg Mirex kg^{-1}. After the 104 day feeding period, the fish were placed on a Mirex-free diet, and the elimination was investigated for 385 days. The rate of accumulation was rapid, however, never reached a plateau level (Fig. 13). After the 385-day period of giving Mirex-free feed, the concentration of Mirex in fish had dropped from an average of 6.3 µg g^{-1} (wet weight) to an average of 2.1 µg g^{-1} (wet weight). The calculated apparent half-life of Mirex in these brook trout is 198 days and the apparent elimination rate 0.0035 day^{-1}. However, during the elimination phase the body weight of the fish increased from an average of 175 g to 571 g. If the average body burden (absolute amount of Mirex per fish) is calculated, this does not appear to be any decline in amount of Mirex per fish during the elimination phase (Fig. 14). The average body burden at the end of Mirex feeding was 1,100 µg per fish and after 385 days of uncontaminated feed, the amount averaged 1,200 µg Mirex per fish. There was no significant reduction of Mirex amount found in fish after 385 days. Thus, the decline in Mirex concentration (see Fig. 13) which was found during the elimination phase was due to **growth dilution** and is not a real depuration [211d]. Furthermore, by this experiment it was confirmed that Mirex stored in fish is not eliminated after 385 days and therefore this super-hydrophobic chemical has an extremely high bioconcentration potential.

8.2.4
Bioconcentration of Polychlorinated Bornanes (Toxaphene)

Polychlorinated bornanes are the main components of Toxaphene which is produced by chlorination of camphene under UV light. Toxaphene is a complex mixture of at least 180 to 190 components, mostly with the formula $C_{10}H_{18-n}Cl_n$ or $C_{10}H_{16-n}Cl_n$ where n is 6–10 [212a, b, c]. Today, more than 60 of these compounds have been identified in their structure. According to Saleh [212d], Toxaphene consists of 76% chlorinated bornanes, 18% chlorobornenes, 2% chlorobornadienes, 1% chlorinated hydrocarbons and 3% nonchlorinated hydrocarbons, although the only unsaturated compounds isolated to date are polychlorinated camphenes. Toxaphene was first produced in the USA and became one of the most heavily used pesticides for several decades. The global usage of Toxaphene from 1950 to 1993 has been estimated to be about 1.33 million tons [213]. This pesticide was banned in many countries (USA, Canada, Western Europe) due to its persistence and biological effects. However, it is still used in other countries, such as Eastern Europe, the former Soviet Unions, India, African countries, Central and South America. These chlorinated bornanes have been globally dispersed largely by atmospheric transport to the same extent as polychlorinated biphenyls (PCBs), DDT, lindane, and other hexachlorocyclohexanes [214,215]. Due to its long-range transport, stability, and high bioaccumulation potential Toxaphene belongs to the persistent organic pollutants (POPs). Especially the polychlorinated bornanes are major contaminants in sediments, fish, marine mammals, human milk, etc.

However, in the course of examining these residues in biota it was noticed that their GC pattern was different compared to Toxaphene standard [216–220]. Differences in GC pattern might be caused by photodegradation, selective bioaccumulation and/or metabolism in aquatic and terrestrial organisms including humans. Due to the differences in the Toxaphene composition in environmental samples a precise Toxaphene residue analysis requires the availability of pure chlorinated bornane indicator congeners. At this time the research group of H. Parlar succeeded in producing the 22 most important single congeners of Toxaphene. Most of them are octa- and nonachlorobornanes [224 a, b] which are commercially available from Ehrenstorfer (Augsburg, Germany) or Promochem (Wesel, Germany). The peak area percentage of all components identified, measured by ECD, amounts to 50% of the total technical Toxaphene. Of these compounds, only about 25 are regularly found in environmental samples. Most of the nona- and decachlorobornanes are normally absent, while many of the hexa- and heptachlorobornanes as well as some of the octa- and nonachlorobornanes are detected in sediments, fish and other biotic samples.

In this context the structure-stability investigations of chlorinated bornanes by Parlar and Fingerling and coworkers [218 a–d] are of great importance. They found that bornane compounds with only a single chlorine atom at each secondary ring atom in alternating orientation, such as Parlar No. 26, 40, and 50, were extremely stable to photodegradation [212 c, 218 a–d]. Studies on the degradation of Toxaphene in soil have shown that higher chlorinated bornanes, such as deca- and nonachlorobornanes, are dechlorinated to lower chlorinated bornanes more easily under anaerobic than under aerobic conditions [212 c, 218 a–d]. Studies on the degradation of single chlorinated bornane congeners within loamy flooded soil have shown that congeners with only one chlorine atom at each C atom in alternating orientation were highly persistent. Congeners with geminal dichloro groups on the ring were rather labile, especially if the dichloro group was located at the C_2 atom [218 a–d].

It is important to note that some polychlorinated bornanes are very persistent to biodegradation and are bioconcentrated in the fatty tissue of fish [227] and other aquatic organisms. In marine fish and other aquatic organisms, such as seals, whales, and penguin 2-endo,3-exo,5-endo,6-exo,8b,8c,10a,10b-octachlorobornane (Parlar No. 26, B8–1413, Tox8) and 2-endo, 3-exo,5-endo,6-exo,8b,8c,9c,10a,10c-nonachlorobornane (Parlar No. 50, B9–1679, Tox9, Toxicant Ac) are dominant [221–223]. Furthermore, other persistent chlorinated bornanes are 2-exo,3-endo,5-exo,9b,9c,10a,10b-heptachlorobornane (Tox7, B7–1453), 2-

Table 13. Chemical name, chemical structure, molecular formula, molecular mass, n-octanol/water partition coefficient (log K_{OW}), predicted bioconcentration factors (BCFs) of 7 persistent **Polychlorinated Bornanes** (**Toxaphene** Components) with high bioconcentration potential, and measured BCFs of **Hexachloronorbornadiene, Heptachloronorbornene** and of **Bromocyclen**

Chemical IUPAC name (Parlar No., No. of Andrews and Vetter, No. of Oehme, and other abbreviations)	Chemical structure	Molecular formula and molecular mass [g mol^{-1}]	log K_{OW}	Bioconcentration factor (BCF) in fish and other aquatic organisms	
				BCF_W[a]	BCF_L[b]
(±)-2-*exo*,3-*endo*,5-*exo*,9b,9c, 10a, 10b-Heptachlorobornane (TOX 7, B7–1457[i])	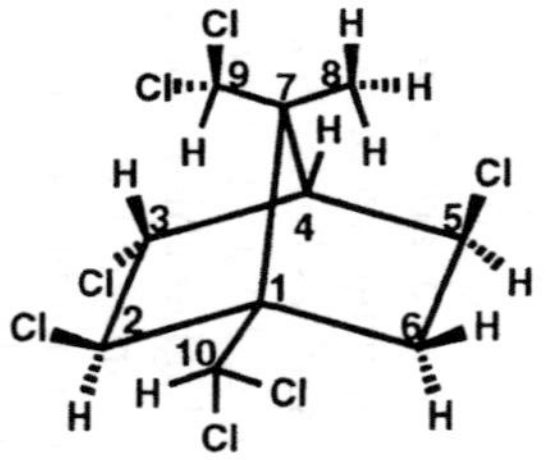	$C_{10}H_{11}Cl_7$ 379.3	5.80[c]	fish (lipid: 5%) 31,500[g]	630,000[g]
			5.93[e]	43,000[g]	850,000[g]
(±)-2-*endo*,3-*exo*,5-*endo*,6-*exo*, 8b, 8c,10a,10c-Octachlorobornane (Parlar No. 26, TOX 8, T 2, B8–1413[i], 169–603[j])	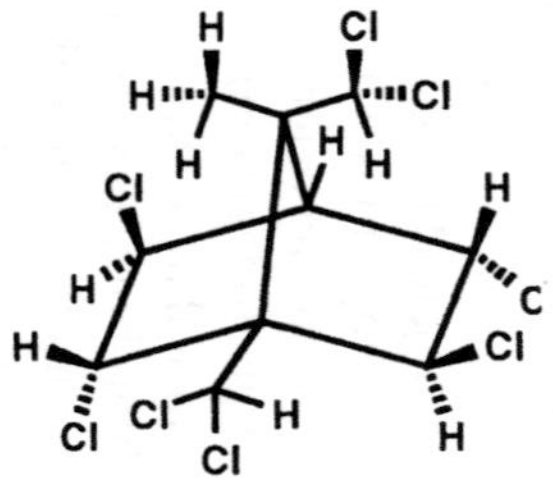	$C_{10}H_{10}Cl_8$ 413.8	5.98[c]	fish (lipid: 5%) 48,000[g]	1,000,000[g]
			6.11[e]	65,000[g]	1,300,000[g]
				zooplankton (1.48%) 163,000[h]	11,000,000[h]
				long-nose sucker (0.96%) 133,000[h]	13,900,000[h]
				lake whitefish (2.69%) 800,000[h]	30,000,000[h]
				lake trout (8.4%) 5,660,000[h]	67,500,000[h]

(±)-2-*endo*,3-*exo*,5-*endo*,6-*exo*, 8b,9c, 10a,10c-Octachlorobornane	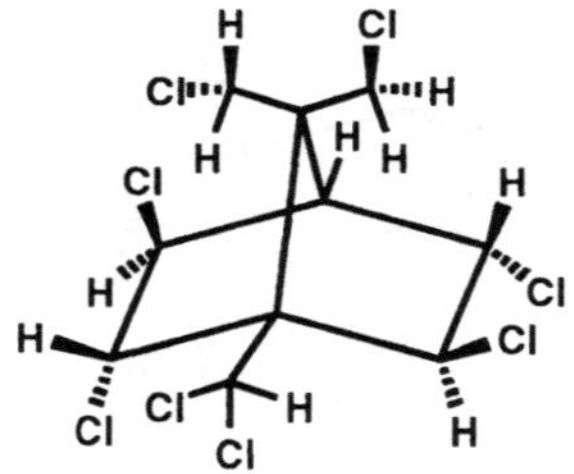	$C_{10}H_{10}Cl_8$ 413.8	6.05[c]	fish (lipid: 5%) 56,000[g]	1,100,000[g]
(Parlar No. 40, B8-1414[i], 297-243[j])			6.18[e]	76,000[g]	1,500,000[g]
(±)-2-*exo*,3-*endo*,5-*exo*,8c,9b, 9c,10a, 10b-Octachlorobornane	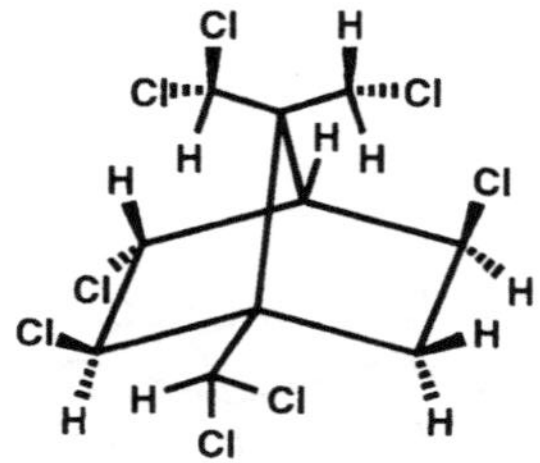	$C_{10}H_{10}Cl_8$ 413.8	6.05[c]	fish (lipid: 5%) 56,000[g]	1,120,000[g]
(Parlar No. 41, B8-1945[i], 41-643[j])			6.18[e]	76,000[g]	1,500,000[g]
(±)-2-*exo*,5,5,8c,9b,9c,10a,10b-Octachlorobornane	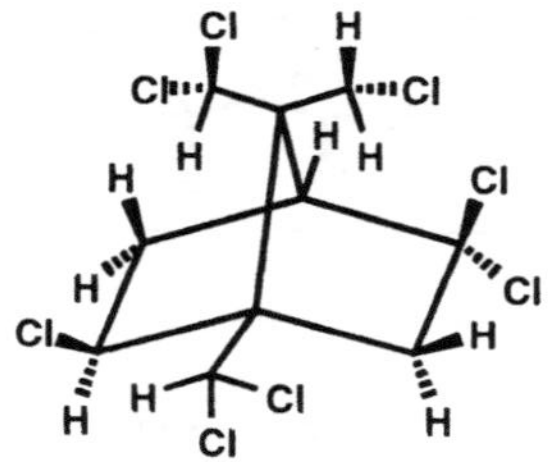	$C_{10}H_{10}Cl_8$ 413.8	6.79[c]	fish (5% lipid) 308,000[g]	6,200,000[g]
(Parlar No. 44, B8-2229[i], 97-643[j])			6.92[e]	420,000[g]	8,300,000[g]

Table 13 (continued)

Chemical IUPAC name (Parlar No., No. of Andrews and Vetter, No. of Oehme, and other abbreviations)	Chemical structure	Molecular formula and molecular mass [g mol^{-1}]	log K_{OW}	Bioconcentration factor (BCF) in fish and other aquatic organisms	
				BCF_W[a]	BCF_L[b]
(±)-2-*endo*,3-*exo*,5-*endo*,6-*exo*, 8b,8c, 9c,10a,10c-Nonachlorobornane (Parlar No. 50, TOX 9, T12, Toxicant Ac, B9-1679[i], 297-303[j])	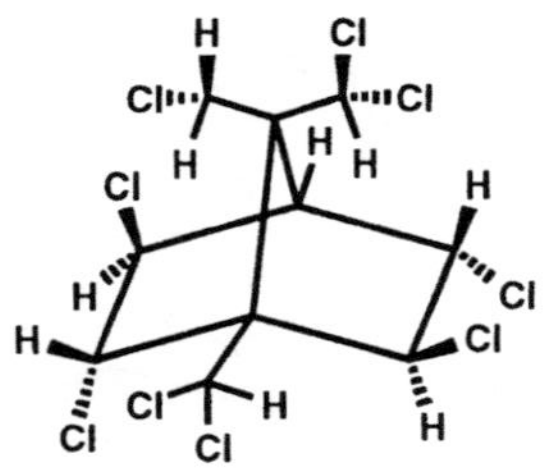	$C_{10}H_9Cl_9$ 448.3	6.23[c] 6.36[e]	fish (5% lipid) 85,000[g] 115,000[g]	 1,700,000[g] 2,300,000[g]
				zooplankton (1.48%) 290,000[h]	20,000,000[h]
				long-nose sucker (0.96%) 100,000[h]	10,000,000[h]
				lake whitefish (2.69%) 680,000[h]	25,000,000[h]
				lake trout (8.4%) 650,000[h]	77,000,000[h]
(±)-2,2,5,5,8c,9b,9c,10a,10b-Nonachlorobornane (Parlar No. 62, B9-1025[i], 99-643[j])	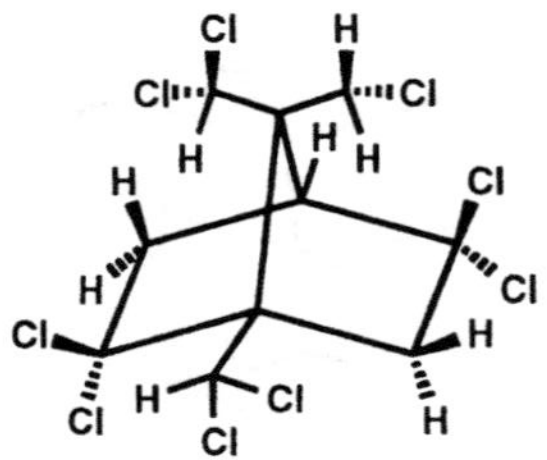	$C_{10}H_9Cl_9$ 448.3	7.72[c] 7.85[e]	fish (lipid: 5%) 2,630,000[g] 3,500,000[g]	 53,000,000[g] 71,000,000[g]

1,2,3,4,7,7-Hexachloro-2,5-nor-bornadiene; 1,2,3,4,7,7-Hexachloro-bicyclo [2,2,1]hepta-2,5-diene (HCND)	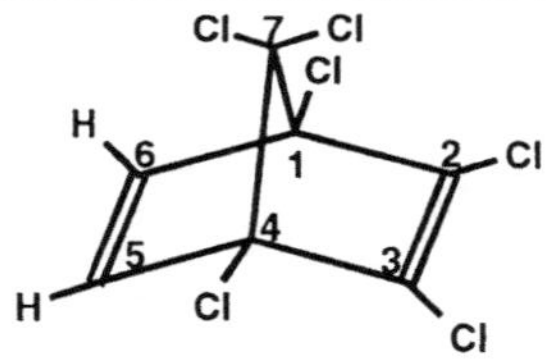	$C_7H_2Cl_6$ 298.8	5.15[c] 5.28[d]	fathead minnow (lipid: 4%) 6,400[f]	160,000
1,2,3,4,5,7,7-Heptachloro-2-nor-bornene; 1,2,3,4,5,7,7-Heptachloro-bicyclo [2,2,1]hept-2-ene (HepCNB)	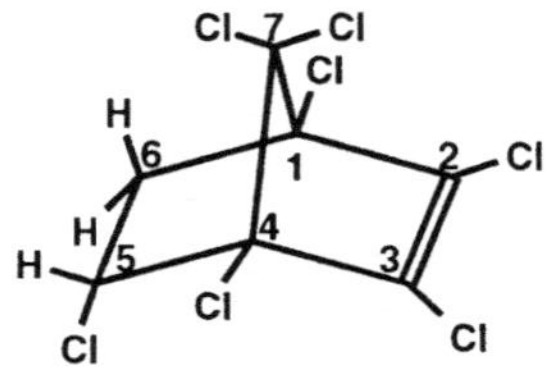	$C_7H_3Cl_7$ 335.3	5.55[c] 5.28[d]	fathead minnow (lipid: 4%) 11,200[f]	280,000
Bromocyclen; Bromodan®; Alugan®; 5-Bromomethyl-1,2,3,4,7,7-hexa-chlorobicyclo[2,2,1]hept-2-ene	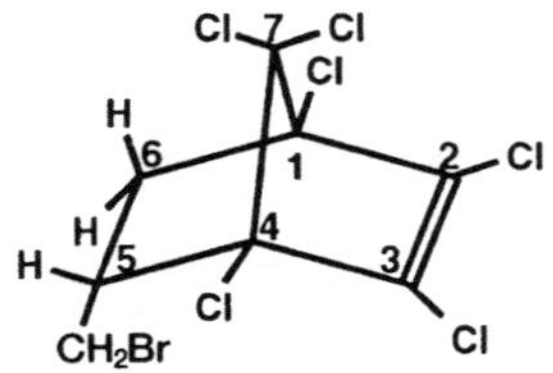	$C_8H_5BrCl_6$ 393.75	5.90	rainbow trout (lipid: 3.7%) 8,700[k]	235,400[k]

[a] BCF_W; Bioconcentration factor on a wet weight basis. [b] BCF_L; Bioconcentration factor on a lipid basis.
[c] The log K_{OW} values were calculated by Andreas Kaune using the LOG KOW Program of Meylan and Howard [232a, b].
[d] The log K_{OW} value was measured by Veith et al. by the HPLC method [62].
[e] The log K_{OW} value was calculated by Andreas Kaune on the basis of the measured log K_{ow} value of 5.28 for 1,2,3,4,7,7-hexachloro-2,5-norbornadiene.
[f] BCF_W value in fathead minnows (0.12 g body weight, 4% lipid) was measured in a 30-day flow-through test by Spehar et al. [231].
[g] BCF value predicted from the log K_{OW} value.
[h] BCF data calculated from the concentration in biota and water of a Canadian fresh water lake (K. Kidd and D. Muir [226]).
[i] Congener No. proposed by Andrews and Vetter [230a]. [j] Congener No. proposed by Oehme [230b].
[k] Flow-through test and kinetic approach (H. Kuhlmann, G. Rimkus and W. Butte (1999) unpublished).

endo,3-exo,5-endo,6-exo,8b,9c,10a,10c-octachlorobornane (Parlar No. 40, B8–1414), 2-exo,3-endo,5-exo,8c,9b,9c,10a,10b-octachlorobornane (Parlar No. 41, B8–1945), 2-exo,5,5,8c,9b,9c,10a,10b-octachlorobornane (Parlar No. 44, B8–2229), and 2,2,5,5,8c,9b,9c,10a,10b-nonachlorobornane (Parlar No. 62, B9–1025). These chlorinated bornane congeners are bioaccumulated in aquatic animals such as fish, seals, dolphins, and whales and also in terrestrial organisms including human [221–223, 227–229].

A recent study by Vetter et al. [225] showed that 11 polychlorinated bornanes were abundant in different seal species. The most important persistent 7 polychlorinated bornanes with their IUPAC name, different abbreviations, chemical structure, molecular formula, n-octanol/water partition coefficient (log K_{OW}) and predicted bioconcentration factors (BCF_W and BCF_L) in fish are compiled in Table 13. The predicted BCF_L values of hepta-, octa- and nonachlorobornanes are between 600,000 and 71,000,000, and the predicted BCF_W values of these congeners in fish with 5% lipid range from ca. 32,000 to 3,500,000. Furthermore, in Table 13 the BCF_W and BCF_L values of two polychlorinated bornane congeners (Parlar No. 26 and No. 50) are included, which were calculated by the authors from the measured concentrations in zooplankton and different fish species and the water of a Canadian fresh water lake [226]. It is obvious that the BCF values of the chlorinated bornanes calculated from concentrations in aquatic organisms and water from the environment are by a factor between 1 and ca. 70 greater than the BCFs predicted from the log K_{ow} values. This can be explained in part by bioaccumulation.

In the future it is now possible to measure experimentally the K_{ow} values and also the BCF values of special chlorinated bornane congeners in aquatic organisms. However, it is necessary to use chlorinated bornane concentrations in the water which are below the water solubility of these very hydrophobic chemicals. Otherwise the BCF values are too low [35–37, 84c, 85–88]. First of all, it is necessary to determine the water solubility of these very hydrophobic chlorobornane congeners before the bioconcentration test is performed. Furthermore, it is necessary to use the flow-through system and the kinetic approach.

It is also known that Toxaphene has low estrogenic activity. However, it is not clear which compounds of the mixture are responsible for this effect. With the pure single chlorinated bornane congeners this is now also possible. It is assumed that some hydroxylated chlorinated bornane metabolites and/or keto metabolites may be responsible for the estrogenic activity.

8.3
Bioconcentration of Polychlorinated Norbornene and Norbornadiene

Heptachloronorbornene (1,2,3,4,5,7,8-heptachlorobicyclo[2,2,1]hept-2-ene) is produced by the Diels-Alder reaction by addition of vinylchloride to cyclopentadiene. This product is converted by dehydrochlorination to hexachloronorbornadiene (1,2,3,4,7,7-hexachlorobicyclo[2,2,1]hepta-2,5-diene) which is an intermediate in the syntheses of the stable chlorinated cyclodiene insecticides. Hexachloronorbornadiene (HCND) and heptachloronorbornene are very hydrophobic chemical intermediates with very high log K_{OW} values of 5.15

and 5.55, respectively. Therefore, it is not surprising that these compounds along with the epoxy metabolite 1,2,3,4,7,7-hexachloro-5,6-endo-epoxy-2-norbornene were identified as contaminants in edible fish from rivers in the USA [233].

Spehar et al. [231] investigated the bioconcentration potential of hexachloronorbornadiene and heptachloronorbornene using 30-day flow-through test with early juvenile fathead minnows (*Pimephales promelas*) with a body weight of 0.12 g and 4% lipid content. The measured mean concentrations of hexachloronorbornadiene (HCND) in water was 20.0 ± 3.9 µg l^{-1} and the water concentration of heptachloronorbornene (HepCNB) was 25.9 ± 3.4 µg l^{-}. The bioconcentration factors on a wet weight basis after 30 days in this fish species were 6,400 and 11,200, respectively. The bioconcentration factors on a lipid basis (BCF_L) of HCND and HepCNB after 30 days were 160,000 and 280,000, respectively (see Table 13).

8.4 Bioconcentration of Tetrachlorobenzyltoluenes (TCBTs)

Tetrachlorobenzyltoluenes (TCBTs) are the main components in products marketed as Ugilec 141. These TCBTs are one group of polychlorinated biphenyls (PCBs) replacements due to their dielectric properties, chemical stability and their good thermal conductivity. TCBTs are used as hydraulic fluid in the underground mining industry, as a dielectric fluid for capacitors, as well as a cooling isolation fluid for transformers [234, 235]. Theoretically, 96 tetrachlorobenzyl toluene isomers are possible. A systematic numbering system was developed on the basis of the IUPAC nomenclature for the numbering of diphenylmethane derivatives [235]. 3,5-Dichlorotoluene is formed only in very small amounts in the technical production of TCBTs, thus reducing the number of the relevant isomers in TCBTs to 70 [235].

TCBTs can enter the environment if they are released in the underground via mine outputs, pit water, or to minor extent via ventilation systems. As a consequence, TCBTs can contaminate river water and especially aquatic organisms, such as algae, mussels, and fish. In Germany near the rivers Ruhr and Lippe, there are areas with extensive underground mining. In fish from these rivers, concentrations of Ugilec 141 ranged from 0.1 to 25 mg/kg based on the edible portion [234, 236–238]. In 1990 Wester and Van der Valk [239] found concentrations up to 4.8 mg/kg (edible portion) in eels of the Dutch rivers Meuse and Rhine, which are connected with the rivers Ruhr and Lippe. These high levels found in fish seem to indicate that TCBTs possess a relatively high bioconcentration potential although the TCBT concentration in the rivers was not measured.

Recently van Haelst et al. [241, 242] have determined the bioconcentration factors of single TCBT isomers in zebra mussels (*Dreissena polymorpha*) and in adult female guppies (*Poecilia reticulata*). The bioconcentration factors on a wet weight basis (BCF_W) and on a lipid basis (BCF_L) of eight tetrachlorobenzyltoluenes (TCBTs) isomers in mussels are compiled in Table 14. The melting points, water solubilities, and the n-octanol/water partition coefficients (log

K_{ow}) of the investigated TCBTs are presented also in Table 14. The BCF_W values of TCBTs isomers in the mussels ranged from ca. 27,000 to 154,900 and the BCF_L values ranged from ca. 620,000 to 3,560,000.

Bouraly and Millisher [240] have determined the bioconcentration factors of the technical mixture Ugilec 141 in zebrafish (*Danio rerio*). After 30 days they found a BCF_W value of 2,300. This value is relatively low as compared to that found in literature for PCBs. Preliminary bioconcentration factors (BCF_W) of six TCBTs in the guppies reported by Van Haelst et al. [242] ranged from ca. 50 to 480, whereas the BCF values on a lipid basis are ranging from ca. 500 to 5,100. The bioconcentration potential of TCBTs in fish, which is some orders of magnitude lower as compared to that of mussels, may be due to biotransformation, as proposed by Bouraly and Millischer [240] and by van Haelst et al. [242]. It is also very likely that in the study of Bouraly and Millischer [240] in 30 days no steady-state concentration in the fish was attained. The BCF was calculated by dividing the concentration of Ugilec 141 in fish after 30 days by the concentration in the water. It is clear that if no equilibrium was reached, the resulting BCF values would be underestimated. However, the main reason for the low BCF value determined by Bouraly and Millischer [240] was the high TCBT concentration in the water (530–810 $\mu g\ l^{-1}$) which was some orders of magnitude above the water solubility of TCBT. Therefore in Table 14 the BCF value of UGILEC 141 is given in brackets. The real BCF values of UGILEC 141 should be some orders of magnitude higher.

Van Haelst et al. [242] determined uptake and elimination of TCBTs in guppies during 15 and 28 days, respectively. The water was continuously loaded with TCBTs during the exposure experiment and was renewed at day 9. The integration method described by Gobas and Zhang [243] was applied to calculate the uptake rate constants, elimination rate constants, and the BCF values. It was stated that the bioconcentration factors derived by the iterative method are independent of the variability in the concentration of TCBT in water and of duration of the exposure experiment. However, for such superhydrophobic chemicals the kinetic method under flow-through conditions is the best method for the determination of BCF values and should be applied. It seems, therefore, necessary to repeat bioconcentration tests of special isomers of TCBTs in fish.

To the best of our knowledge, up to now no metabolites of TCBTs were identified in fish or mammals. But the authors of this paper predicted some TCBT metabolites (see Fig. 15) which could show estrogenic or other hormonal effects in fish and other aquatic organisms as well as in terrestrial organisms including human. Recently, Körner et al. [246] have shown that tetrabromo bisphenol-A [2,2-bis-(3,5-dibromo-4-hydroxyphenyl)-propane], which has structural similarities with the predicted TCBT metabolites in Fig. 15, has weak estrogenic potency in the proliferation assay with the human breast cancer (MCF-7) cell line.

Table 14. Physico-chemical properties and bioconcentration factors on a wet weight basis (BCF_W) and on a lipid basis (BCF_L) of **Tetrachlorobenzyl-toluenes (TCBTs)** of zebra mussels (*Dreissena polymorpha*) and/or fish (*Danio rerio*)

Tetrachlorobenzyl-toluene isomer (TCBT)	TCBT No.[a]	Chemical structure	Melting Point (°C)	Water solubility[b] at 25 °C (µg l^{-1})	log K_{OW}[c]	Bioconcentration factor (BCF)	
						BCF_W	BCF_L
2,2′,4,6′-Tetrachloro-5-methyl-diphenylmethane	28	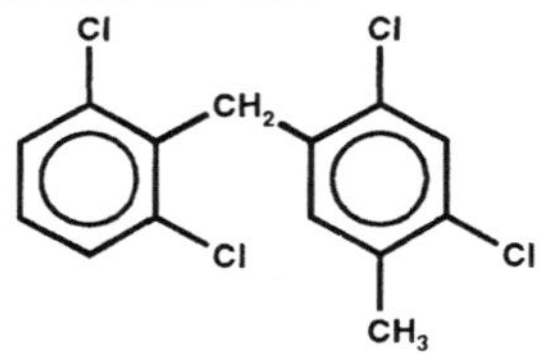	114	1.4	6.73	mussel: 67,610[d]	1,554,100[e]
2,2′,4,5′-Tetrachloro-5-methyl-diphenylmethane	25	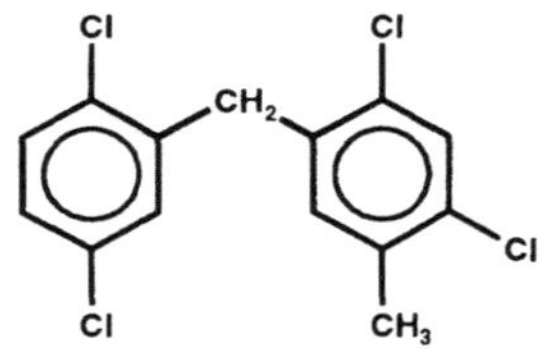	79–80.5	2.8	7.54	mussel: 51,300[d]	1,179,000[e]
2,2′,4,4′-Tetrachloro-5-methyl-diphenylmethane	22	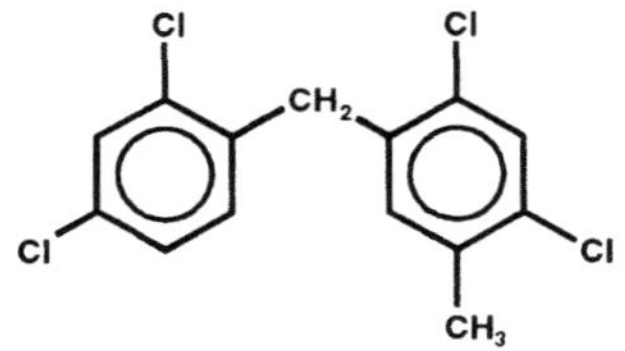	62–64	10.4	7.43	mussel: 74,130[d]	1,704,200[e]

2,2′,4,6′-Tetrachloro-3-methyl-diphenylmethane	27	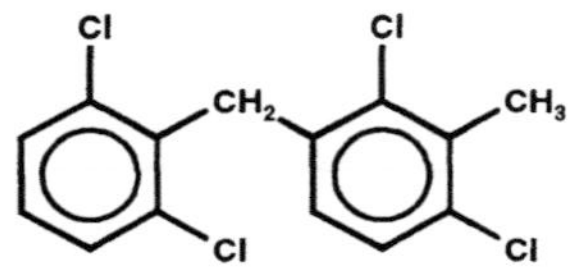	87	3.3	7.20	mussel: 154,900[d]	3,560,500[e]
2′,3,4,6′-Tetrachloro-6-methyl-diphenylmethane	80	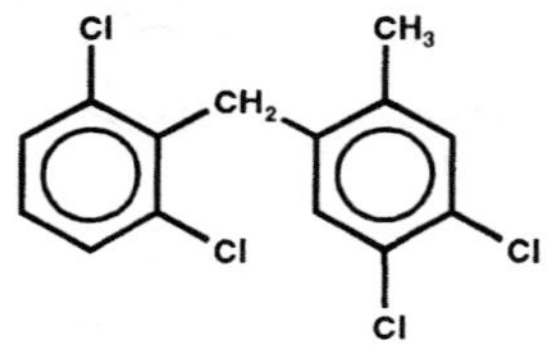	107	18.3	7.15	mussel: 26,920[d]	618,740[e]
2,2′,4,4′-Tetrachloro-3-methyl-diphenylmethane	21	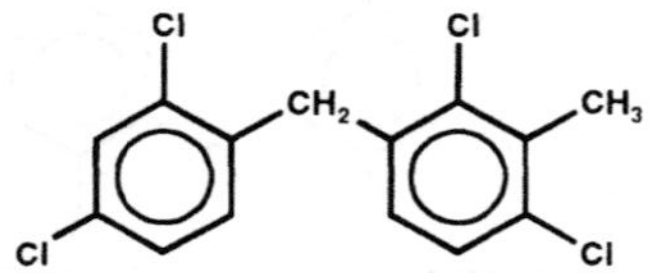	76	12.8	7.20	mussel: 38,020[d]	874,000[e]
2,3′,4,4′-Tetrachloro-5-methyl-diphenylmethane	52	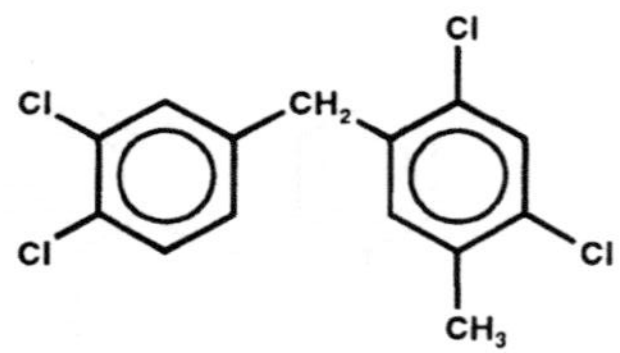	75	2.1	7.26	mussel: 33,880[d]	779,000[e]
2′,3,4,4′-Tetrachloro-6-methyl-diphenylmethane	74	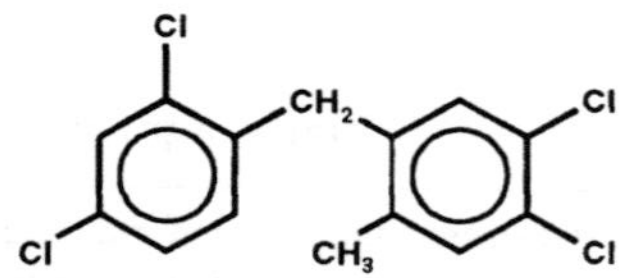	83	11.6	7.41	mussel: 45,710[d]	1,050,700[e]

Table 14 (continued)

Tetrachlorobenzyl-toluene isomer (TCBT)	TCBT No.[a]	Chemical structure	Melting Point (°C)	Water solubility[b] at 25 °C (μg l⁻¹)	log K_{OW}[c]	Bioconcentration factor (BCF)	
						BCF_W	BCF_L
Tetrachlorobenzyltoluene (UGILEC 141)	–	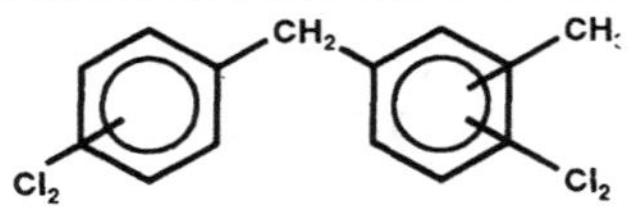			6.2	fish: (2,300)[f]	(35,000)[f]

[a] Nomenclature and numbering according to Ehmann and Ballschmiter [235] and in order of elution.
[b] Water solubility determined by the generator column method by van Haelst et al. [244].
[c] n-Octanol/water partition coefficients determined by Van Haelst et al. using the slow stirring method [245].
[d] Mean steady-state bioconcentration factors on a wet weight basis of two independent tests with zebra mussels from Van Haelst et al. [242].
[e] Mean steady-state bioconcentration factors on a lipid weight basis calculated by the authors using the mean lipid content of 4.35% given by Van Haelst et al. [242].
[f] Bioconcentration factor of the technical mixture of TCBTs (UGILEC 141) in zebrafish (*Danio rerio*) (160–170 mg body weight, lipid content ≈ 6.5%) determined by Bouraly and Millischer [240] using the flow-through system (uptake period = 30 days, elimination period = 30 days). However, the concentration of TCBT in water was 530–810 μg l^{-1} (suspension of carboxymethylcellulose) and the mortality rate was 12%. Therefore this is not a "real" BCF value and is given in parentheses.

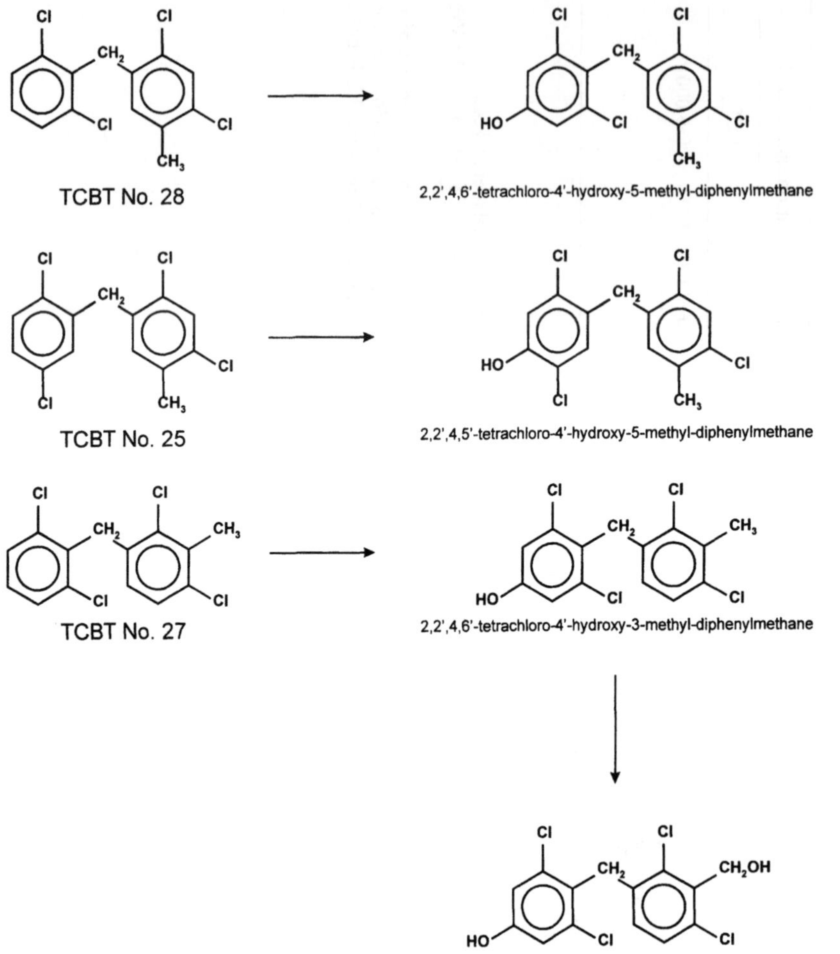

Fig. 15. Predicted metabolites of some tetrachlorobenzyltoluene (TCBT) isomers with supposed estrogenic and/or other endocrine-disrupting effects in fish and mammals including human

8.5
Bioconcentration of Polybrominated Benzenes (PBBz) and Polybrominated Biphenyls (PBBs)

Polybrominated benzenes (PBBz) and polybrominated biphenyls (PBBs) have been widely used as flame or fire retardants. The use of flame retardants is recommended or required in diverse areas such as synthetic polymers which are used in building materials, textiles, packing materials, electric applications, automobile manufacturing etc. to protect the public from fire accidents [247, 248]. PBBs were introduced as flame retardants in the early 1970s. In Japan e.g. the annual consumption of polybrominated compounds used as fire retardants in 1987 has increased by about nine times that of 1975 [247]. However, PBBs have never been produced in Japan, but, up to 1978, some were imported [248]. The estimated production of PBBs in the USA between 1970 and 1976 was 6000 tons. Octabromobiphenyl and decabromobiphenyl were produced in the USA until 1979 [248].

Polybrominated benzenes and polybrominated biphenyls can enter the aquatic and terrestrial environment during production by losses through waste waters, emission into the air, or during handling and shipping or as a result of the incineration of materials containing PBBz or PBBs. These brominated aromatic chemicals, depending on the structure and degree of bromination, have a relatively high n-octanol/water partition coefficient and therefore it is not surprising that polybrominated benzenes and polybrominated biphenyls are detected in sediments, mussels, fish, and human fat (Table 15). Recently de Boer et al. [272b] determined polybrominated biphenyls (PBBs: No. 15, 49, 52, 101, 153 and 169) in mackerels, harbor seals, minke whales, sperm whales, and whitebeaked dolphins from the Atlantic Ocean and Dutch coastal seas. The total PBB concentrations in sperm whale blubber was around 2 μg kg^{-1}. The presence of PBBs in sperm whales indicates that this group of chemicals has reached deep ocean waters, as sperm whales are not usually found in shelf seas. They hunt in waters of depths 400 to 1200 m or more. The environmental occurrence, toxicity, and analysis of polybrominated biphenyls was reviewed by Pijnenburg et al. [249]. In the following section the uptake and bioconcentration of PBBzs and PBBs in aquatic organisms, especially fish, is critically reviewed.

The uptake of some brominated benzenes and polybrominated biphenyls in fish (Atlantic salmon) was investigated in 1976 for the first time in a static test by Zitko and Hutzinger [250]. After 96 hours, they found no uptake of hexabromobenzene (HBB) from water or food in *Salmo salar*. The fish contained 2.3% hexane-extractable fat. The polybrominated biphenyls were accumulated to a lower extent than the polychlorinated biphenyls. Zitko and Hutzinger [250] suggested that HBB with a high molecular weight of 552 can not be taken up by fish. They concluded also that it is possible that HBB can be taken up by fish but that this compound may be converted into a non-extractable form in fish. Experiments with ^{14}C-labeled HBB are required to determine its fate. But as far as we know, no bioconcentration experiments with ^{14}C-labeled HBB and fish were conducted.

However, in 1985 Oliver and Niimi [251] investigated the bioconcentration of hexabromobenzene (HBB) and other highly lipophilic organic chemicals in a flow-through test in rainbow trout (*Oncorhynchus mykiss*). After 96 days, they found a bioconcentration factor (BCF_W) on a wet weight basis of 1400 (lipid = 7.6%) and a BCF_L value of 18,000. But these authors stated also that no steady-state after 96 days was reached. However, the metabolism of HBB was not investigated. Nevertheless, these experiments by Oliver and Niimi [251] have clearly shown that hexabromobenzene can be taken up from water by fish and is bioconcentrated to a relatively high amount. However, the BCF_L value of HBB is lower than predicted from its K_{OW} value.

Butte et al. [253] investigated in a flow-through test system the bioconcentration of hexabromobenzene (HBB) beside other chemicals in zebrafish

(*Danio rerio*) in the laboratory and under outdoor-conditions. By means of the kinetic method they calculated for HBB a BCF_W value of 166 and a BCF_L value of 2,188. However, the concentration of HBB in the water (mean 1.72 µg l^{-1}) was not constant. Under outdoor-conditions no bioconcentration factor could be determined because most of the HBB in the water was bound to particles (ca. 96%) and the concentration in the fish showed a relatively great variation.

Gobas et al. [252] investigated the bioconcentration of some other polybrominated benzenes and polybrominated biphenyls in fish. These BCF_L data together with molecular mass, log K_{ow} data, and occurrence in sediments, mussels, fish, and human fat are compiled in Table 15. The experiments of Gobas et al. [252], Butte et al. [253], and Oliver and Niimi [251] have clearly shown that super-hydrophobic polybrominated benzenes and polybrominated biphenyls can be taken up from water and are bioconcentrated in fish, mussels, and other aquatic organisms. Gobas et al. [252] showed also the importance of bioavailability for the bioconcentration of such super-hydrophobic compounds. From the BCF data presented in Table 15, it can also be concluded that the bioconcentration potential of PBBz and PBBs is increasing with their lipophilicity (log K_{ow}). However, the BCF_L values of these brominated aromatics are lower than predicted from their log K_{ow} values. The authors concluded that the reason for this phenomenon may be metabolism by which hydroxylated and/or debrominated compounds are formed. These metabolites are eliminated faster than the parent compound. Nevertheless, these superhydrophobic brominated compounds are able to cross membranes. They are then transported by the circulating blood to liver and to the lipid depots and bioaccumulated there. The elimination from the fat depot is very slow. These considerations are in agreement with the results of Watanabe and Tatsukawa [247] who found tribromo-, tetrabromo-, pentabromo-, and hexabromobenzene in fish and shellfish in Japan.

The abundant use of brominated organic compounds, like polybrominated benzenes and polybrominated biphenyls, as flame retardants may lead to serious environmental pollution and hazards both for wildlife and human. Beside the high bioconcentration potential, the formation of polybrominated dibenzofurans (PBDFs) and/or polybrominated dibenzo-*p*-dioxins (PBDDs) during thermolysis or combustion is another disadvantage of polybrominated biphenyls and/or benzenes. The thermolysis of the commercial fire retardant Fire Master BP-6, which contained 54–68% 2,2′,4,4′,5,5′-hexabromobiphenyl and 2–17% heptabromobiphenyl [254], was studied by Buser et al. [255]. In the presence of air at 600 °C, the highly toxic 2,3,7,8-tetrabromodibenzofuran (TBDF) was formed in the percentage range. At 400 °C, the yield of conversion was in the mg kg^{-1} range [255].

It is very likely that the polybrominated biphenyls are metabolized in fish, sea mammals, and human to debrominated and hydroxylated polybrominated biphenyls. It is suggested by the authors that some of these polybrominated metabolites can bind to the estrogen receptor and act like xenoestrogens, as was shown for hydroxy polychlorinated biphenyls (HO-PCBs) by Korach et al. [256], McKinney and Waller [257], and Waller et al. [258]. Especially the chemical structure of *para*-substituted hydroxylated metabolites of polybrominated biphenyls would be similar to that of estradiol, preferably in the presence of

Table 15. Chemical name, trade name, CAS No., chemical structure, molecular formula, molecular weight, n-octanol/water partition coefficient (K_{OW}), bioconcentration factors on a lipid basis (BCF_L) in fish, and residues found in environmental samples (sediments, mussels, fish) and human fat of **Polybrominated Benzenes (PBBzs), Polybrominated Biphenyls (PBBs), and Polybrominated Diphenylethers (PBDEs)**

Chemical Name (abbreviation, PBB No. or PBDE No. and/or trade name of flame retardant)	CAS No.	Chemical structure	Molecular formula	Molecular mass [g mol^{-1}]	log K_{OW}	BCF in fish. Detected in water (W), sediments (S), mussels (M), fish (F), whales (Wh) and/or human fat (H)
Polybrominated Benzenes (PBBzs)						
1,4-Dibromobenzene (1,4-DBBz)	106–37–6		$C_6H_4Br_2$	235.92	3.89 4.07	BCF_L:1,413[b]
1,3-Dibromobenzene (1,3-DBBz)	108–36–1	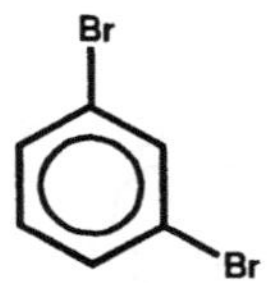	$C_6H_4Br_2$	235.92	4.02	BCF_L(90 d): 8,040[c]
1,2,4-Tribromobenzene (1,2,4-TBBz)	615–54–3	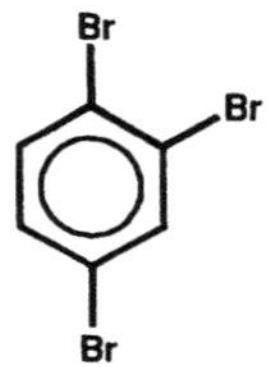	$C_6H_3Br_3$	314.8	4.73	BCF_L(90 d): 40,100[c]

Table 15 (continued)

Chemical Name (abbreviation, PBB No. or PBDE No. and/or trade name of flame retardant)	CAS No.	Chemical structure	Molecular formula	Molecular mass [g mol⁻¹]	log K_{OW}	BCF in fish. Detected in water (W), sediments (*S*), mussels (M), fish (F), whales (Wh) and/or human fat (H)
1,3,5-Tribromobenzene (1,3,5-TBBz)	626–39–1	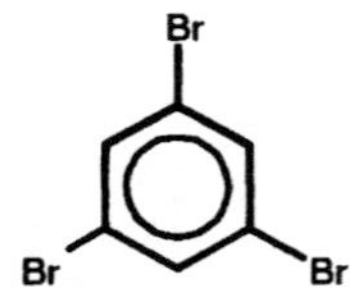	$C_6H_3Br_3$	314.8	4.98	BCF_L: 26,300[b] BCF_L(90 d): 45,800[c] S: + M: + F: + H: – (< 0.1 ppb fat)
1,2,4,5-Tetrabromo-benzene (1,2,4,5-TeBBz)	636–28–2	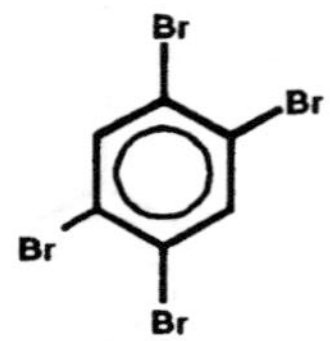	$C_6H_2Br_4$	393.7	5.45	BCF_L(90 d): 52,700[c,d] S: + M: + F: + H: +
Pentabromobenzene (PeBBz)	608–90–2	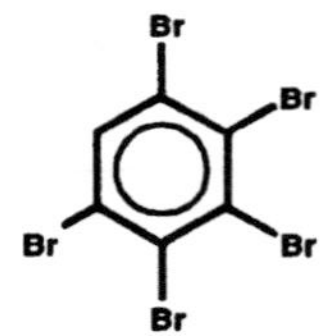	C_6HBr_5	472.6	6.21	S: + M: + F: + H: +
Hexabromobenzene (HBBz)	87–82–1	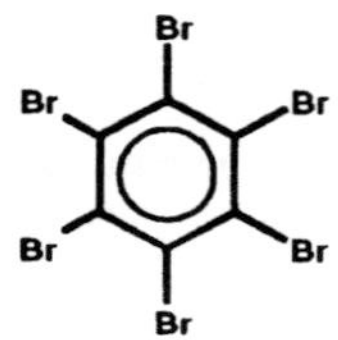	C_6Br_6	551.5	6.80	BCF_L (96 d):18,000[c,d] S: + M: + F: + H: +

Polybrominated Biphenyls (PBBs)						
4,4′-Dibromobiphenyl (PBB # 15)	92–86–4	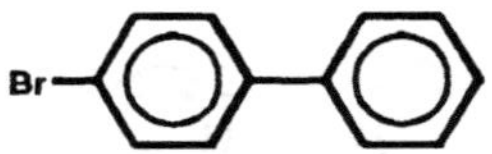	$C_{12}H_8Br_2$	312.01	5.72	BCF_L: 269,000[b] F: + W: +
2,4,6-Tribromobiphenyl (PBB # 30)	59080–33–0	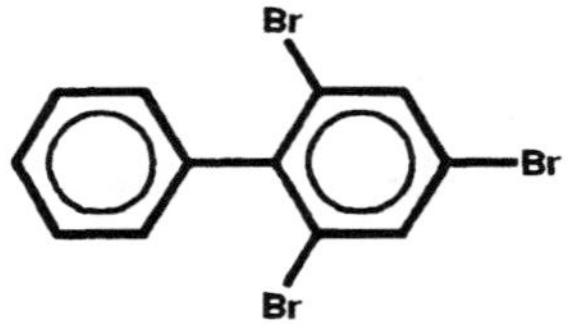	$C_{12}H_7Br_3$	390.90	6.03	BCF_L: 114,800[b]
2,2′,5,5′-Tetrabromo-biphenyl (PBB # 52)	59080–37–4	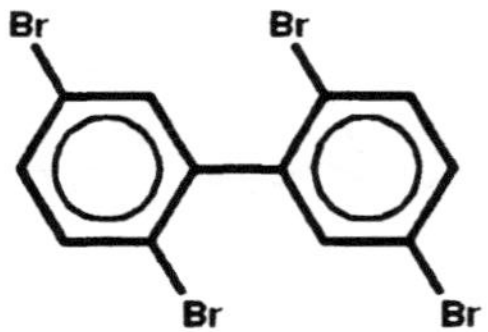	$C_{12}H_6Br_4$	469.80	6.50	BCF_L: 2,042,000[b] F: + W: +
2,2′,4,4′,5,5′-Hexabromo-biphenyl (PBB # 153) and other isomers (Fire-Master BP-6)	59080–40–9	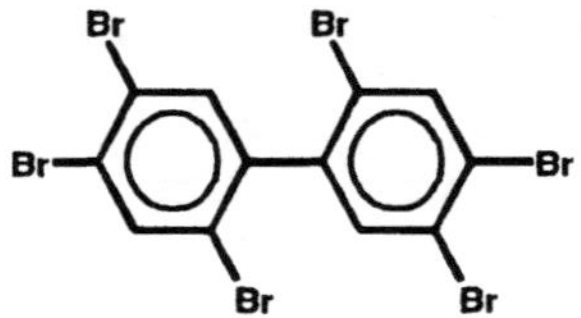	$C_{12}H_4Br_6$	627.6	7.50	S: + F: + W: + H: +
2,2′,4,4′,6,6′-Hexabromo-biphenyl (HBB) (PBB # 155)	59261–08–4	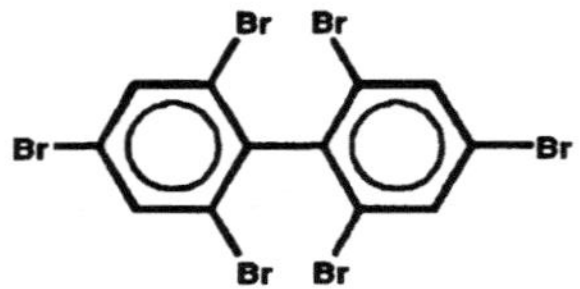	$C_{12}H_4Br_6$	627.6	7.20	BCF_L: 2,820,000[b]

Table 15 (continued)

Chemical Name (abbreviation, PBB No. or PBDE No. and/or trade name of flame retardant)	CAS No.	Chemical structure	Molecular formula	Molecular mass [g mol^{-1}]	log K_{OW}	BCF in fish. Detected in water (W), sediments (*S*), mussels (M), fish (F), whales (Wh) and/or human fat (H)
Decabromobiphenyl (DeBB) (PBB # 209)	13654–09–6	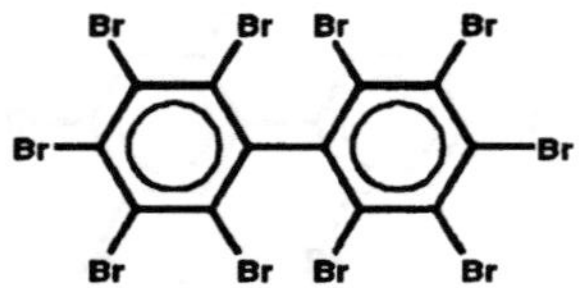	$C_{12}Br_{10}$	943.2	8.60	not yet found in aquatic organisms
Polybrominated Diphenyl Ethers (PBDEs)						
4-Bromodiphenylether (PBDE #3) (MBDE) and other isomers	101–55–3	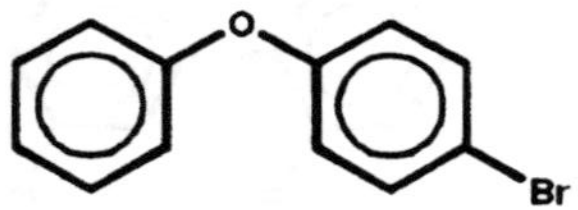	$C_{12}H_9BrO$	249.03	4.28 (4.08 – 4.98)	W: + S: + Aquatic organisms: +
4,4′-Dibromodiphenyl-ether (DiBDE) (PBDE # 15)	2050–47–7	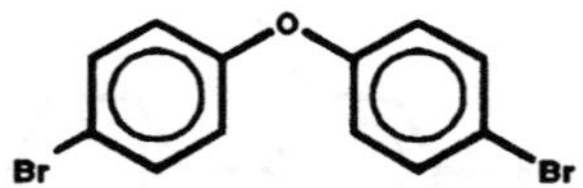	$C_{12}H_8Br_2O$	327.9	5.03 [a]	W: – (< 10–30 ng l^{-1}) S: – (< 0.05 μg kg^{-1} dry wt.)
2,4,4′-Tribromodiphenyl-ether (TrBDE) (PBDE # 28)	49690–94–0	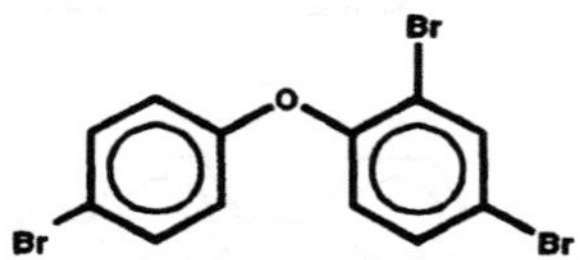	$C_{12}H_7Br_3O$	406.8	5.53 (5.47 – 5.58 [a])	M: + F: not detected (< 0.2 ppb wet weight)

2,2′,4,4′-Tetrabromodiphenylether (TeBDE) (PBDE # 47)	40088–47–9	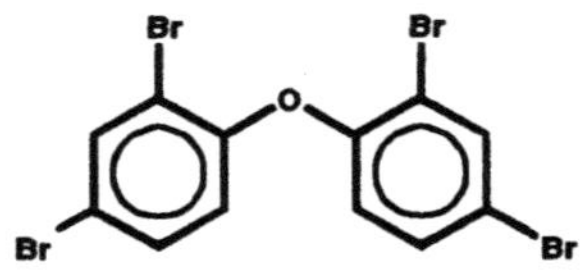	$C_{12}H_6Br_4O$	485.8	6.04 (5.87 –6.16)[a]	BCF in mussel[g,h] BCF_W : 144,400 BCF_L : 13,000,000 S: +; M: +; F: +; Wh: +; H: +
2,2′,4,4′,5-Pentabromodiphenylether (PBDE # 99) and other isomers (PeBDE) (Great Lakes DE-71; Bromkal 70-5-DE)	32534–81–9	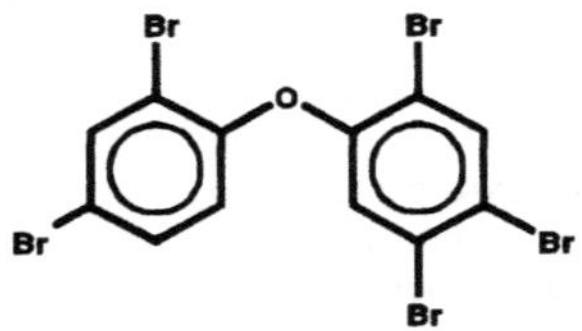	$C_{12}H_5Br_5O$	564.7	6.84 (6.64 – 6.97)[a]	BCF in carp (3.4% lipid) (8 weeks uptake; c_W =10 μg l^{-1}) BCF_W (> 10,000)[d,f] BCF_L (> 290,000)[d,f] BCF in mussel[g,i] BCF_W : 156,000 BCF_L : 15,400,000 S: +; M: +; F: +; Wh: +; H: +
2,2′,4,4′,5,5′-Hexabromodiphenylether (HxBDE) (PBDE # 153) and other isomers (BR 33 N)	36483–60–0	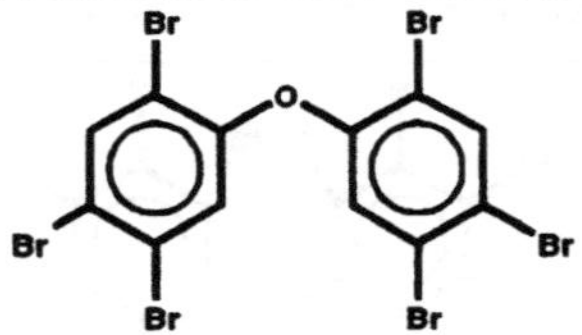	$C_{12}H_4Br_6O$	643.6	7.66 (6.86 – 7.92)[a]	BCF in mussel[g,j] BCF_W : 24,400 BCF_L : 2,200,000 F: + H: + S: +
2,2′,3,3′,4,4′,5,5′-Octabromodiphenylether (OBDE) (PBDE # 194) and other isomers (Great Lakes DE-79; DOW FR-1208HM)	32536–52–0	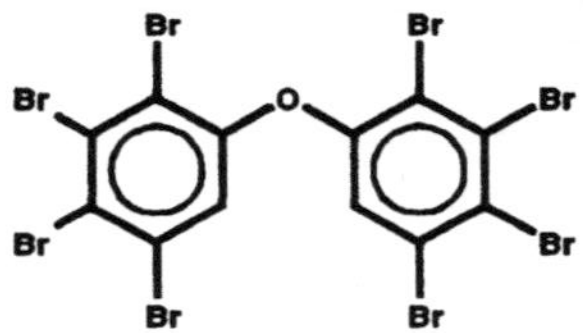	$C_{12}H_2Br_8O$	801.4	8.71 (8.35 – 8.90)[a]	W: – (< 70 ng l^{-1}) S: + H: +

Table 15 (continued)

Chemical Name (abbreviation, PBB No. or PBDE No. and/or trade name of flame retardant)	CAS No.	Chemical structure	Molecular formula	Molecular mass [g mol^{-1}]	log K_{OW}	BCF in fish. Detected in water (W), sediments (S), mussels (M), fish (F), whales (Wh) and/or human fat (H)
Decabromodiphenylether (DeBDE) (PBDE # 209) (Great Lakes DE-83; DOW FR - 300 BA)	1163–19–5	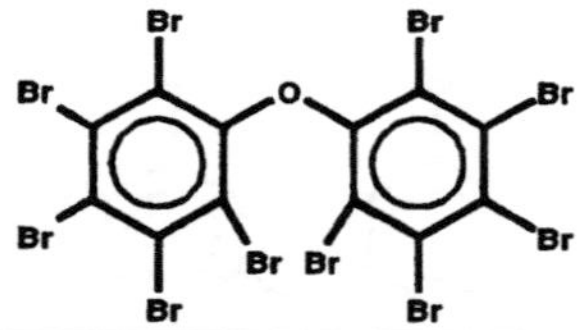	$C_{12}Br_{10}O$	959.2	9.97 [a]	S: + M: + F: + H: +

[a] The log K_{OW} values were determined by Watanabe and Tatsukawa [268] using the HPLC method, as otherwise cited.
[b] BCF_L: Bioavailability-corrected bioconcentration factor on a lipid basis (kinetic method) [252].
[c] The bioaccumulation factors were determined by Oliver and Niimi [251] in rainbow trout using a flow-through test.
[d] No steady-state was reached.
[e] The main compounds of the flame retardant Bromcal 70–5 DE contained tetrabromodiphenylether (41%) and 2 pentabromodiphenylethers (45% and 7%) as was found by Sundström and Hutzinger [263]. However, the Chemische Fabrik Kalk GmbH, Germany, stopped the production in 1985
[f] Water solubility 0.9 ng l^{-1} [259]. The BCF value was determined by the Chemical Inspection and Testing Institute, Tokyo. However, this BCF value is too low, because after 8 weeks no steady-state is reached and because the concentration in the water was 10 µg l^{-1}. That is more than 10,000 times higher than the water solubility.
[g] Kinetic apprach Ref. [401] and personal communication from Michael Gilek to H. J. Geyer.
[h] Concentration of PBDE #47 in water: 0.31 ng l^{-1}.
[i] Concentration of PBDE #99 in water: 0.07 ng l^{-1}.
[j] Concentration of PBDE #153 in water: 0.086±0.11 ng l^{-1}.

ortho-located bromine groups of PBBs. These ortho bromines are able to sterically hinder any rotation and it is suggested that these metabolites have a relative high binding affinity for the estrogen receptor. However, it is important to note that the binding affinity is not necessarily indicative of the biological activity of this chemical in an organism and/or that the activity agrees with the order of binding affinity.

8.6
Bioconcentration of Polybrominated Diphenyl Ethers (PBDEs)

Polybrominated diphenyl ethers (PBDEs) are widely used as additive flame retardants in polymers, especially in electric devices, TV sets, computers, building materials, resins, paints, and textiles [259]. PBDEs are added to these materials at levels up to 10–20%. The use of flame retardants has increased due to stricter fire regulations in many countries [260]. In Sweden e.g. the consumption of PBDEs varies between 1,400 and 2,200 tons per year. In the Netherlands, the annual consumption of these chemicals is estimated to be 2,500 tons [260]. The annual consumption of decabromodiphenyl ether was 4,000 tons, of octabromodiphenyl ether 1,000 tons and of tetrabromodiphenyl ether 1,000 tons during 1987 in Japan [268]. PBDEs are also produced in France (1500 tons), Israel, and the USA. According to OECD, in 1992 world-wide 600,000 tons of flame retardants were used [261]. 150,000 tons were brominated chemicals and 40,000 tons were PBDEs [261]. PBDEs are commercially produced via direct bromination of diphenyl ether with bromine in the presence of a catalyst. The technical products are generally mixtures of isomers and congeners. Commercial PBDEs can be classified into three groups, based on the degree of bromine substitution [262]:

(1) low brominated products which are mixtures of tetra-, penta-, and hexabrominated diphenyl ethers (e.g. Great Lakes DE-71 and Bromkal 70–5-DE),
(2) octabrominated diphenyl ethers (e.g. Great Lakes DE-79, Dow FR-1208 HM), and
(3) decabrominated diphenyl ethers (e.g. Great Lakes DE-83, Dow FR-300BA).

Sundström and Hutzinger [263] found that the main components of the low brominated flame retardant Bromkal 70–5-DE were tetrabromo diphenyl ethers (41%) and two pentabromo diphenyl ethers (45 and 7%). De Boer and Dao [260] analyzed Bromkal 70–5-DE again and compared it with pure standards of 2,2′,4,4′-tetrabromodiphenyl ether (TBDE) and 2,2′,4,4′,5-pentabromodiphenyl ether (PeBDE). They found that the technical mixture contained 36.1% TBDE and 35.5% PeBDE. It has to be mentioned that in 1985 the Chemische Fabrik Kalk GmbH, Germany, discontinued the production of these polybrominated flame retardants. However, these polybrominated aromatic compounds are still produced in Japan, USA, Canada, Sweden, Netherlands, and other industrialized countries.

In 1981, the presence of PBDEs in fish (*Esox lucius*) from Swedish rivers was first reported by Anderson and Blomkvist [264]. The highest PBDE concentrations were found in fish caught in a locally contaminated Swedish river. These environmental chemicals were also detected in sediment, sludge, mussels, fish, and other biological samples collected in the North Sea, Baltic Sea, Arctic Ocean, in Japan, and the USA [265–272a]. De Boer et al. [272b] found polybrominated diphenyl ethers (PBDEs, 2,2′4,4′-tetrabromodiphenyl ether, PBDE #47; 2,2′4,4′,5-pentabromodiphenyl ether, PBDE # 99, and another pentabrominated diphenyl ether with unknown structure) in 13 marine animals of four species (mackerels, harbor seals, minke whales, sperm whales, and whitebeaked dolphins) from the Atlantic Ocean and Dutch coastal seas. The presence of PBDEs in sperm whale blubber (ca. 100 $\mu g\ kg^{-1}$) indicates that these compounds have reached deep ocean waters, as sperm whales are not usually found in shelf seas. Males occur as far north as northern Norway, Iceland, and Greenland. At this latitude, sperm whales hunt in waters of depths 400 to 1200 m or more. The PBDEs were also found in human adipose tissue [273, 274, 286b].

It is important to note that 2,2′3,3′,4,4′,5,5′,6,6′-decabromodiphenyl ether (DeBDE; PBDE # 209) was found in sludge, sediments, and mussels. Recently, it was also detected in fish (pike) samples that were just above the detection limit of about 100 $ng\ g^{-1}$ lipid [259a]. This high detection limit is due to a broad, late-eluting chromatographic peak. In the gas chromatography column, some thermal degradation of PBDE # 209 to lower brominated diphenyl ethers, such as hepta- to nona-BDEs, also occurs, both in standards and fish samples. The PBDE # 209 could not be quantitated in the fish species pike. Nevertheless, the super-hydrophobic chemical with a log K_{ow} of 9.97 can pass the membranes and is bioconcentrated in fish. The super-hydrophobicity of DeBDE may hinder its release from sediment and other particles in the water. Therefore, the real dissolved concentration in the water is very low and so only a very small fraction is bioavailable to fish, mussels, and other gill-breathing organisms.

From their high n-octanol/water partition coefficient (K_{ow}) (see Table 15) it can be assumed that PBDEs could be bioconcentrated to a high extent in fish and other aquatic organisms. Recently the analysis, environmental fate, toxicokinetics, biotransformation, bioaccumulation, toxicity, and environmental occurrence was reviewed by Pijnenburg et al. [249]. In the following part the bioconcentration of PBDEs in aquatic organisms, especially fish, is critically reviewed. Some information on endocrine disrupting properties of PBDEs is also presented.

There is only little information available on bioconcentration using PBDEs in fish and other aquatic organisms. In 1982 the Chemicals Inspection and Testing Institute in Tokyo, Japan, investigated the bioconcentration of pentabromobiphenyl ether in carp [259, 293]. The fish were exposed for 8 weeks to commercial pentabromobiphenyl ether at a concentration of 10 and 100 $\mu g\ l^{-1}$. The bioconcentration factors (on a wet wt. basis) were more than 10,000. However, it is clear that for such super-hydrophobic chemicals with log K_{ow} values of 6.64–6.97 no steady-state is reached after 8 weeks. The second reason that this BCF value was underestimated is that the test chemicals concentration in the water was ca. four to five orders of magnitude higher than the water solubility. The water solubility of pentabromobiphenyl ether (PBDE) is $9 \times 10^{-7}\ mg\ l^{-1} = 0.9\ ng\ l^{-1}$ at 20°C [259].

It is further reported that octabromodiphenylether (OBDE) and decabromodiphenylether (DeBDE) are expected not to bioconcentrate in aquatic organisms [259]. However, this may be due to the experimental conditions. The "real" BCF values of these super-hydrophobic chemicals should be determined by using the kinetic approach and chemical concentrations below the water solubility. Recently in the laboratory of Bo Jansson, Stockholm University, Sweden, the uptake of decabromodiphenyl ether (DeBDE) and other PBDEs from food by rainbow trout (*Oncorhynchus mykiss*) was studied [275]. The rainbow trout were fed with either clean or DeBDE prepared food (7.5–10 mg $kg^{-1}day^{-1}$). Muscle and liver samples were collected for analysis after 0, 16, 49, and 120 days of exposure. A depuration group was fed clean food for 71 days after 49 days of exposure. It was found that the levels of DeBDE and 2,2′,4,4′,5,5′-hexabromo diphenyl ether (HexBDE) increased with time span of exposure. However, after 49 days no steady state was reached. The concentration of a number of brominated organic compounds corresponding to retention time intervals for hexa- to nonabrominated diphenyl ethers also increased with exposure time. Kierkegaard et al. [275] came to the conclusion that DeBDE may be metabolized to lower brominated PBDEs and possibly hydroxylated brominated organic compounds.

From these experiments, from their high lipophilicity, and from the bioconcentration experiments with polychlorinated diphenyl ethers (see Sect. 8.7), it can be concluded that the higher brominated diphenyl ethers possess also a high bioconcentration potential. However, it is very likely that the BCF values of the PBDEs are somewhat lower in comparison to the isosteric polychlorinated diphenyl ether (PCDE) congeners. The reason could be that the brominated aromatic compounds are dehalogenated and/or hydroxylated easier than the chlorinated compounds. Another concern in relation to PBDEs, beside their high bioaccumulation potential, is the formation of toxic polybrominated dibenzofurans (PBDFs) and polybrominated dibenzo-*p*-dioxins (PBDDs) by photolysis, accidental burning, incineration or thermolysis [267, 276–284]. Toxicity studies with PBDDs and PBDFs in rats, mice, monkeys as well as in cell cultures have shown that these compounds exhibit biological and toxic effects (hepatic microsomal AHH and EROD induction, thymic atrophy, body weight loss, and LD_{50}) which are often similar to, although a little less potent, than those of their chlorinated analogues [287–292]. These results have lead to proposals for legislative actions for use of PBDEs in Germany, and other countries [285]. In Germany, since 1989 the chemical industry and the plastic manufacturers renounce voluntarily the use of PBDEs. Nevertheless, PBDEs are found in plastics and will be found in the next years especially in polymers and electronic scrap.

From the high production volume and application, environmental persistence, and their high n-octanol/water partition coefficients (K_{ow}) it can be concluded that the higher brominated diphenyl ethers are bioconcentrated to a high amount in algae, mussels, fish, and other aquatic organisms. The PBDEs may be considered to be a potential threat for aquatic mammals and human health, especially through fish consumption.

Recently it was shown by Darnerud and Sinjari [286a] that 2,2′,4,4′-tetrabromodiphenyl ether decreased the total thyroxine plasma levels in rats and mice after 14 days. It is very likely that PBDEs are metabolized to hydroxylated compounds in mammals including human. In this context it is interesting to note that Haglund et al. [286b] identified and quantified methoxy-polybrominated diphenyl ethers (MeO-PBDEs) beside PBDEs in fish, gray seal, and human adipose tissue. It is suggested by the authors that some of these metabolites can act as endocrine-disrupters or like hormones, such as thyroxine. It may be pointed out, that hydroxylated polybrominated diphenyl ethers (HO-PBDEs) have been detected by Asplund et al. [286c] in fish from the Baltic Sea. Some HO-PBDEs show close structural resemblance with the thyroid hormones 3,3′,5,5′-tetraiodo-L-thyronine (T4) and 3,3′,5-triiodo-L-thyronine (T3). It is therefore not surprising that some hydroxylated PBDEs are bound with different affinity to the thyroid hormone receptors THR-α and THR-β [286d]. The finding that some HO-PBDEs have significant affinity for the thyroid hormone receptors may have far-reaching implications. Furthermore, it was found by Meerts et al. [286e] that there are clear indications that hydroxylated metabolites of PBDEs are potent competitors for thyroxine-binding to the human plasma thyroid hormone transport protein, transthyretin (TTR). It is also possible that these compounds induce the enzymatic conjugation and excretion of thyroxine and thus behave like endocrine-disrupting chemicals. Therefore, it is important to assess the occurrence of PBDEs in the environment to investigate their metabolisms, and to assess the thyromimetic potency of these chemicals, in order to clarify their role as endocrine disrupters.

8.7
Bioconcentration of Polychlorinated Diphenyl Ethers (PCDEs)

Polychlorinated diphenyl ethers (PCDEs) form a group of 209 congeners with physico-chemical properties similar to those of polychlorinated biphenyls (PCBs) [295, 296]. The chlorine substitution on the diphenyl ring and numbering for PCDE congeners are the same as for PCBs [295, 296]. The synthesis of PCDE congeners was performed and described for the first time by Sundström and Hutzinger in 1976 [297]. Up to now over one hundred PCDE congeners have been synthesized by Paasivirta and Koistinen [298] and by Kurz and Ballschmiter [296]. The syntheses, structure verification, and gaschromatographic retention times of PCDEs were recently described by Nevalainen et al. [299].

PCDEs are used as intermediates in chemical syntheses, e.g. production of herbicides chloroxuron, 2,4-dichlorophenyl-*p*-nitrophenyl ether (nitrofen) and binofex [300]. Because PCDEs have physico-chemical properties like PCBs, they are or were used also as heat exchangers. One product is Dowtherm A, a PCDE – PCB mixture, with heat transfer applications. However, in general their use pattern is unknown [301]. Beside the use as heat-exchange fluids, the lower chlorinated PCDEs have been used as flame retardants. It must be also stated that the PCDEs were never directly produced in large quantities. Nevertheless, these aromatic chlorinated compounds have been identified as widespread environmental contaminants. The widespread appearance of PCDEs in the aquatic and terrestrial environment could be most likely due to their presence as impurities in chlorophenols [295, 302]. Chlorophenols or their sodium salts in the past have been widely used as fungicides, bactericides, slimicides, herbicides, and wood preservatives. PCDEs are also found as impurities in commercial preparations of chlorinated phenoxy acetic acids, such as 2,4-dichlorophenoxy acetic acid (2,4-D) and 2,4,5-trichlorophenoxy acetic acid (2,4,5-T) which are produced from chlorinated phenols. The concentrations of PCDEs in chlorophenol preparations vary depending on the used production method [302]. Tri- and tetrachlorophenols have been manufactured by the chlorination of phenol and pentachlorophenol (PCP) by treatment of hexachlorobenzene (HCB) with alkali [302]. Nilson and Renberg [302] found between 100 and 1,000 mg kg^{-1} tetra- to octachlorinated diphenyl ethers in trichlorophenol formulations. The dominating PCDEs in a technical 2,3,4,6-tetrachlorophenol formulation were hexachlorinated diphenyl ethers [303]. PCDEs can also be formed during combustion and therefore are found in fly ash from municipal waste incinerators [295, 296, 304].

In 1976 Sundström and Hutzinger [297] have suggested that leakage of PCDEs into the biosphere may cause bioaccumulation problems similar to those caused by PCB because of the similarity of their physico-chemical properties. It was also shown that PCDEs are relatively stable in the environment [305]. Therefore, it is not surprising that PCDEs are widespread in the environment and are found as environmental contaminants in sediments [303, 306], mussels [306], lobster [306], fish [307, 308], seals [286b, 303], and in human [286b, 309–311].

Neely et al. [312] studied the uptake, elimination, half-life, and bioconcentration of a tetrachlorinated diphenyl ether in trout muscle. Using the kinetic approach, they calculated a BCF_W value of 12,590 and a half-life of 29 days. Zitko and Carson [313] studied the uptake, distribution, and elimination of the 3 chlorinated diphenyl ethers 2,4,4′-trichlorodiphenyl ether (TCDE), 2,3′,4,4′-tetrachlorodiphenyl ether (TeCDE), and 2,2′,4,4′,5-pentachlorodiphenyl ether (PeCDE) in juvenile Atlantic salmon (3.5% lipid) using a static test system. The uptake and excretion of the 3 PCDEs resembled those of the corresponding PCBs. The biological half-lives ranged from 15 to 55 days. The biological half-lives ($t_{1/2}$) of 16 PCDEs were determined by Niimi [314] in rainbow trout (325 g body weight) at 13 °C. The $t_{1/2}$ values ranged from 63 days for trichlorodiphenyl ether to 167 days for 2,2′,4,4′,5,5′-hexachlorodiphenyl ether.

More recently Chui et al. [315] measured the bioconcentration, uptake and elimination kinetics of 4-chlorodiphenyl ether, 2,4-dichlorodiphenyl ether, 2,4,4′-trichlorodiphenyl ether, and 2,4,4′,5-tetrachlorodiphenylether in brook trout (*Salvelinus fontinalis*) of 4–8 g body weight in a flow-through system at 14 °C. The half-lives ranged from 4 to 63 days. The bioconcentration factors on a wet weight basis (BCF_W) ranged from 1570 for 4-chlorodiphenyl ether to 15,700 for 2,4,4′-trichlorodiphenyl ether. The bioconcentration factors calculated on a lipid basis (BCF_L) varied from 28,500 to 285,000. These experimentally determined and some other predicted BCF values of PCDEs in fish and mussels are compiled with their n-octanol/water partition coefficients (log K_{ow}) in Table 16. It can be concluded that the bioconcentration potential of PCDEs is relatively high and is increasing with their increasing lipophilicity (K_{ow} value).

Table 16. Chemical name, chemical structure, molecular formula, molecular mass, water solubility (WS), n-octanol/water partition coefficient (log K_{OW}), and estimated or predicted bioconcentration factors on a wet weight basis (BCF_W) and on a lipid basis (BCF_L) of **Polychlorinated Diphenyl Ethers** (PCDEs) in fish and mussel

Polychlorinated Diphenyl Ethers (PCDE No.)	Chemical structure	Molecular formula	Molecular mass (g mol^{-1})	WS[a] (ng l^{-1})	log K_{OW}[b]	Lipid content of fish, and/or mussel (%)	Bioconcentration factor (BCF)	
							BCF_W	BCF_L
4-Chloro diphenyl ether (3)	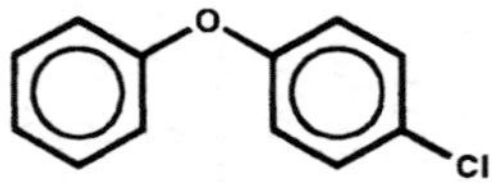	$C_{12}H_9ClO$	204.66	$9.8 \cdot 10^6$	4.70	fish (5.5)	1,570[c]	28,500
						mussel (1)	5,000[d]	50,000[d]
2,4-Dichloro diphenyl ether (7)	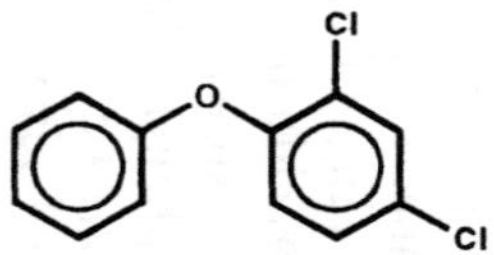	$C_{12}H_8Cl_2O$	239.1	$5.6 \cdot 10^6$	4.93	fish (5.5)	3,670[c]	66,800
						fish (11.6)	9,360[e]	80,700
						mussel (1)	850[d]	85,000[d]
2,4,4′-Trichloro diphenyl ether (28)	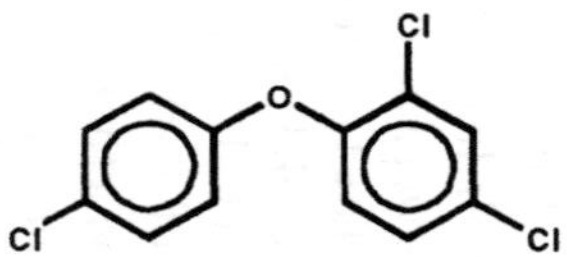	$C_{12}H_7Cl_3O$	273.5	$1.6 \cdot 10^5$	5.53	fish (5.5)	15,690[c]	285,000
						mussel (1)	3,390[d]	339,000[d]
2,4,5-Trichloro diphenyl ether (29)	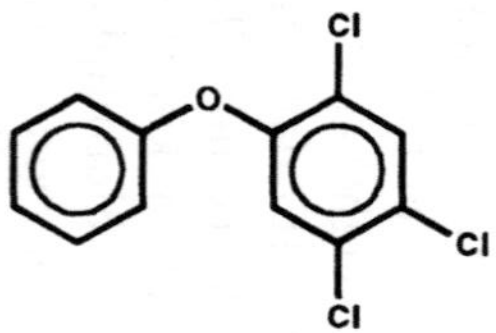	$C_{12}H_7Cl_3O$	273.5	$7.2 \cdot 10^4$	5.58	fish (5.36)	15,000[h]	280,000
						mussel (1)	3,800[d]	380,000[d]
						fish (5)	19,000[f]	380,000[f]
						fish (10)	38,000[f]	

2,2′,4,4′-Tetrachloro diphenyl ether (47)	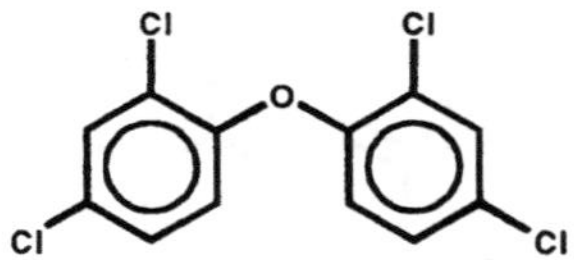	$C_{12}H_6Cl_4O$	308.0	$4.7 \cdot 10^4$	5.95	fish (N.R.)[i]	12,590[g]	
						mussel (1)	8,900[d]	890,000[d]
						fish (5)	44,500[f]	
						fish (10)	89,000[f]	890,000[f]
2,4,4′,5-Tetrachloro diphenyl ether (74)	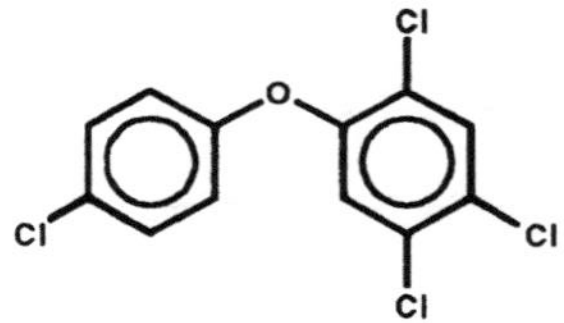	$C_{12}H_6Cl_4O$	308.0	$2.8 \cdot 10^4$	5.99	fish (5.5)	10,940[c]	199,000
							23,900[j]	434,500
						mussel (1)	9,800[d]	980,000[d]
						fish (5)	49,000[f]	
						fish (10)	98,000[f]	980,000[f]
3,3′,4,4′-Tetrachloro diphenyl ether (77)	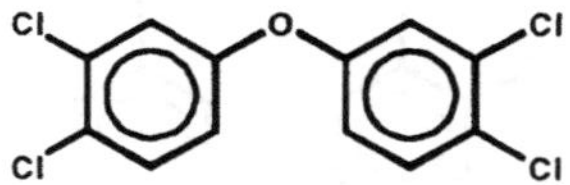	$C_{12}H_6Cl_4O$	308.0	$3.2 \cdot 10^4$	6.36	fish (5.36)	32,000[h]	597,000
						mussel (1)	23,000[d]	
						fish (5)	115,000[f]	$2.3 \cdot 10^{6}$[d,f]
						fish (10)	23,000[f]	
2,2′,4,4′,5-Pentachloro diphenyl ether (99)	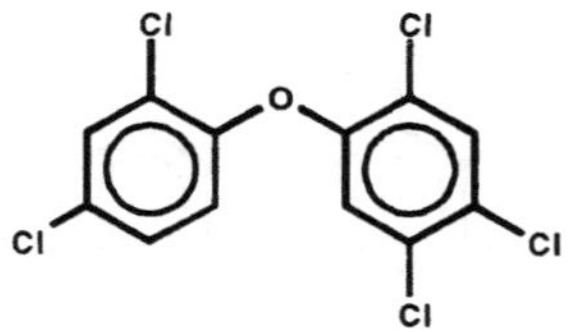	$C_{12}H_5Cl_5O$	342.4	$8.4 \cdot 10^3$	6.38	mussel (1)	24,000[d]	$2.4 \cdot 10^{6}$[d]
						fish (5)	120,000[f]	$2.4 \cdot 10^{6}$[f]
						fish (10)	240,000[f]	
2,2′,3,4′,5,5′-Hexa-chloro diphenyl ether (146)	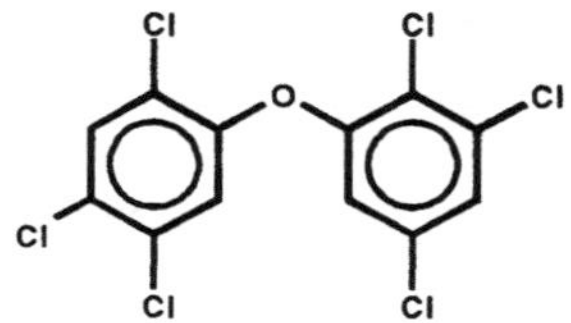	$C_{12}H_4Cl_6O$	376.88	1,500	6.76	mussel (1)	57,500[d]	$5.75 \cdot 10^{6}$[d]
						fish (5)	288,000[f]	$5.75 \cdot 10^{6}$[f]
						fish (10)	575,000[f]	

Table 16 (continued)

Polychlorinated Diphenyl Ethers (PCDE No.)	Chemical structure	Molecular formula	Molecular mass (g mol^{-1})	WS[a] (ng l^{-1})	log K_{OW}[b]	Lipid content of fish, and/or mussel (%)	Bioconcentration factor (BCF)	
							BCF_W	BCF_L
2,3,3′,4,4′,5-Hexachloro diphenyl ether (156)	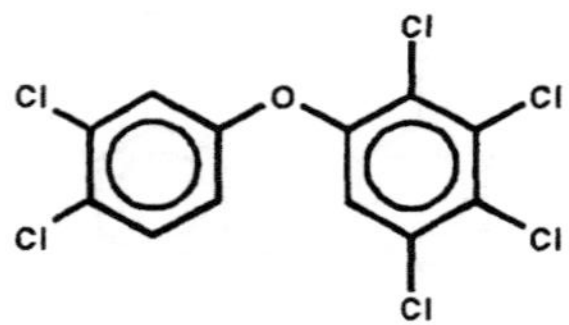	$C_{12}H_4Cl_6O$	376.88	626	7.07	mussel (1)	117,000[d]	$11.7 \cdot 10^{6}$[d]
						fish (5)	585,000[f]	$11.7 \cdot 10^{6}$[f]
						fish (10)	$1.17 \cdot 10^{6}$[f]	
2,2′,3,4,4′,5,5′-Heptachloro diphenyl ether (180)	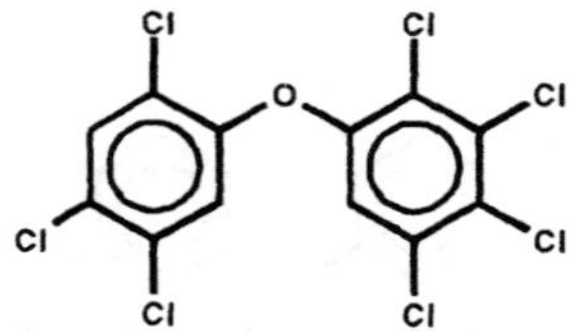	$C_{12}H_3Cl_7O$	411.3	130	7.46	mussel (1)	290,000[d]	$28.8 \cdot 10^{6}$[d]
						fish (5)	$1.44 \cdot 10^{6}$[f]	$28.8 \cdot 10^{6}$[f]
						fish (10)	$2.88 \cdot 10^{6}$[f]	
2,2′,3,3′,4,4′,5,6-Octachloro diphenyl ether (195)	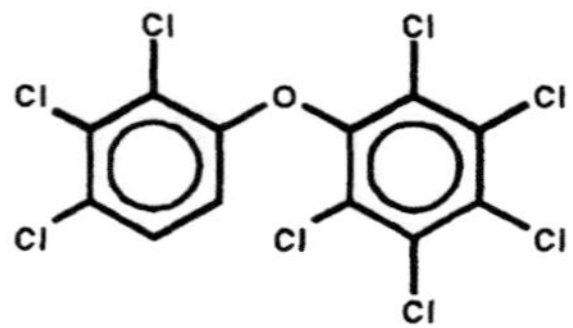	$C_{12}H_2Cl_8O$	445.77	13	7.84	mussel (1)	690,000[d]	$69.0 \cdot 10^{6}$[d]
						fish (5)	$3.45 \cdot 10^{6}$[f]	$69.0 \cdot 10^{6}$[f]
						fish (10)	$6.90 \cdot 10^{6}$[f]	
2,2′,3,4,4′,5,5′,6-Octachloro diphenyl ether (203)	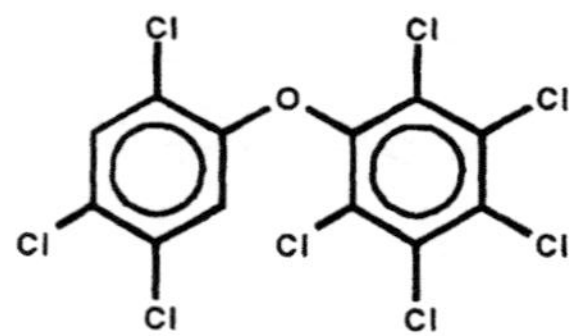	$C_{12}H_2Cl_8O$	445.77	32	7.81	mussel (1)	650,000[d]	$64.6 \cdot 10^{6}$[d]
						fish (5)	$3.23 \cdot 10^{6}$[f]	$64.6 \cdot 10^{6}$[f]
						fish (10)	$6.46 \cdot 10^{6}$[f]	

2,2′,3,3′,4,4′,5,5′,6,6′-Decachloro diphenyl ether (209)	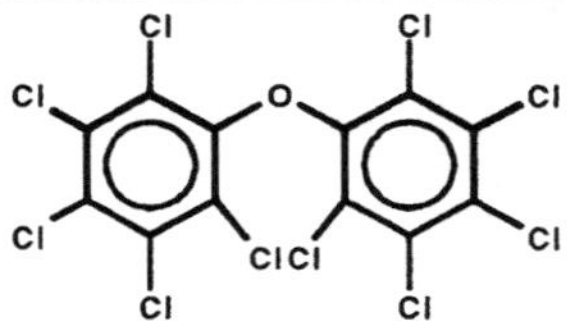	$C_{12}Cl_{10}O$	514.66	0.058	8.16	mussel (1)	$1.45 \cdot 10^{6}$ [d]	$145 \cdot 10^{6}$ [d]
						fish (5)	$7.25 \cdot 10^{6}$ [f]	$145 \cdot 10^{6}$ [f]
						fish (10)	$14.5 \cdot 10^{6}$ [f]	

[a] WS: The water solubility was estimated by Kurz [296] from the relationship of WS and the retention time of test chemicals in reverse-phase high-performance liquid chromatography (RP-HPLC method).
[b] The log K_{OW} values were estimated by Kurz [296] (RP-HPLC method).
[c] BCF value determined by Chui et al. [315].
[d] BCF value predicted in mussel from the n-octanol/water partition coefficient if the original compound is not metabolized or only to a minor extent.
[e] BCF value determined by Oliver and Niimi [325b].
[f] BCF value predicted in fish from the K_{OW} value if the original compound is not metabolized or only to a minor extent.
[g] BCF value determined by Neely [312].
[h] BCF value determined by Opperhuizen and Voors [325a].
[i] N.R.: Not Reported.
[j] Recalculated by Dr. Xiulin Wang from K_1 and K_2.

PCDEs have been reported to elicit biochemical and toxic responses similar to those reported for PCBs and related aromatic hydrocarbons [315–317]. Beside the high bioconcentration potential which is mainly due to their high persistence and long half-life in aquatic organisms, the conversion to polychlorinated dibenzofurans (PCDFs) and polychlorinated dibenzo-*p*-dioxins (PCDDs) during industrial processes and thermal and photolytic reactions is another disadvantage of PCDEs (for review see [316]). It is also known that PCDEs, such as 4-chlorodiphenylether, 2,4-dichlorodiphenylether, 2,4,4′-trichlorodiphenylether, and 2,2′,4,4′,5-pentachlorodiphenylether can be metabolized to hydroxylated products in fish and rats [318, 319]. The parent compounds are hydroxylated primarily at the 4′ position. However, if the 4 and 4′ positions in the PCDE molecule were occupied, their corresponding metabolic rates were slower and *ortho*-hydroxylated metabolites were observed. The monohydroxylated metabolites predominated among the metabolic compounds [320].

The possibility of deleterious health effects from low level exposure to environmental chemicals, especially with regard to endocrine disruption, is of great interest. It is known that PCDEs induce cytochrome P-450 1A1 mediated enzyme activities, and they therefore should bind to the Ah (dioxin) receptor. Because PCDEs and thyroid hormones, such as L-thyroxine (L-3,3′,5,5′-tetraiodothyronine, T_4) and L–3,3′5-triiodothyronine (T_3), show structural similarities it is conceivable that these halogenated diphenylethers could interfere with the thyroid receptor binding and/or thyroid hormone metabolism. Especially the non-planar hydroxylated PCDEs should bind with high affinity to the thyroid receptor and/or transthyretin (TTR) and thus disrupt the thyroid hormone transport. This can be concluded also from the investigation of Lans et al. [321]. They found that 4-hydroxy-3,3′5,5′-tetrachlorobiphenyl, a major metabolite of 3,3′5,5′-tetrachlorobiphenyl, selectively inhibited the binding of T_4 to transthyretin in plasma of rats. The binding strength of 4-hydroxy-3,3′5,5′-tetrachlorobiphenyl is 4 times greater to TTR than T_4. This binding is due to the structural resemblance of the hydroxy-ring and the diiodophenyl-ring of the thyroid hormone [322, 324]. This competitive binding to TTR by the hydroxylated PCBs causes increased glucuronidation and biliary excretion of thyroxin (T_4) resulting in decreased T_4 plasma levels [321, 323]. The same phenomenon may occur with hydroxylated PCDE metabolites.

8.8 Bioconcentration of Nitro Musk Compounds (NMCs)

Nitro musks, especially musk xylene and musk ketone (for their structures see Table 17) have been used since many years in large amounts as fragrances in the industrial production of soaps, cosmetics, and laundry detergents [326–328]. Their world-wide production was numbered 1987 about 2,500 tons per year [326]. Mainly China and India are producing nitro musks for the world market [329, 330]. In 1991, an increase in the Chinese musk xylene production of about 29% was reported [329]. Recently Geyer et al. [331] and Rimkus and Brunn [332] reviewed the significance of nitro musk compounds (NMCs) in the aquatic environment. Some ecotoxicological data, such as the acute toxicity of the nitro musk compounds to bacteria, algae, and *Daphnia*, were investigated and summarized by Schramm et al. [333].

Yamagishi et al. [334] identified 1981 for the first time musk xylene and musk ketone in the aquatic ecosystem. They analyzed these nitro aromatics in freshwater fish collected from a Japanese river. Additionally, in a further study [335] fish, mussels, river water, and waste water from this area were investigated to identify the routes and extent of contamination. About ten years later, Rimkus and Wolf [336, 337] analyzed nitro musks in fish, mussels, and shrimps from various locations and started a broad discussion and many activities in this field. They found the highest concentrations in some samples of rainbow trout (*Oncorhynchus mykiss*) from Danish and Spanish aquacultures [336–338]. In the meantime these results have been confirmed in general by other studies carried out in Germany, Switzerland, Denmark, and the Netherlands (summarized in [331, 332]). The highest musk xylene and musk ketone residue levels reported in literature till now were found by Eschke et al. [338] in some eels from a pond which received the water from a municipal sewage treatment plant. In this context it is important to note that the Spanish regulations allow fish farmers to take up to 75% of total river flow for their fish ponds [347]. In Denmark the fish farmers use also river water for their aquacultures. Therefore it was suggested that the fish were contaminated by NMCs by uptake and bioconcentration of these compounds from the water.

Up to now, there are some data on the bioconcentration (uptake from water) and bioaccumulation or biomagnification (uptake from food) of nitro musk compounds in fish. Even in the first environmental studies of Yamagishi et al. [335], relatively high bioconcentration factors on a wet weight basis (BCF_W) of 4,100 and 1,100 for musk xylene and musk ketone, respectively, were reported. These BCF_W values were estimated semi-quantitatively as ratios between the average analyzed concentrations in muscle of fish and in river water. BCF_W values of 640 –5,820 (10 µg musk xylene l^{-1}) and 1,440 –6,740 (1 µg musk xylene l^{-1}) for musk xylene were found in an experiment of 10 weeks with Japanese carps [340]. But there is not enough information about the exact parameters of this fish test and the reasons for the relatively broad range of values. Recently in a long-term bioconcentration study rainbow trout (*Oncorhynchus mykiss*) were exposed in a flow-through test system for several months to musk xylene at relatively low water concentrations (in average, 22.5 ng musk xylene l^{-1}) [341, 346]. A fast and high bioconcentration in fish muscle was observed, with an estimated BCF_W of about 4,400 for musk xylene. Further calculations resulted in BCF_W values between 4,200 and 5,100 as well as 115,000–122,000 for bioconcentration factors on a lipid basis (BCF_L) respectively, depending on the mathematical model applied to describe the data [342, 343].

In the MITI list for musk xylene a log K_{ow} value of 5.20 was published [340]. Rimkus et al. [344] determined by reversed-phase HPLC a log K_{ow} value of 4.90. From this log K_{ow} value we estimated by means of a Quantitative Structure-Activity Relationship (QSAR) of Mackay [345] a BCF_W for musk xylene of about 3,800 and a BCF_L of 79,200 for this compound in fish [346]. The predicted BCF_W of musk ketone (log K_{ow} = 4.20 [344]) was 760 and the BCF_L 15,800 [346]. These data, the chemical structures, n-octanol/water partition coefficients (log K_{ow} s)

Table 17. Trivial name, CAS number, chemical name, chemical structure, molecular formula, molecular mass, n-octanol/water partition coefficient (log Kow), and bioconcentration factors on a wet weight basis (BCF_W) and/or on a lipid basis (BCF_L) of **Nitro Musk Compounds (NMCs)** in mussel and fish, which were detected in an aquatic environment, and/or in human milk and adipose tissue

Trivial name (abbreviation)	CAS No.	Chemical name	Chemical structure
Musk Xylene (MX)	81–15–2	1-*tert*-Butyl-3,5-dimethyl-2,4,6-trinitrobenzene; 2,4,6-Trinitro-1,3-dimethyl-5-*tert*-butylbenzene; 2,4,6-Trinitro-5-*tert*-butyl-1,3-xylene	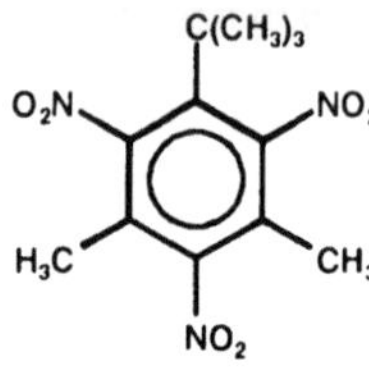
Musk Ketone (MK)	81–14–1	1-*tert*-Butyl-3,5-dimetyl-2,6-dinitro-4-acetylbenzene; 4-*tert*-Butyl-3,5-dinitro-2,6-dimethyl-acetophenone	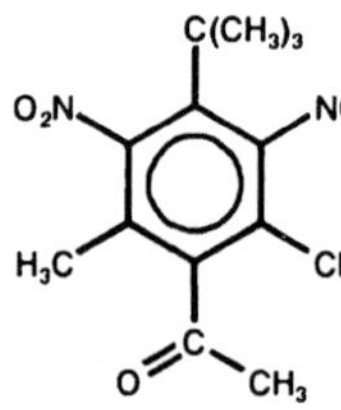
Musk Ambrette (MA)	83–66–9	1-*tert*-Butyl-2-methoxy-4-methyl-3,5-dinitrobenzene; 4-*tert*-Butyl-2,6-dinitro-3-methoxytoluene	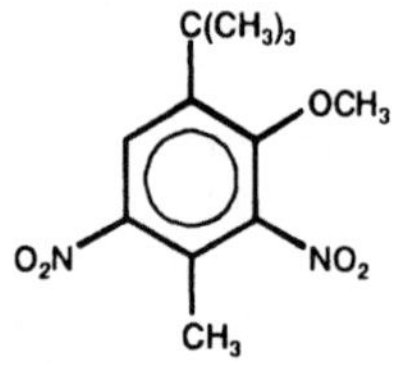
Musk Tibetene (MT)	145–39–1	1-*tert*-Butyl-3,4,5-trimethyl-2,6-dinitrobenzene	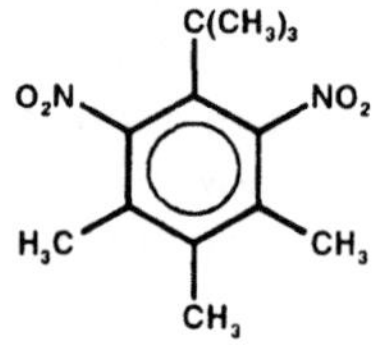
Toluene Musk (TM)[h]	547–94–4	1-*tert*-Butyl-3-methyl-2,4,6-trinitrobenzene; 2,4,6-Trinitro-3-*tert*-butyltoluene; 2-*tert*-Butyl-4-methyl-1,3,5-trinitrotoluene	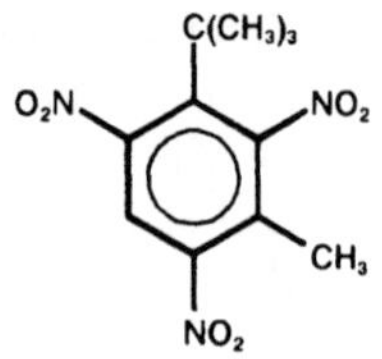

Molecular formula	Molecular mass [g mol^{-1}]	log Kow[a]	Bioconcentration factor		Detected[b] in rivers (R), waste water (W), mussel (M), fish (F) and/or human (H)
			BCF_W (Lipid %)	BCF_L	
$C_{12}H_{15}N_3O_6$	297.3	4.90	1) Carp (3.4%) 1,440–6,740[c]	42,400 – 198,200[c]	R: +
			2) Carp (3.4%) 640–5,820[c]	18,800– 171,200[c]	W: + M: + F: +
			Rainbow trout (3.7%) 4,400[i]	118,900[i]	H: +
			mussel (1%) 790[e]	79,000	
$C_{14}H_{18}N_2O_5$	294.3	4.20	1,100[d]		R: +
			mussel (1%) 160[e]	15,800[e]	W: + M: +
			fish (5%) 790[e]	15,800[e]	F: + H: +
			fish (10%) 1,580[e]		
$C_{12}H_{16}N_2O_5$	268.2	4.44	mussel (1%) 275[e]	27,500[e]	M: + F: +
			fish (5%) 1,370[e]	27,500[e]	H: +
			fish (10%) 2,750[e]		
$C_{13}H_{18}N_2O_4$	266.3	5.01	mussel (1%) 1,020[e]	102,000[e]	N. D.[g]
			fish (5%) 5,100[e]	102,000[e]	
			fish (10%) 10,200[e]		
$C_{11}H_{13}N_3O_6$	283.2	4.34[f]	mussel (1%) 219[e]	21,900[e]	N. D.[g]
			fish (5%) 1,095[e]	21,900[e]	
			fish (10%) 2,190[e]		

Table 17 (continued)

Trivial name (abbreviation)	CAS No.	Chemical name	Chemical structure
Musk Moskene (MM)	116-66-5	1,1,3,3,5-Pentamethyl-4,6-dinitroindane; 2,3-Dihydro-1,1,3,3,5-pentamethyl-4,6-dinitroindane	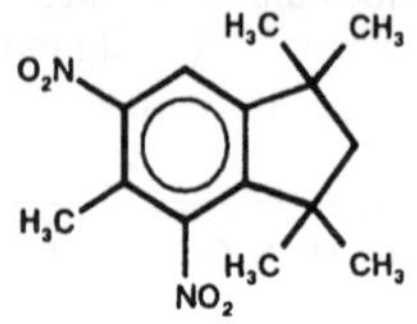

Source: Adopted with modifications from Geyer et al. [331], Eschke et al. [339], and Rimkus and Brunn [332].

[a] The K_{OW} values were estimated by Rimkus et al. from the relationship of log K_{OW} and the retention time of test chemicals in reverse-phase high-performance liquid chromatography (RP-HPLC).

[b] Data from Rimkus and Brunn [332] and Eschke et al. [339].

[c] Bioconcentration test with carp (3.4% lipid), flow-through tests for 10 weeks. MX concentration in water: 1) 1 μg l^{-1}. 2) MX concentration in water: 10 μg l^{-1} [340].

[d] BCF value was calculated by Yamagishi et al. from the concentration of MK in freshwater fish from the environment and the concentration in water [334].

and other information as well as the predicted BCF_W and BCF_L values of NMCs in fish and mussels are compiled in Table 17.

On the other hand, bioaccumulation fish tests with spiked feed (1 and 10 μg musk xylene/kg feed, respectively) resulted after 140 days in non-detectable residues in the fish [341]. That means that no biomagnification and no bioaccumulation occurs and that the residues in fish and may be also in other aquatic gill-breathing organisms can be explained by the uptake from water alone.

In summary, there is a relatively good conformity of all these BCF data of musk xylene (MX) in fish (Table 17). However, Boleas et al. [348] found very low BCF_W values between 10 and 60 of MX in edible portion of rainbow trout (body weight: 44.2 ± 2.8 g). The test was performed under static conditions with daily water renewals. However, for such highly lipophilic compounds this method is not suitable. Therefore these BCF values can not be accepted especially because the water concentration was not measured and the analytical method for the determination in fish and water is questionable [346]. All the other BCF values document the relatively high bioconcentration potential of these lipophilic substances, which is comparable to some other typical pollutants such as some organochlorine pesticides, chlorinated benzenes, and lower chlorinated PCBs etc. Due to the large world-wide production and use as well as their persistence and the high bioconcentration potential of these lipophilic substances, the nitro musk compounds are apparently ubiquitously distributed in the aquatic environment and, therefore, are found in fish, mussels, and shrimps [332, 337]. Thus, nitro musk compounds represent a new class of environmental contaminants of high relevance and priority in aquatic ecosystems.

Molecular formula	Molecular mass [g mol^{-1}]	log Kow[a]	Bioconcentration factor		Detected[b] in rivers (R), waste water (W), mussel (M), fish (F) and/or human (H)
			BCF$_W$ (Lipid %)	BCF$_L$	
$C_{14}H_{18}N_2O_4$	278.3	5.29	mussel (1%) 1,950[e] fish (5%) 9,750[e] fish (10%) 19,500[e]	195,000[e] 195,000[e]	F: + H: +

[e] BCF values of NMCs in mussels and fish predicted from the n-octanol/water partition coefficient (K_{OW}) if the original compound is not or only slowly metabolized.
[f] The log K_{OW} value was calculated by Kaune on the basis of log K_{OW} of musk xylene (4.90).
[g] N. D.: Not detected.
[h] TM has no relevance of industrial production.
[i] Flow-through test (concentration of MX in water 22.5 ng l^{-1}). For more information see Ref. [341] and [346].

It is also important to note that some nitromusk compounds were also found in human fat and milk. It is generally proven and accepted that the main route of uptake of persistent lipophilic chemicals, such as DDT, PCBs, PCDDs and PCDFs, is performed to more than 95% by food, especially of animal origin, such as fish, cow's milk, cheese, eggs, and meat of pigs, cattle, etc. However, because NMCs were found only in aquatic organisms and not in other food of terrestrial animal origin, the oral uptake of these NMCs by human is negligible. These compounds are mainly taken up by human via dermal absorption due to their frequent and intense dermal contact as fragrances in cosmetics and washed textiles [349–351 a].

Furthermore, it has to be noted that the nitro groups in the NMCs are metabolized by microorganisms and animals such as fish and rats. It is known that aromatic amines (substituted anilines) are acetylated to acetanilides. Some of these compounds possess anti-androgenic properties [351 b, c, d]. It is supposed that some *N*-acetylated metabolites of NMCs, e.g. 2-methyl-3-nitro-4-methoxy-5-*tert*-butyl-acetanilide (metabolite of musk ambrette) and 4-*tert*-butyl-2,6-dimethyl-3,5-dinitro-acetanilide (metabolite of musk xylene) are bound to the androgen receptor (AR) and may act as weak anti-androgens [351 e].

8.9 Bioconcentration of Polycyclic Musk Fragrances (PMFs)

Polycyclic musk fragrances (PMFs) are indane and tetraline derivatives with different substituents [361 a]. These chemicals with strong musk odor are used as fragrances in cosmetics and laundry detergents and are of great industrial importance. According to a study of the fragrance industry in 1987 the world-wide production of polycyclic musk fragrances was 4,300 metric tons [352]. In the year 1996 ca. 5,600 tons of PMFs were used world-wide [361 b]. The world-wide production e.g. of HHCB (e.g. Galaxolide) has been reported to be 1,000 tons per year [353]. The chemical structures of PMFs, together with their abbreviations, chemical names, CAS numbers, trade names, molecular mass, and n-octanol/water partition coefficients are presented in Table 18.

The state of the art of the polycyclic musk fragrances was recently reviewed by Rimkus and Brunn [362, 363]. Eschke and coworkers [354, 355] for the first time found some of these PMFs in surface waters, waste waters, and fish in Germany. All fishes from the river Ruhr contained HHCB (e.g. Galaxolide) and AHTN (e.g. Tonalide). Adult eels contained the highest concentrations of these PMFs because this fish species had also the highest lipid content. Rimkus and Wolf [356] investigated eels and pike-perches from the river Elbe (Germany), rainbow trout from Danish aquacultures, different fish species from the German River Stör, mussels and crabs from the North Sea, as well as shrimps from Asia for these PMFs. In nearly all these aquatic gill-breathing organisms HHCB (e.g. Galaxolide) and AHTN (e.g. Tonalide) were found. It was obvious that the concentrations of HHCB and AHTN were higher than the levels of nitro musk compounds. This could be due to higher concentrations in the water, caused by higher production and usage rates and/or to the higher n-octanol/water partition coefficients (log K_{ow}) and thus higher bioconcentration potential of the PMFs in comparison to the nitro musk compounds (see Tables 17 and 18). We predicted the BCF values of these compounds in mussels and fish from their n-octanol/water partition coefficients (K_{ow}). The K_{ow} values were determined by reversed-phase HPLC method by Eschke et al. [357] and are compiled together with measured and/or predicted BCF_W and BCF_L values in Table 18.

The bioconcentration of ^{14}C labeled HHCB and AHTN in bluegill sunfish (*Lepomis macrochirus*) has been tested using two concentrations in a flow-through test according to the OECD guideline 305 E [367 a, b]. In both tests dimethylformamide or Tween 80 were used as solubilizers of HHCB and AHTN. While HHCB was radiochemically pure (three isomer groups), AHTN was only 78.8% radiochemically pure.

(a) BCF of HHCB: In fish the concentration of HHCB reached plateau levels after 3–7 days. However, no uptake rate could be determined. The elimination of HHCB from fish followed first-order kinetics. The elimination half-lives were 2–3 days. The bioconcentration factors (BCF_W) were calculated from the plateau level in fish after 28 days and the overall mean ^{14}C-HHCB concentrations in water ($0.91 \pm 0.10\ \mu g\ l^{-1}$). The BCF_W value of HHCB based on total radioactivity in whole fish was 1624 while the BCF_W based on the parent chemical was 1584 [367 b].

(b) BCF of AHTN: The concentration of AHTN reached the plateau level after 3–7 days if the concentration of AHTN in the water was $0.99 \pm 0.12\ \mu g\ l^{-1}$. Due to a rapid stabilization of the AHTN concentration in fish no uptake rate could be determined. The elimination half-lives were 0.8–2.1 days. The BCF_W value was calculated from the total radioactivity in fish after plateau was reached and the mean concentration in water. The BCF_W value in whole fish for ^{14}C-AHTN was 1320. However, in fish a very polar metabolite fraction was found in the same or higher amounts as the parent compound. If the BCF_W value for the whole fish was calculated on actual concentration of parent compound at plateau level in fish and in the exposure water, a BCF_W value of 597 was obtained [337b].

These BCF_W values of these very lipophilic polycyclic musk fragrances are relatively low compared to the predicted BCF values calculated by means of Eq. (26). At this time no exact explanation for this phenomenon can be given. It is known that the parent chemicals HHCB and AHTN are metabolized in the fish to more polar compounds that will be eliminated at a higher rate. It is also possible that the low BCF value of ^{14}C-AHTN may be due to the low radiochemical purity of 78.8%. It seems therefore necessary to perform bioconcentration tests with PMFs of high purity in the absence of a solubilizer and to use water concentrations of these very lipophilic PMFs in the lower ng l^{-1} range, which are found in fresh water systems [362], and to use the kinetic approach. At this time no exact water solubility data are available.

Some of these polycyclic musk fragrances were also found in humans [358–360]. Consumption of fish and other food from aquatic ecosystems contaminated with PMFs can not explain the concentration in human. It is assumed that the occurrence of these lipophilic compounds in human adipose tissue or mother's milk is mainly due to dermal sorption from cosmetics and detergents [349–351]. In the future the production and use of polycyclic musk fragrances will still increase and the nitro musk compounds will be replaced by the PMFs.

It is assumed that some compounds of this group (AHTN and ATTN) may bind to the retinoid acid receptor (RAR) or retinoid X receptor (RXR) because their structure shows some similarity with synthetic RXR ligands [364, 365]. The RAR and RXR belong to the steroid/thyroid hormone nuclear receptor super family. They play a central role in the regulation of many intracellular receptor pathways [366]. However, all these assumptions and predictions, especially the predicted high bioconcentration potential of the PMFs, have to be investigated experimentally.

8.10
Bioconcentration of Sunscreen Agents (SSAs)

Sunscreen agents (SSAs), called also UV filter substances, are preferably utilized in the production of sun protective agents. Moreover, these chemicals are used partly for means of preservation in many other cosmetic products, such as shampoos, hair cosmetics, fragrance waters, and foam [368]. In 1993, the amount of sun protective products in Germany was 8000 metric tons. These products can contain up to 10% sunscreen agents. The production of sunscreen agents (UV filter substances) in 1993 in Germany amounted to approximately 1000 metric tons [369]. In 1993/94 in Germany 23 sunscreen agents were allowed in cosmetics [368]. In Table 19, the most relevant sunscreen agents with their chemical name, CAS registration number, chemical structure, molecular formula, and molecular mass are presented.

Table 18. Chemical name, trade name, CAS number, chemical structure, molecular formula, molecular mass, n-octanol/water partition coefficient (log K_{ow}), measured and/or predicted bioconcentration factors on a wet weight basis (BCF_W) and on a lipid basis (BCF_L) in mussel or fish, and occurrence of **Polycyclic Musk Fragrances (PMFs)** in aquatic environment and/or human milk and adipose tissue

Chemical name (abbreviation)	Trade name(s)	CAS No.	Chemical structure
1,3,4,6,7,8-Hexahydro-4,6,6,7,8,8-hexamethyl-cylopenta-(γ)-2-benzopyran (HHCB)	Galaxolide Abbalide Pearlide	1222–05–5	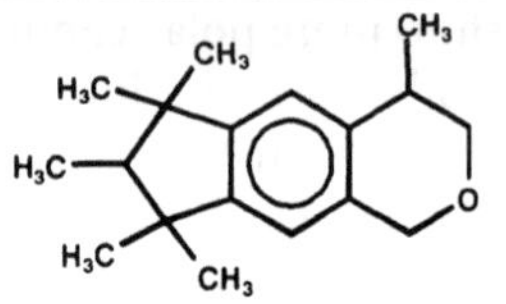
7-Acetyl-1,1,3,4,4,6-hexa-methyl-1,2,3,4-tetrahydro-naphthalene (AHTN)	Tonalide Fixolide	1506–02–1	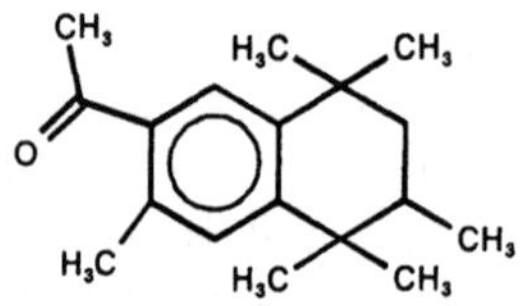
4-Acetyl-1,1-dimethyl-6-*tert*-butylindan (ADBI)	Celestolide Crysolide	13171–00–1	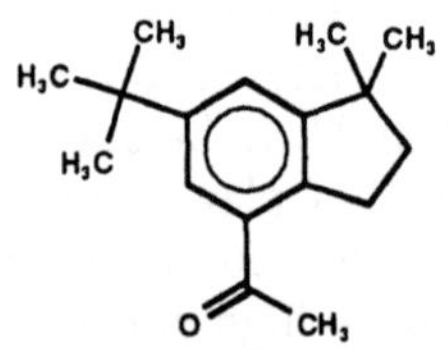
6-Acetyl-1,1,2,3,3,5-hexa-methylindan (AHMI)	Phantolide	15323–35–0	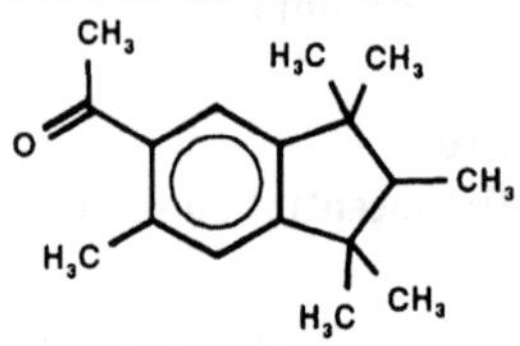
6,7-Dihydro-1,1,2,3,3-penta-methyl-4(5H)-indanone (DPMI)	Cashmeran	33704–61–9	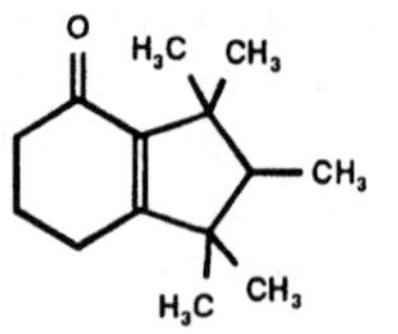

Molecular formula	Molecular mass [g mol⁻¹]	log Kow[a]	Bioconcentration factor		Detected in rivers (R), waste water (W), mussel (M), fish (F) and/or human (H)
			BCF_W (Lipid %)	BCF_L	
$C_{18}H_{26}O$	258.40	5.9	mussel (1.4%): 620[i] fish (NR): 1,624[e] 1,584[g] 624[g,h]	 44,400[i] 33,200[g,h]	R: + W: + F: + H: +
$C_{18}H_{26}O$	258.40	5.8	mussel (1.4%): 560[i] fish (NR): 1,320[f] 597[g] 600[g,h]	 40,100[i] 33,700[g,h]	R: + W: + F: + H: +
$C_{17}H_{24}O$	244.38	5.4	fish (5%): 670[b]	 13,300[b]	R: + W: + F: + H: +
$C_{17}H_{24}O$	244.38	5.8	fish (5%): 1,670[b]	 33,400[b]	R: + W: + F: + H: +
$C_{14}H_{22}O$	206.33	4.5	fish (5%): 84[b]	 1,680[b]	N.D.[c]

Table 18. (continued)

Chemical name (abbreviation)	Trade name(s)	CAS No.	Chemical structure
5-Acetyl-1,1,2,6-tetramethyl-3-isopropylindan (ATII)	Traseolide	68140-48-7	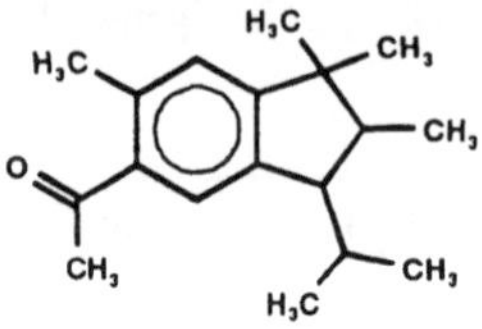
7-Acetyl-1,1,4,4-tetramethyl-6-ethyl-1,2,3,4-tetrahydro-naphthalene (ATTN)	Versalide[d]	88-29-9	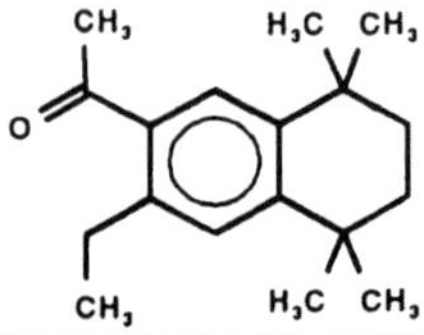

Source: Adapted with modifications and extensions from Eschke et al. [354, 355] and Rimkus and Wolf [356].

[a] The n-octanol/water partition coefficients were determined by the RP-HPLC method by Eschke [357].

[b] Predicted bioconcentration factors of PMFs in fish from the n-octanol/water partition coefficient **under consideration of metabolism.**

[c] N. D.: Not detected.

[d] Versalide has neurotoxic effects and is therefore no longer produced since 1980.

Analytical methods for the determination of sunscreen agents by gas chromatography-mass spectrometry (GC/MS) were published by Ternes et al. [374], Kazuo et al. [375] and Ro et al. [376]. The photostability and photoreactivity of 4-isopropyldibenzoylmethane (IDBM) and 4-*tert*.butyl-4′-methoxy-dibenzoylmethane (TDM) was recently investigated by Schwack and Rudolph [377].

It is interesting to note that recently Hany and Nagel [373] determined benzophenone-3 (BP-3) and octyl methoxycinnamate (OMC) in German human breast milk samples at a concentration range between 16 and 417 $\mu g\ kg^{-1}$ (on a fat basis). According to manufacturers information, up to 2% of the applied sunscreen agents can be absorbed via skin. Some cases of contact and photocontact allergies to certain sunscreen agents have been reported in clinical studies. Therefore, in the European Union 4-isopropyl dibenzoylmethane (IDBM) is no longer allowed as a sunscreen agent in sun protective products.

In 1993, Ternes [370] for the first time identified and quantified the sunscreen agents 3-(4′-methyl benzyliden)-camphor [MBC] and *p*-dimethylamino benzoicisooctylester [DABI] in fish from five different lakes of Germany. The contamination of water and fish of some lakes in Germany was further investigated in the years 1991 and 1993 by Nagtegaal et al. [371]. These scientists

Molecular formula	Molecular mass [g mol^{-1}]	log Kow[a]	Bioconcentration factor		Detected in rivers (R), waste water (W), mussel (M), fish (F) and/or human (H)
			BCF_W (Lipid %)	BCF_L	
$C_{18}H_{26}O$	258.40				F: + H: +
$C_{18}H_{26}O$	258.40				N.D.

[e] Estimated BCF_W value of ^{14}C-labeled compound in bluegill sunfish (0.35 g initial weight) after 28 d [367b].
[f] Estimated BCF_W value of ^{14}C-labeled compound in bluegill sunfish (1.2–1.4 g initial weight) after 28 d [367b].
[g] BCF based on the parent compound.
[h] Estimated BCF value in zebrafish from Ewald [367c].
[i] BCF estimated by Gattermann et al. in mussels (*Mytilus edulis*) from a pond of a sewage treatment plant [403].

detected and quantified six different sunscreen agents (see Table 19) in the fish species perch (*Perca fluviatilis*) and roach (*Rutilus rutilus L.*) of the lake Meerfelder Maar/Eifel in Germany. Both fish species were contaminated in the same range with sunscreen agents and organochlorinated chemicals, such as polychlorinated biphenyls (PCBs) and DDT. In the lake water, the concentrations of sunscreen agents were mostly below the detection limits. However, in a lake the concentration of E-3-(4′-methyl benzylidene)-camphor (MBC) was 4 ng l^{-}. The bioconcentration factor on a wet weight basis (BCF_w) of this chemical in perch with 2.24% lipid was calculated by Nagtegaal et al. [371] to be 5,400. The bioconcentration factor on a lipid basis (BCF_L) in fish is 240,000. This BCF_L value of E-3-(4'methylbenzylidene)-camphor is in excellent agreement with the n-octanol/water partition coefficient (log K_{ow}: 5.4) of this chemical [372]. At this time, to the best of our knowledge, no bioconcentration factors in fish or K_{ow} values of other sunscreen agents had been published.

The investigations and results by Ternes [370], Nagtegaal et al. [371], and Hany and Nagel [372] indicate that some sunscreen agents have probably to be counted as a new group of environmental chemicals which are relatively lipophilic and are therefore bioconcentrated in aquatic organisms, such as algae, *Daphnia*, mussels, and fish. It is important to protect the human skin against ultraviolet radiation of sunlight to prevent sunburn and especially skin cancer.

Table 19. Trivial name, chemical name, abbreviation, CAS No., chemical structure, molecular formula, and molecular mass of **Sunscreen Agents** (SSA) which were identified and quantified in fish and/or human

Trivial name or synonym, chemical name (abbreviation)	CAS No.	Chemical structure	Molecular formula	Molecular mass [g mol^{-1}]	Bioconcentrated or detected in fish and/or human[a]
4-Isopropyldibenzoylmethane; 1-(4-Isopropylphenyl)-3-phenyl-1,3-propanedione **(IDBM)**[b]	63250–25–9	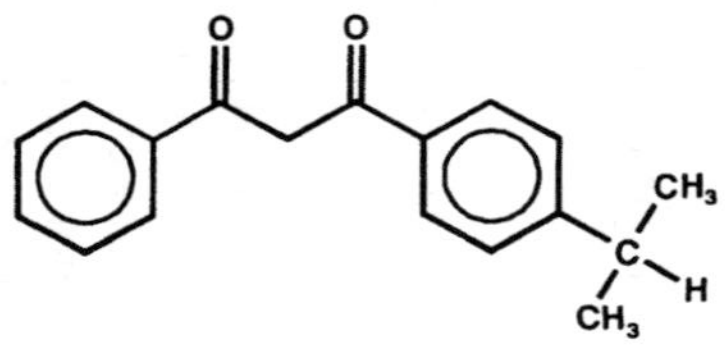	$C_{18}H_{18}O_2$	266.37	fish: +
Butyl methoxydibenzoyl-methane; 4-*tert*-Butyl-4′-methoxy-dibenzoylmethane **(TDM)**	70356–03–1	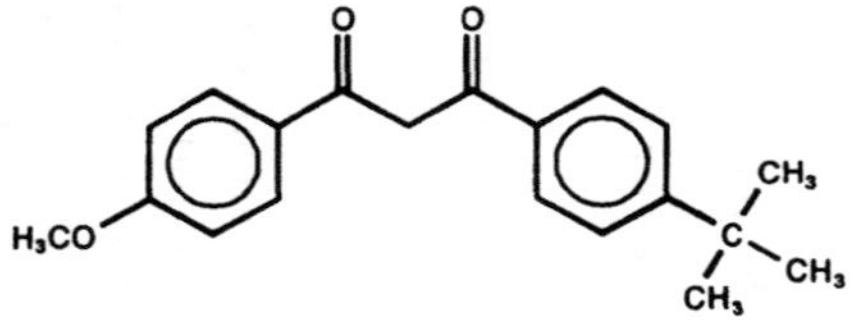	$C_{20}H_{22}O_3$	310.39	fish: +
4-Methylbenzylidenecamphor; Bicyclo[2,2,1]heptan-2-one-1,7,7-trimethyl-3-(4′-methyl-benzylidene); 3-(4′-Methylbenzylidene)-bornan-2-one **(MBC)**	38102–62–4	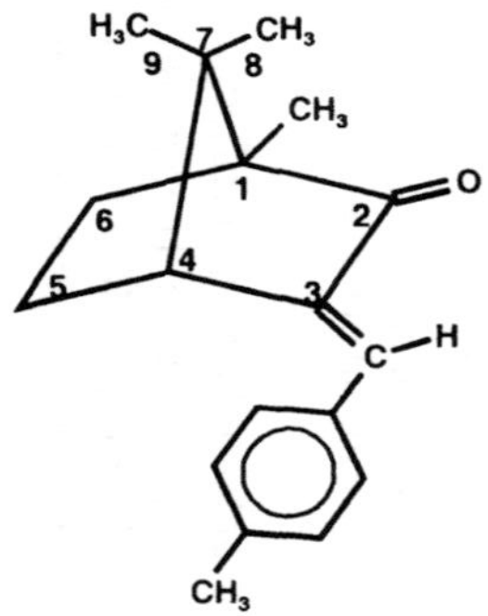	$C_{18}H_{22}O$	254.37	fish: + log K_{OW}: 5.4 bioconcentration factor in fish (lipid: 2.24%): BCF_W: 5,400[c] BCF_L: 241,000[c]

Benzophenone-3; 2-Hydroxy-4-methoxybenzo-phenone **(BP-3)**	131–57–7	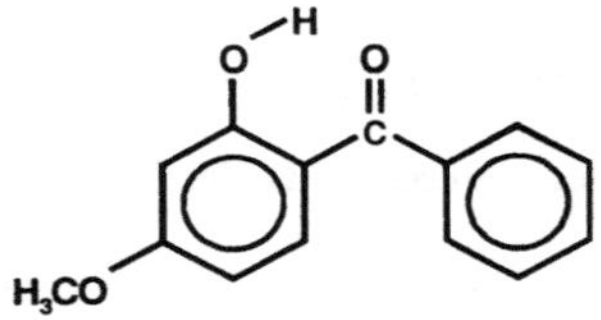	$C_{14}H_{12}O_3$	228.26	fish: + human: +
Mexenone; 2-Hydroxy-4-methoxy-4′-methylbenzophenone; Benzophenone-10 **(BP-10)**	1641–17–4	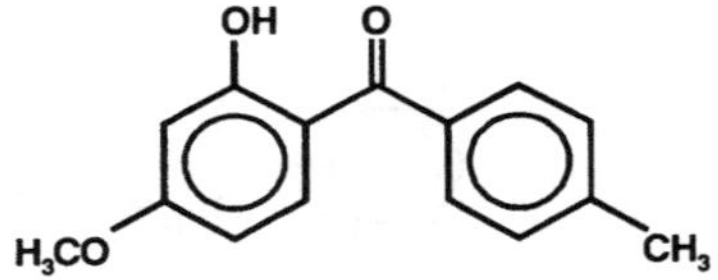	$C_{15}H_{14}O_3$	242.27	
Homosalate; Homomenthyl salicylate; 3,3,5-Trimethylcyclohexyl (2-hydroxy)-benzoate **(HMS)**	118–56–9	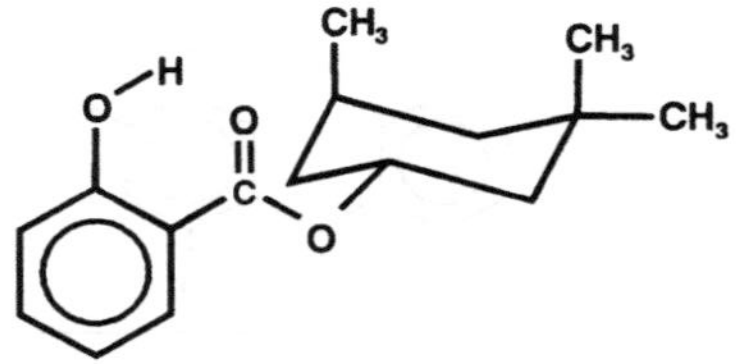	$C_{16}H_{22}O_3$	262.35	fish: +
Isoamyl-*p*-methoxycinnamate; 3-(4-Methoxyphenyl)-2-propenoic acid 3-methylbutyl ester **(IMC)**	71617–10–2	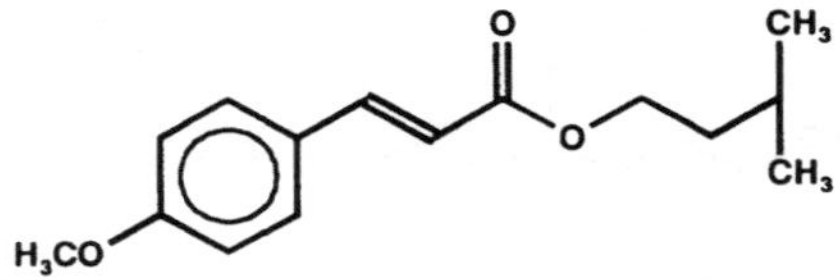	$C_{15}H_{20}O_3$	248.34	Not detected
Octylmethoxycinnamate; 3-(4-Methoxyphenyl)-2-propenoic acid 2-ethylhexyl ester **(OMC)**	5466–77–3	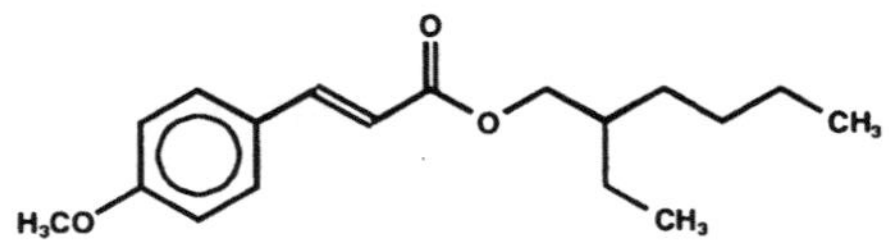	$C_{18}H_{26}O_3$	290.40	fish: + human: +

Table 17. (continued)

Trivial name or synonym, chemical name (abbreviation)	CAS No.	Chemical structure	Molecular formula	Molecular mass [g mol^{-1}]	Bioconcentrated or detected in fish and/or human[a]
p-Dimethylaminobenzoic acid isooctylester; *N*,*N*-Dimethyl-4-amino-benzoic acid-2-ethylhexyl ester **(DABI)**	21245-02-3	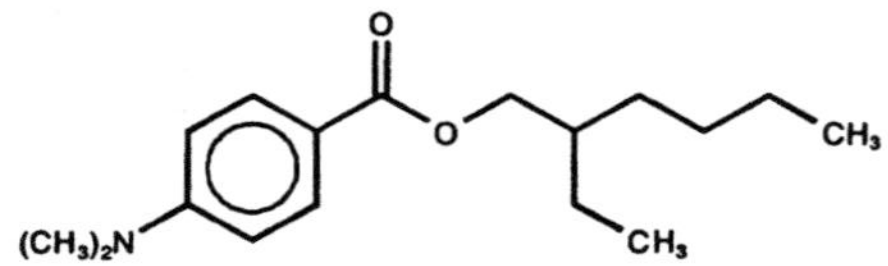	$C_{17}H_{27}NO_2$	277.40	Not detected
Sulisobenzone; 2-Hydroxy-4-methoxybenzo-phenone-5-sulfonic acid; Benzophenone-4 **(BP-4)**	4065-45-6	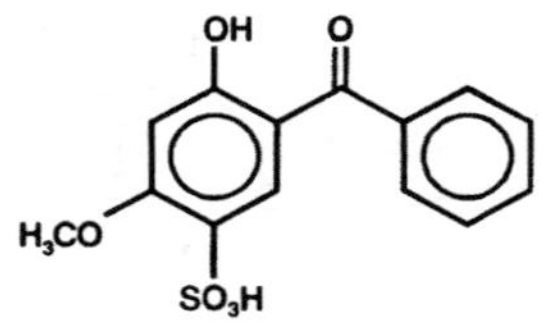	$C_{14}H_{12}O_6S$	308.31	
Drometrizole; 2-(2′-Hydroxy-5′-methylphenyl) benzotriazole	2440-22-4	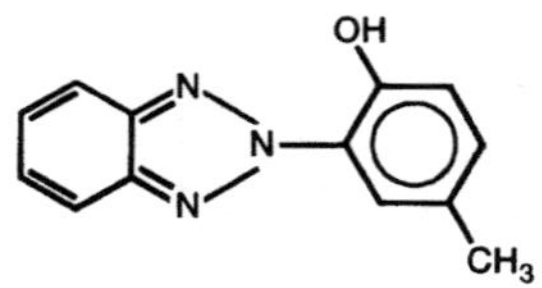	$C_{13}H_{11}N_3O$	225.25	

Source: Adapted with modifications from Hany and Nagel [373], Nagtegal et al. [371], and The MERCK Index [152].
[a] For more information see Nagtegal et al. [371].
[b] This chemical in some cases has photocontact allergic properties and therefore the permission as a sunscreen agent was cancelled for the European Community.
[c] BCF calculated from measured fish and water concentrations of MBC in a lake [371].

However, more research on biodegradation, photostability, physico-chemical properties, toxicity, metabolism, bioconcentration in aquatic organisms, and especially an advantage-disadvantage and ecological hazard/risk assessment of these sunscreen agents is necessary. In this context it is also necessary to test the estrogenic activity of the sunscreen agents and their metabolites and/or degradation products. The authors of this paper and Ternes [372] suggest from structure-activity relationship (SAR) that *o*-hydroxy benzophenone, benzophenone-3 (BP-3), the demethylated metabolite of BP-3, particularly the hydroxylated metabolites 2-hydroxy-4-methoxy-4′-hydroxy-benzophenone, and 2,4,4′-trihydroxy-benzophenone possess weak estrogenic activity.

9 New Aspects and Considerations on Bioconcentration of Chemicals with High Molecular Size and/or Cross-Section

It is important to note that some chemicals with a cross-section greater than 9.5 Å are able to cross the membranes of the gills (may be slowly) and can be bioconcentrated in aquatic organisms to a high extent, which is in agreement with the predicted BCF_L values from their n-octanol/water partition coefficient (K_{OW}). Examples for such super-hydrophobic chemicals are octachlorodibenzo-*p*-dioxin (OCDD) and Mirex. Because these chemicals were tested at concentrations some orders of magnitude higher than their water solubility, relatively low BCF values were found. However, because only the truly dissolved chemical can be taken up by fish etc., the bioconcentration potential of a chemical in aquatic organisms has to be tested below its water solubility. Because the super-hydrophobic compounds are stored in the lipids of the organisms, it is necessary to measure the elimination for a long time (some months) and to measure also the growth rate. In agreement with our conclusions that chemicals with cross-sections > 9.5 Å are able to cross membranes are the experimental results of Belfroid et al. [381]. They found that octachloronaphthalene (OCN) and hexabromobenzene (HBB) are taken up in earthworms (*Eisenia andrei*) and their elimination was slow [381].

More than 20 years ago Zitko [387] came to the conclusion that for compounds with a molecular mass greater than 600, uptake through biological membranes decreases exponentially with increasing molecular mass. Zitko [387] stated that chemicals with molecular masses of 1,000 or greater are only insignificantly absorbed by aquatic organisms. However, these statements seem not generally held for all chemicals. Exceptions from these rules may be avermectin B_{1a} and ivermectin.

Recently, van den Heuvel et al. [378a] studied the bioconcentration of [^{3}H]avermectin B_{1a} in an 28-d uptake flow-through test with bluegill sunfish (*Lepomis macrochirus*). Avermectin B_{1a} (see Fig. 16), the major component of abamectin, possesses a molecular mass of 872. The molecular dimensions are 17.0 × 18.7 × 18.4 Å and were determined by Nachbar (cited in [378]) by finding the smallest parallelepiped whose faces were centered on the inertial axes of the molecule and would enclose the van der Waals surface of the molecule. A van der Waals radius of 1.2 Å for hydrogen was used and the atomic coordin-

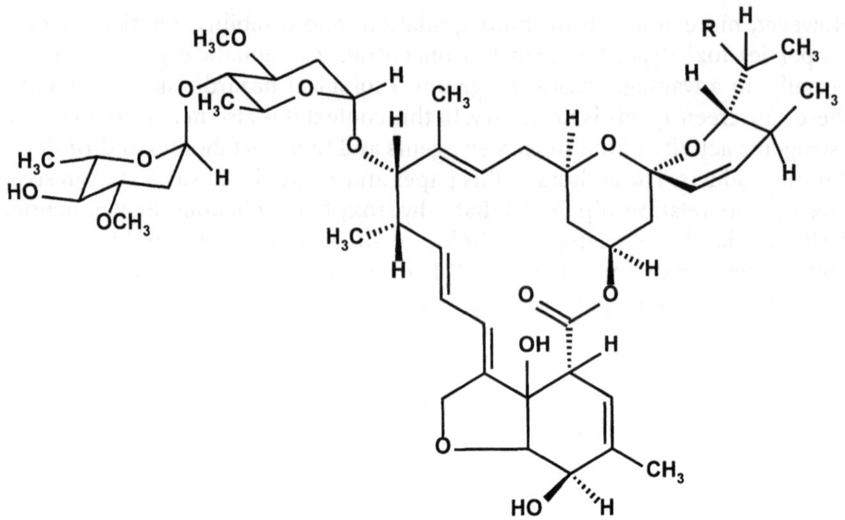

Fig. 16. Chemical structure of abamectin: avermectin B_{1a}, $R = C_2H_5$, and avermectin B_{1b}, $R = CH_3$. (Tritium label at the 5 position)

ates were taken from the crystal structure [379]. This chemical has a cross-section of ca. 25 Å. However, van den Heuvel et al. [378] calculated BCF_W values from the steady-state concentrations (a) in whole fish: 56, (b) viscera: 84, and (c) filet: 28, respectively. The lipid content of the bluegill sunfish with a body weight of 6.2 g and a length of 55 mm is ca. 3%, then the BCF_L value for the whole fish would be 1900. This BCF value is in satisfactory agreement with the BCF_L value of 9,000 predicted from the log K_{OW} value of 3.996 for avermectin B_{1a}. It is possible that this compound is metabolized in the fish, and therefore the BCF_L value is ca. 5 times lower than predicted from its K_{OW} value.

Davies et al. [378b] studied the bioconcentration of ivermectin (22,23-dihydroavermectin B_1) in mussels (*Mytilus edulis*). Ivermectin has been proposed as a chemotherapeutant for the treatment of farmed salmon infected with sea lice. The commercial ivermectin contains two avermectin derivatives: at least 80% of 22,23-dihydroavermectin B_{1a} ($C_{48}H_{74}O_{14}$; molecular mass 874.5 g mol^{-1}) and not more than 20% of 22,23-dihydroavermectin B_{1b} ($C_{47}H_{72}O_{14}$, molecular mass 860.5 g mol^{-1}). Both compounds possess nearly the same molecular dimensions with the same cross-section of ca. 25 Å as avermectin B_{1a}. The water solubility of ivermectin is low, between 6 and 9 µg l^{-1}. For comparison, the solubility of hexachlorobenzene (HCB) in water is 5 µg l^{-1}.

The mussels bioconcentrated ivermectin from water at 6.9 µg l^{-1} for 6 days under semi-static conditions by a factor on a wet weight basis (BCF_W) of 750 (confidence limits 720–790). The lipid content of *Mytilus edulis* is between 1 and 2%. The bioconcentration factor an a lipid basis (BCF_L) of ivermectin in mussels is therefore between 37,500 and 75,000. That means that the bioconcentration potential of ivermectin is very high and that ivermectin B_{1a} and ivermectin B_{1b} are able to cross membranes of gill-breathing organisms although the cross-section is much bigger than 9.5 Å.

In this context some dinoflagellate toxins and other marine toxins are of great interest. Examples are brevetoxin-B (BTX-B, $C_{50}H_{70}O_{14}$, molecular mass: 895 g mol^{-1}) [382], ciguatoxin (CTX, $C_{60}H_{86}O_{19}$, molecular mass: 1110 g mol^{-1}) [384], palytoxin (PTX, $C_{129}H_{223}O_{54}N_3$, molecular mass: 2677 g mol^{-1}) [384], maitotoxin (MTX, $C_{164}H_{256}O_{68}S_2Na_2$, molecular mass: 3422 g mol^{-1}) [383] and other marine toxins (for review see reference [384]). These natural compounds have a very high molecular mass and nevertheless they can be detected in mussels, oysters, crabs, and fish. All these natural chemicals are very toxic to mice, rats, and other mammals including humans and were frequently involved in fatal seafood (mussels, clams, and/or fish) poisoning and intoxication in human [384, 389]. It is important to note that brevetoxins are very toxic to fish while ciguatoxin, palytoxin, and maitotoxin are not. Mussels are also very resistant to these marine toxins. BTX-B and MTX are highly polar polycyclic ethers which are well soluble in water while ciguatoxin is a lipophilic substance. The molecular mass of maitotoxin (MTX) exceeds that of any other natural products. The cross-section of maitotoxin calculated by Bräse [388] yielded 15.8 Å. Gusovsky and Daly [385] expected that the highly polar MTX would not cross membrane lipid bilayers.

It seems interesting to study the mechanism of uptake, bioaccumulation, and toxicity of these organic compounds. Furthermore, it would be interesting to investigate if these chemicals with such high molecular mass can go or can not go through membranes of gill-breathing organisms.

Another example of an organic compound with a high molecular mass of 1355.4 g mol^{-1}, a cross-section of >9.5 Å, and a high water solubility is vitamin B_{12} (cyanocobalamin). It is known that this big molecule is transported by a protein through the membranes of the intestine of mammals including human. However, it is not known if the transport of vitamin B_{12} through the gills of aquatic organisms, such as fish, mussels, and *Daphnia* is also possible. Therefore, it would be interesting to study the kinetics of bioconcentration of this compound in aquatic organisms. However, it is necessary to use ^{14}C, ^{3}H, or ^{60}Co labeled vitamin B_{12} to differentiate between the natural occurring vitamin B_{12} and this compound which may be taken up from water through the gills of fish etc.

On the other side it has also to be noted that chemicals with very large cross sections, very high molecular mass, and extremely high hydrophobicity can not go through membranes and are not bioconcentrated in fish and other gill breathing organisms. However, the threshold value of membrane permeability (if indeed it exists) is above the cross section of 9.5 Å. Examples may be some super-hydrophobic pigments, silicone oils, and paraffins with very high molecular mass. The totally chlorine substituted copper phthalocyanine, which is called hexadecachloro phthalocyanato copper (II) ($C_{32}Cl_{16}CuN_8$, molecular mass: 1127.2 g mol^{-1}), is an example for an extremely hydrophobic pigment. It is assumed that this organic compound is not bioconcentrated in fish, mussels, *Daphnia*, and other gill breathing organisms because its cross section is 17.5 Å and the molecular mass greater than 1,000. The dimensions of this compound were determined by Uyeda et al. [380] by means of high voltage electron microscopy. The chemical structure of this pigment is shown in Fig. 17.

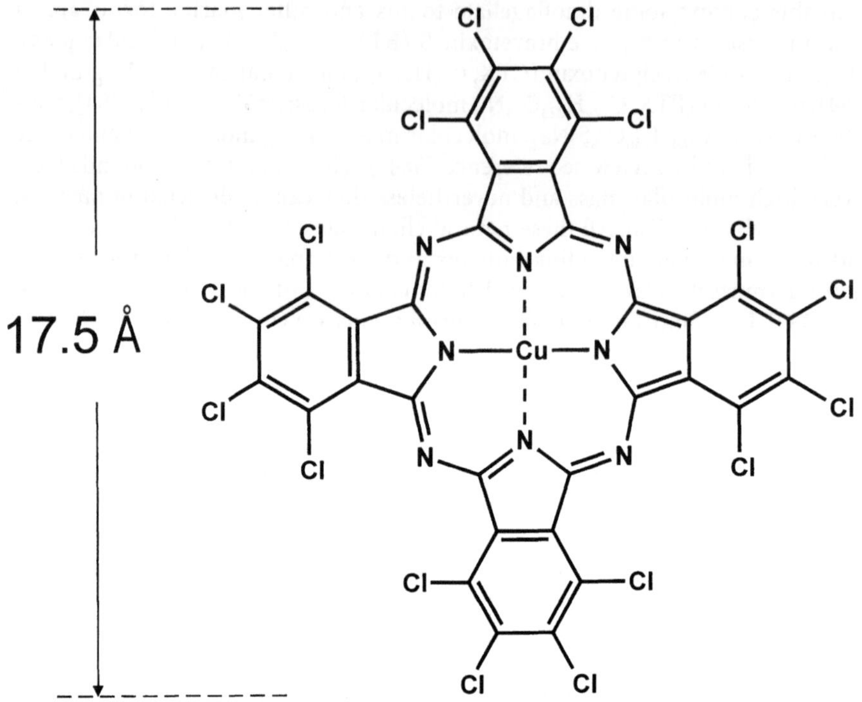

Fig. 17. Chemical structure of copper(II) hexadecachloro phthalocyanine (cross-section: 17.5 Å)

10
Discussion and General Conclusions

It is known and accepted that very lipophilic persistent organic pollutants (POPs) [159, 393], such as the following 12 chemicals or group of substances DDT, aldrin, dieldrin, endrin, hexachlorobenzene (HCB), Mirex, chlordane, heptachlor, Toxaphene, highly polychlorinated biphenyls (PCBs), highly polychlorinated (especially 2,3,7,8-chlorinated) dibenzo-*p*-dioxins (PCDDs) and dibenzofurans (PCDFs) which are metabolized and excreted by aquatic organisms only to a small extent or not at all, can bioconcentrate in organisms to a very high extent. For hazard-assessment of chemicals the bioconcentration factor (BCF) is used. Because the BCF value on a wet weight basis (BCF_W) of a chemical is dependent on the lipid content (L% on a wet weight basis) of the organisms, it is necessary to refer the BCF value on the lipid content. Otherwise the BCF_W value refers only to this specific organism with its lipid content, and is not comparable to the BCF values of the same chemical in other aquatic organisms. In Table 20, a classification scheme for organic chemicals by their bioconcentration potential is presented. It is clear that a chemical is classified in a lower bioconcentration potential group if the BCF value was determined in a

Table 20. Classification Scheme for Organic Chemicals by their Hydrophobicity or Lipophilicity (log K_{OW}) and by their "reasonable worst-case" Bioconcentration Factors

Chemical		Properties			Bioconcentration factor (BCF_L or BCF_W)				
Group	Examples of chemicals	Bioconcentration potential (BP)	Hydrophobicity or Lipophilicity	log K_{ow}	BCF_L [a]	Bioconcentration factor (BCF_W) [b]			
					fish 100%	adult eel 20% [c]	Fathead minnows 10% [c]	Guppy 5% [c]	Mussel, *Daphnia* 1% [c]
1	Nitrobenzene, Aniline	very low	very low hydrophobic/ or lipophilic	≤2	<100	<20	<10	<5	<1
2	Monochlorobenzene, Chloronitrobenzene	low	low hydrophobic/ or lipophilic	>2–3	>10–1,000	>20–200	>10–100	>5–50	>1–10
3	Dichlorobenzene, Biphenyl, Pentachlorophenol (PCP)	moderately	moderately hydrophobic/ or lipophilic	>3–4	>1,000–10,000	>200–2,000	>100–1,000	>50–500	>10–100
4	Trichlorobenzene, Musk xylene (MX)	high	highly hydrophobic/ or lipophilic	>4–5	>10,000–100,000	>2,000 –20,000	>1,000–10,000	>500–5,000	>100–1,000
5	Pentachlorobenzene (PeCB), Hexachlorobenzene (HCB), Dieldrin, Kepone	very high	very highly hydrophobic/ or lipophilic	>5–6	>100,000–1,000,000	>20,000–200,000	>10,00–100,000	>5,000–50,000	>1,000–10,000
6	TCDD, OCDD, Mirex, penta-, hexa-, hepta-, octa-, nona-, deca-chlorobiphenyl, p,p′-DDT, p,p′-DDE, p,p′-DDD	extremely high	super-hydrophobic/ or super- lipophilic	>6–9	>1,000,000	>200,000	>100,000	>50,000	>10,000

[a] BCF_L: Worst-case steady-state biconcentration factor on a lipid basis of a chemical which is not metabolized or only to a low extent and which gives no bound residues.
[b] BCF_W: Worst-case steady-state biconcentration factor on a wet weight basis of a chemical which is not metabolized or only to a low extent.
[c] Assumed lipid content (%) on a wet weight basis.

gill-breathing organism, such as fish, mussel, or *Daphnia* with a low lipid content. This means this chemical seems to be more harmless to the aquatic environment than it really is.

In this chapter BCF values of some other chemicals or groups of substances have been presented, which show that these can also be classified to and named persistent organic pollutants (POPs) or persistent environmental pollutants (PEPs). Beside the "dirty dozen" POPs cited above, the following chemicals or groups of substances may also be considered as potential POPs: the highly polybrominated biphenyls (PBBs), polybrominated diphenlyethers (PBDEs), polychlorinated diphenlyethers (PCDEs), and tetrachlorobenzyltoluenes (PCBTs). There are indications that nitro musk compounds (NMCs) and some sun screen agents (SSAs) are relatively persistent to total biodegradation and can be considered as new environmental contaminants. Furthermore we conclude that polychlorinated terphenyls (PCTs), tin organic compounds (e.g. tributyl tin, triphenyl tin etc.), tris(*p*-chlorophenyl)methane, tris(*p*-chlorophenyl)methanol, chlorinated paraffins (CPs) [390–395], and polychlorinated naphthalenes (PCNs) [396–398] may also be considered as potential POPs. However, these chemicals/or groups of substances could not be considered in this review.

It was found that with increasing lipophilicity (log K_{OW} value) of the chemicals steady-state conditions are not achieved within some days or few weeks, but in many instances only after many months. One consequence is that the BCF values of super-hydrophobic chemicals can be evaluated only under flow-through conditions using the "kinetic method". It appears self-evident that aquatic organisms should be exposed only to concentrations below water solubility. This is also valid for aquatic toxicity tests with these organisms. Otherwise these BCF and lethal concentration (LC_{50} in $\mu g\, l^{-1}$) or effect concentration (EC_{50}) data are meaningless. However, to fulfill both experimental conditions with super-hydrophobic compounds, severe practical problems emerge. In the past all bioconcentration experiments with super-hydrophobic chemicals such as OCDD, Mirex, pentabromo diphenyl ether (PBDE), technical mixture of tetrachlorobenzyltoluene (UGILEC 141) etc. failed these requirements, resulting in BCF values which were some orders of magnitude too low. The only exception is 2,3,7,8-tetrachlorodibenzo-*p*-dioxin (TCDD) for which recently a correct bioconcentration factor was determined [28].

11 Recommendations

We recommend for BCF evaluations of hydrophobic chemicals in aquatic organisms such as fish, mussels, etc.:

(1) The flow-through systems according to the "kinetic method" (OECD guideline) should be applied.
(2) The ambient chemical concentrations in the water must be below their water solubility and should be measured during the uptake phase.

(3) During the uptake and especially during the elimination phase no or only minimal toxic effects of the test organisms should occur.
(4) To obtain reliable values of k_2 or $t_{1/2}$ it is necessary to measure the elimination of super-hydrophobic compounds from organisms for a long time (some months).
(5) For fish whose growth is fast or if the bioaccumulation and elimination takes a long time the specific growth rate (k_G) must be considered for calculation of the BCF value.
(6) The lipid content of the organism is a critical controlling factor of body residues of organic chemicals. Bioconcentration studies often provide lipid-corrected results to compensate for this. Therefore, the lipid content of organisms used in bioassays should be reported routinely in all aquatic bioassays, such as bioconcentration, bioaccumulation, biomagnification, and toxicity studies with organic chemicals.
(7) There is a need to make measurements of the resistance to transport of larger, high molecular weight chemicals across gill and gut membranes to ascertain if there is a size dependence or "cut off".
(8) It is important to investigate the bioconcentration potential of natural hormones, such as 17β-estradiol, estrone (using ^{14}C or tritium labeled compounds) and synthetic hormones (e.g., mestranol, diethylstilbestrol etc.). However, the concentration in the water should be at environmental relevant concentrations (< 10 ng l^{-1}).
(9) In the future it is also necessary to test all new substances if they have endocrine disrupting effects. If their log K_{OW} is 3.0 or greater a bioconcentration test with fish should be performed.
(10) The relationship being found between endocrine system, the nervous system, and immune system will make these endpoints prime areas for further development of chronic toxicity test methods for aquatic organisms and should be considered for ecological and hazard risk assessments of chemicals [7, 386].

The freshwater and marine environment are not only important and precious habitats for fishes, oysters, mussels, lobsters, squids, octopus, cuttlefish, etc., but also for many other aquatic organisms, such as sea-mammals (whales, sea lions, seals, dolphins etc.). The first class and partly the second class are important sources of protein and fat for many humans. It is now established that fish and fish products constitute an important route of human exposure to polychlorinated biphenyls (PCBs), polychlorinated dibenzo-*p*-dioxins (PCDDs), dibenzofurans (PCDFs), DDT, DDE, and other persistent organic chemicals. A number of studies have shown that fish samples from various waterways, including those lying in prestine areas, are contaminated at various levels. Thus the dietary intake of organic chemicals with bioaccumulation potential in aquatic organisms can be governed to a large extent by the quantity of fish consumed by individuals. Furthermore, it is important to note that marine organisms, such as sponges (*Pseudaxinyssa cantharella, Halichondria okodai, Luffariella variabilis*), sea hairs (*Dolabella auricularia*), sea squirt (*Trididemnun solidum*), fan corals (*Pseudopterogorgia elisabethae*), algae, bacteria, marine bryozoans

(*Bugula neritina*) and related organisms produce substances with antibacterial, antitumor (e.g. bryostatins, didemnin B, dolastatin, girodazol, halichondrin B), anti-inflammatory (e.g. pseudopterosin E, manoalide derivatives), antifungal, antiviral, or immuno-suppressive (e.g. microcolin A and B) activity [399, 400]. These compounds and/or their synthetic derivatives may be important novel bioactive pharmaceutical substances. It is also very likely that some new natural marine substances or their derivatives can be used as antifouling compounds, insecticides, or fungicides.

It is desirable that industrial and domestic wastes be treated or eliminated to ensure that lakes, rivers, and oceans are not contaminated by persistent bioaccumulating and toxic substances including those which are endocrine disrupters. Atmospheric inputs must also be considered and reduced when necessary. The incentive for achieving this is both the protection of the population of freshwater and marine organisms from toxic effects but also the protection of the human and wildlife population which consumes these organisms. It is thus critically important that there be reliable quantification of the phenomena of bioconcentration, bioaccumulation, and biomagnification in any ecological, hazard or risk assessment of chemicals. This chapter has sought the state of the art in this task. The old and now discredited view that "dilution is the solution to pollution" is clearly misguided given the magnitude of the concentration increases of factors of millions or more by which pollutants can achieve as a result of bioconcentration.

Acknowledgement. The authors are grateful to J. Altschuh, Drs. B. Beek, J. de Boer, L. Brooke, W. Butte, P. Cook, B. Danzo, C. Franke, C. Gammerl, M. Gilek, F. Gobas, D.W. Hawker, W.L. Hayton, O. Hutzinger, U. Irmer, D.E. Kime, R. Länge, G. Lien, H. Loonen, M. Mansour, M. Matthies, L.S. McCarty, J.M. McKim, J.A. McLachlan, M. Nendza, A. Niimi, J. Oehlmann, H. Parlar, J. Petty, S. Safe, D. Sijm, J. Schmitzer, S. Schwartz, H. Schweinfurth, S. Trapp, G.D. Veith, Patricia Schmieder and Andrea Wenzel for helpful discussions and for providing data and literature. We thank Dr. H. Seibert and Dr. H. Gülden for valuable information on endocrine active environmental chemicals. Special thanks are due to Dr. Kurt Bunzl and Dr. Rainer Brüggemann for regression analysis, to Dr. Xiulin Wang for recalculating some BCF values, and to Marcus Rummler for his skillful preparation of the graphics and drawing of the chemical structures.

References

1. Beek B (1991) In: Nagel R, Loskill R (eds) Bioaccumulation in Aquatic Systems. Contribution to the Assessment. Proceedings of an International Workshop, Berlin 1990. VCH, Weinheim New York Basel Cambridge, pp 1–5
2. a. Nagel R, Loskill R (eds) (1991) Bioaccumulation in Aquatic Systems. Proceedings of an International Workshop. VCH, Weinheim New York Basel Cambridge. b. Connell DW (1990) Bioaccumulation of Xenobiotic Compounds. CRC Press, Boca Raton, Florida. c. Bartell SM, LaKind JS, Moore J, Anderson P (1998) Int J Environ Pollution 9(1):3
3. Franke C, Studinger G, Berger G, Böhling S, Bruckmann U, Cohors-Fresenborg D, Jöhnke U (1994) Chemosphere 29(7):1501
4. Steinberg C, Kern J, Pitzen G, Traunspurger W, Geyer H (1992) Biomonitoring in Binnengewässern. Grundlagen der biologischen Überwachung organischer Schadstoffe für die Praxis des Gewässerschutzes. Ecomed, Landsberg/Lech, ISBN 3-609-65560-7
5. Franke C (1996) Chemosphere 32(10):1897

6. Commission of the European Communities (1996) Expended Scheme for Harmonization of Transport and Supply and Use Classification Schemes for Dangers to the Aquatic Environment Proposed by European Commission. Directorate-General XI, EU, Brussels, April 19, 1996
7. Zeeman M (1997) Aquatic toxicology and ecological risk assessment: US-EPA/OPPT perspective and OECD interactions. In: Zelikoff JT, Lynch J, Schepers J (eds) Ecotoxicology: Responses, Biomarkers, and Risk Assessment. Organization for Economic Cooperation & Development, Paris, published for the OECD by SOS Publications, Fair Haven, N.J., pp 89–108
8. Environment Canada (1995) Toxic Substances Management Policy – Persistence and Bioaccumulation. Environment Canada, Ottawa, Canada
9. Zeeman M (1995) Ecotoxicity testing and estimation methods developed under Sect. 5 of the Toxic Substances Control Act (TSCA). Chap. 23, In: Rand G (ed) Fundamentals of Aquatic Toxicology: Effects, Environmental Fate, and Risk Assessment, 2nd edn. Taylor & Francis, Washington, D.C., pp 703–715
10. Bruggemann WA (1982) In: O Hutzinger (ed) The Handbook of Environmental Chemistry, vol 2/Part B. Springer, Berlin Heidelberg New York, pp 83–103
11. Spacie A, Hamelink JL (1985) In: Rand GM, Petrocelli R (eds) Fundamentals of Aquatic Toxicology: Methods and Applications, Chapter 13. Hemisphere Publishing Corporation, McGraw-Hill Internat. Book Comp, Washington New York London, pp 495–525
12. Streit B (1992) Experientia 48:955
13. a. Mackay D (1982) Environ Sci Technol 16: 274. b. Geyer HJ, Muir DCG, Scheunert I, Steinberg CEW, Kettrup A (1994) ESPR-Environ Sci Pollut Res 1(2):75
14. Oliver BG, Niimi AJ (1985) Environ Sci Technol 19 (9):842
15. Opperhuizen A, v.de Velde EW, Gobas FAPC, Liem DAK, Steen JMDvd, Hutzinger O (1985) Chemosphere 14 (11/12):1871
16. Chapman PM, Allen HE, Godtfredsen K, Z'Graggen MNC (1996) Environ Sci Technol 30(10):448 A
17. Geyer HJ, Scheunert I, Korte F (1985) Chemosphere 14:545
18. Tadokoro H, Tomito Y (1987) In: QSAR in Environmental Toxicology – II. D. Reidel, Dordrecht Boston Lancaster Tokyo, pp 363–373
19. a. Ernst W (1985) In: Sheehan P, Korte F, Klein W, Bourdeau P (eds) Appraisal of Tests to Predict the Environmental Behaviour of Chemicals, part 4.4. Wiley, New York, pp 243–255. b. Suedel BC, Boraczek JA, Peddicord RK, Clifford PA, Dillon TM (1994) Rev Environ Contam Toxicol 136:21. c. Opperhuizen A (1991): In: Angeletti G, Bjorseth A (eds) Organic pollutants in the aquatic environment. Proceedings of the 6th European SETAC symposium, Lisbon, Portugal, 22–24 May 1990. d. Thomann RV (1989): Environ Sci Technol 23:699
20. a. Sijm DTH, Seinen W, Opperhuizen A (1992) Environ Sci Technol 26:2162. b. Leblance GA (1995) Environ Sci Technol 29:154. c. Randall DJ, Connell DW, Yang R, Wu SS (1998): Chemosphere 37(7):1263
21. OECD (1993) OECD Guidelines for Testing of Chemicals. Organization for Economic Co-Operation and Development, Paris
22. OECD (1996) OECD Guidelines for Testing of Chemicals. Proposal for Updating Guideline 305. Bioconcentration: Flow-through fish test. Updated Guideline, June 1996 OECD, Paris
23. Wang X, Harada S, Watanab M, Koshikava H, Geyer HJ (1996) Chemosphere:32(9):1783
24. Butte W (1991) In: Nagel R, Loskill R (eds) Bioaccumulation in Aquatic Systems, contribution to the Assessment. Proceedings of an International Workshop, Berlin 1990. VCH Weinheim New York Basel Cambridge, pp 29
25. a. Niimi AJ (1987) Rev Environ Contam Toxicol 99:1. b. Niimi AJ, Palazzo V (1985) Water Res 19(2):205
26. Bruggeman WA, Marton L, Koiman D, Hutzinger O (1981) Chemosphere 10:811
27. Lodge K (1989) Chemosphere 18:933

28. Schmieder P, Lothenbach D, Tietge J, Erickson R, Johnson R (1995) Environ Toxicol Chem 14(10):1735
29. a. Geyer HJ, Scheunert I, Brüggemann R, Zitko V, Steinberg C, Pruell R, Kettrup A, Olson J, Schmieder P, Muir DCG (1995) In: Birnbaum L, Clement R, Fingerhut M, Ramamoorthy S, Safe S (eds) Organohalogen Compounds. Dioxin 95, vol 25, pp 273–278. b. Schultz IR, Hayton WL (1997) Environ Toxicol Chem 16(5):997
30. Gobas, FAPC, Clark K, Shiu W, Mackay D (1989) Environ Toxicol Chem 8:231
31. Servos MR, Muir DCG (1989) Environ Toxicol Chem 8:141
32. Servos MR, Muir DCG, Webster GRB (1989) Aquatic Toxicol 14:169
33. Schrap SM, Opperhuizen A (1990) Environ Toxicol Chem 9:715
34. Delbeke K, Zhu L, Mugtadir M, Khaled M (1995) Int J Environ Anal Chem 58:131
35. Geyer HJ, Muir DCG, Scheunert I, Steinberg C, Kettrup A (1992) Chemosphere 25:1257
36. Geyer HJ, Muir DCG, Scheunert I, Steinberg C, Kettrup A (1994) ESPR-Environ Sci Pollut Res 1(2):75
37. Geyer HJ, Muir DCG (1993) In: Mansour M (ed) Fate and Prediction of Environmental Chemical in Soils, Plants, and Aquatic Systems, chapter 18. Lewis, Boca Raton Ann Arbor London Tokyo, pp 185–197
38. Geyer HJ, Scheunert I, Korte F (1985) Chemosphere 14:545
39. Geyer HJ (1994) Unpublished results
40. Geyer HJ, Scheunert I, Brüggemann R, Langer D, Korte F, Kettrup A, Mansour M, Steinberg C, Nyholm N, Muir DCG (1997) Chemosphere 35:343
41. Kristensen P, Nyholm N (1987) Ringtest Programme 1984–1985. Bioaccumulation of Chemical Substances in Fish, Flow-Through Method. Commission of the European Community. Degradation Accumulation Subgroup. March 1987 (Final Report)
42. de Boer J (1988) Chemosphere 17(9):1803
43. a. Randall RC, Lee II H, Ozretich RJ, Lake JL, Pruell RJ (1991) Environ Toxicol Chem 10:1431. b. Randall RC, Young DR, Lee H II, Echols SF (1998) Environ Toxicol Chem 17(5):788
44. Bligh EG, Dyer WJ (1959) Can J Biochem Physiol 37:911
45. Ernst W, Schaefer RG, Goerke H, Eder G (1974) Z Anal Chem 272:358
46. Beck H, Mathar W (1988) Bundesgesundheitsblatt 28(1):1
47. Schmitt CJ, Zajicek JL, Ribick MA (1985) Arch Environ Contam Toxicol 14:225
48. Geyer HJ, Scheunert I, Brüggemann R, Matthies M, Steinberg C, Zitko V, Kettrup A, Garrison W (1994) Ecotoxicol Environ Saf 28:53
49. Veith GD, Comstock VM (1975) J Fish Res Board Can 32:1849
50. a. Opperhuizen A (1986) In: Aquatic Toxicology and Environmental Fate. Eds. TM Poston, R Purdy. ASTM STP 921; American Society for Testing and Materials: Philadelphia, vol 9, pp 304–315. b. Opperhuizen A, Schrapp SM (1987) Environ Toxicol Chem 6:335
51. Hamelink J (1980) In: Maki AW, Dickson KL, Cairns J (eds) Biotransformation and Fate of Chemicals in the Environment. American Society of Microbiology Washington D.C., pp 56–62
52. a. Hamelink JL, Landrum PF, Bergman HL, Benson WH (eds) (1992) "Bioavailability. Physical, Chemical, and Biological Interactions". Proceedings of the 13th Pellston Workshop. Pellston, Michigan, August 17–22, 1992. SETAC Special Publications Series. Series Editors TW La Point, BT Walton, CH Ward. Lewis Publishers, Boca Raton, Ann Arbor, London, Tokyo. b. Gobas FAPC, Russell RW (1991) Comp Biochem Physiol 100C(1/2):17
53. Schrap SM (1991) Comp Biochem Physiol 100C(1/2):13
54. Wang X, Ma Y, Yu W, Geyer HJ (1997) Chemosphere 35(8):1781
55. Haitzer M, Hoess S, Traunspurger W, Steinberg C (1998) Chemosphere 37(7):1335
56. Kaiser KLE, Valdmanis I (1982) Can J Chem 60:2104
57. Stehly GR, Hayton WL (1990) Arch Environ Contam Toxicol 19:464
58. Bude A (1981) Bestimmung des quantitativen Einflusses von biologischen, physikalischen und chemischen Faktoren auf die Bioakkumulation in Fischen am Beispiel von

2,5,4′-Trichlorbiphenyl und Pentachlorphenol. Doctoral Thesis, Technical University Munich, Weihenstephan, Germany
59. McKim JM, Schmieder PK, Erickson RJ (1986) Aquatic Toxicol 9:59
60. Schwarzenbach RP, Gschwend PM, Imboden DM (1993) Environmental Organic Chemistry. Wiley, New York Chichester Brisbane Toronto Singapure
61. Jafvert CT, Westall JC, Gruder E, Schwarzenbach RP (1990) Environ Sci Technol 24:1795
62. Veith DG, DeFoe DL, Bergstedt BV (1979) J Fish Res Board Can 36:1040
63. Kishino T, Kobayashi K (1994) Water Res 28(7):1547
64. Kishino T, Kobayashi K (1995) Water Res 29(2):431
65. Fent K (1996) Crit Rev Toxicol 26(1):1
66. Arnold CG, Weidenhaupt A, David M, Müller S, Haderlein S, Schwarzenbach R (1997) Environ Sci Technol 31:2596
67. Johnson CA, Westall JC (1990) Environ Sci Technol 24:1869
68. Zeeman M, Auer CM, Clements RG, Nabholz JV, Boethling RS (1995) US EPA regulatory perspectives on the use of QSAR for new and existing chemical evaluations. SAR & QSAR in Environ Res 3(3):179
69. Mackay D, Shiu WY, Ma KC (1992) Illustrated Handbook of Physical-Chemical Properties and Environmental Fate for Organic Chemicals. Vol I: Monoaromatic Hydrocarbons, Chlorobenzens, and PCBs. Vol II: Polynuclear Aromatic Hydrocarbons, Polychlorinated Dioxins, and Dibenzofurans. Vol III: Volatile Organic Chemicals. Lewis, Boca Raton Ann Arbor London
70. Johnston PA, Stringer RL, Santillo D (1996) Toxicol Ecotoxicol News (TEN) 6(4):115
71. Greim H, Ahlers J, Bias R, Broecker B, Gamer AO, Gelbke H-P, Haltrich WG, Klimisch H-J, Mangelsdorf I, Schön N, Stropp G, Vogel R, Welter G, Bayer E (1997) Chemosphere 26(9):1653
72. Johnson J (1996) Environ Sci Technol 30(II):476A
73. OECD (1991) The State of the Environment. OECD Paris
74. Geyer HJ, Politzki G, Freitag D (1984) Chemosphere 13:269
75. Geyer HJ, Scheunert I, Brüggemann R, Steinberg C, Korte F, Kettrup A (1991) Sci Total Environ 109/110:387
76. Geyer HJ, Sheehan P, Kotzias D, Freitag D, Korte F (1982) Chemosphere 11:1121
77. a. Mackay D (1982) Environ Sci Technol 16:274. b. Halfon E (1985) Environ Sci Technol 19:747
78. a. Connell DW (1988) Rev Environ Contam Toxicol 101:117. b. Isnard P, Lambert S (1988) Chemosphere 17:21
79. Connell DW (1990) In: Connell DW (ed) Bioaccumulation of Xenobiotic Compounds, chapter 6. CRC Press, Boca Raton, Florida, pp 97–144
80. Nendza M (1991) In: Nagel R, Loskill R (eds) Bioaccumulation in Aquatic Systems, chapter 5. VCH, Weinheim, pp 43–66
81. Connell DW, Hawker DW (1988) Ecotoxicol Environ Saf 16:242
82. Binstein S, Devillers J, Karcher W (1993) SAR QSAR Environ Res 1:29
83. Govers HAJ, Loonen H, Parsons JR (1996) SAR QSAR Environ Res 5:63
84. a. Devillers J, Binstein S, Domine D (1996) Chemosphere 33(6):1047. b. Jager DT, Hamers T (1997) Estimation Methods for Bioaccumulation in Risk Assessment of Organic Chemicals. Report No. 679102013. National Institute of Public Health and the Environment (RIVM), Bilthoven, The Netherlands. c. Schwartz S (1997) In: Matthies M (ed) Organische Schadstoffe in der Nahrungskette – Vorstudie zur Validierung von Expositionsmodellen, Beitrag Nr. 5. Beiträge des Instituts für Umweltsystemforschung der Universität Osnabrück, ISSN 1433–3805. d. EU (1996) Technical Guidance Document in Support of the Commission Directive 93/67/EEC on Risk Assessment for New Notified Substances and the Commission Regulation (EC) 1488/94 on Risk Assessment for Existing Substances, Part I-IV Office for Official Publications of the European Communities, Luxembourg
85. Yen C-PC, Perenich TA, Banghman GL (1989) Environ Toxicol Chem 8:981
86. Gobas FAPC, Schrap SM (1990) Chemosphere 20:495

87. Geyer HJ, Muir DC, Scheunert I, Steinberg CEW, Kettrup A (1992) Chemosphere 25:1257
88. Geyer HJ, Muir DCG, Scheunert I, Steinberg CEW, Kettrup A (1994) ESPR-Environ Sci Pollut Res 1(2):75
89. Servos MR, Muir DCG (1989) Environ Toxicol Chem 8:141
90. Black MC, McCarthy JF (1988) Environ Toxicol Chem 7:593
91. Gobas FAPC, Clark KE, Shiu WY, Mackay D (1989) Environ Toxicol Chem 8:231
92. De Wolf W, De Bruijn JHM, Seinen W, Hermens JLM (1993) Environ Sci Technol 26:1197
93. Muir DCG, Hobden BR, Servos MR (1994) Aquatic Toxicol 29:223
94. Halbach S (1985) Arch Toxicol 57:139
95. Reinert RE, Stone LJ, Wilford WA (1974) J Fish Res Bd Can 31:1649
96. Niimi AJ (1987) Biological half-lives of chemicals in fishes. Rev Environ Contam Toxicol 99:1–46
97. World Health Organization (WHO) (1990) Environmental Health Criteria #101: Methylmercury. Geneva, Switzerland
98. Galassi S, Calamari D (1983) Chemosphere 22:1599
99. Vigano L, Galassi S, Gatto M (1992) Environ Toxicol Chem 11:535
100. Ando K (1962) Bull Jpn Soc Sci Fish 28 (1):73
101. Crisp TM, Clegg ED, Cooper RL, Anderson DG, Baetcke KP, Hoffmann JL, Morrow MS, Rodier DJ, Schaeffer JE, Touart LW, Zeeman MG, Patel YM (1997) Special Report on Environmental Endocrine Disruption: An Effects Assessment and Analysis. US Environmental Protection Agency, Office of Research and Development, Risk Assessment Forum, Washington, DC, EPA/630/R-96/012
102. Colborn T, vom Saal FS, Soto MA (1993) Environ Health Perspect 101:378
103. Schlumpf M, Lichtensteiger W (Eds.) (1996) Sinkt die Fertilität? Daten zur Wirkung von Umweltchemikalien auf Fortpflanzungsprozesse bei Mensch, Wirbeltieren und wirbellosen Tieren, vol 4, Kind und Umwelt, Zürich, ISBN 3-9520483-2-1
104. Schäfer WR, Zahradnik HP, Frijus-Plessen N, Schneider K (1996) Umweltmedizin 1(1):35
105. Gies A, Wenzel A (1996) Endocrinically Active Chemicals in the Environment. Expert Round, 9–10 March 1995, Texte Series 3, vol 36, Umweltbundesamt Berlin, Germany
106. Seibert H (1996) UWSF-Z Umweltchem Ökotoxicol 8(5):275
107. Kavlock R et al. (1996) Environm. Health Perspect 104[Suppl 4]:715
108. Beato M (1989) Cell 56:335
109. Danzo BJ (1997) Environ Health Perspect 105:294
110. Nagel SC, von Saal FS, Thayer KA, Dhar MG, Boechler M, Welshons WV (1997) Environ Health Perspect 105(1):70
111. McDonnell DP, Clemm DL, Hermann T, Goldman ME, Pike JW (1995) Mol Endocrinol 9:659
112. Kelce WR, Monosson E, Gamsick MP, Laws SC, Gray LE (1994) Toxicol Appl Pharmacol 126:276
113. Gray LE, Ostby JS, Kelce WR (1994) Toxicol Appl Pharmacol 129:46
114. Oehlmann J, Fioroni P, Stroben E, Markert B (1996) Sci Total Environ 188:205
115. a. Kime DE (1995) Rev Fish Biol Fisheries 5:52. b. Ronis MJJ, Mason AZ (1996) Marine Environ Res 42:61. c. Parks LG, LeBlanc GA (1996) Aquat Toxicol 34:291. d. LeBlanc GA, Bain LJ (1997) Environ Health Perspect 105[Suppl 1]:65. e. LeBlanc GA, Bain LJ, Wilson VS (1997) Mol Cell Endocrinol 126:1. f. Dickerson RL, Kendall RJ (Eds) (1998) "Endocrine Disruptors". Annual Review Issue. Environ Toxicol Chem 17(1):1–135
116. Umbreit TH, Gallo MA (1988) Toxicol Lett 42:5
117. Safe S, Astroff B, Harries M (1991) Pharmacol Toxicol 69:400
118. Safe S, Krishnan V (1995) Toxicol Lett 82/83:731
119. McKinney JD, Waller CL (1994) Environ Health Perspect 102:209
120. a. Krishnan V, Safe S (1993) Toxicol Appl Pharmacol 120:55. b. Santodonato J (1997) Chemosphere 34:835
121. Sumpter JP (1995) Toxicol Lett 82/83:737
122. Jobling S, Reynolds T, White R, Parker MG, Sumpter JP (1995) Environ Health Perspect 103:582

123. Hammond B, Katzenellenbogen B, Krauthammer N, McConnell J (1979) Proc Natl Acad Sci 76(12):6641
124. a. vom Saal F (1995) Hum Ecol Risk Assess 1:3. b. Laws SC, Carey SA, Kelce WR (1995) Toxicologist 15:294
125. a. Kelce WR, Stone CR, Laws SC, Gray LE, Kemppainen JA, Wilson EM (1995) Nature 375:581. b. Waller CL, Juma BW, Gray LE, Kelce WR (1996) Toxicol Appl Pharmacol 137:219. c. Cook JC, Mullin LS, Frame SR, Biegel LB (1993) Toxicol Appl Pharmacol 119:195. d. Bauer ERS, Meyer HHD, Allner B, Sauerwein H (1998) Hormonal wirksame Substanzen in Umwelt und Lebensmitteln: Vorkommen, Wirkung und toxikologische Bewertung. 8. Seminar für Toxikologie. 17. und 18. April 1998, Graz, pp 103–104. e. Allner B (1997) Dissertation University Mainz. f. Stahlschmidt-Allner P, Allner B, Römbke J, Knacker T (1997) ESPR – Environ Sci & Poll Res 4 (3):155. g. Zock G, Görge G, Kalsch W Nagel R (1991) Sci Total Environ 109/110:411. h. Call DJ, Brooke LT, Kent RJ, Knuth ML, Poirier SH, Huot JM, Lima AR (1987) Arch Environ Toxicol 16:607. i. Spanggord RJ, Gordon GR, Schockan MJ, Starr RJ (1996) Environ Toxicol Chem 15(10):1655. j. Gaido KW, Leonard LS, Lovell S, Gould JC, Babai D, Portier CJ, McDonnell DP (1997): Toxicol Appl Pharmacol 143:205
126. Bitman J, Cecil HC (1970) J Agric Food Chem 18(6):1108
127. Nelson JA, Struck RF, James R (1978) J Toxicol Environ Health 4:325
128. Jobling S, Sumpter JP(1993) Aquatic Toxicol 27:361
129. Routledge EJ, Sumpter JP (1996) Environ Toxicol Chem 15 (3):241
130. Geyer H, Sheehan P, Kotzias D, Korte F (1982) Chemosphere 11:1121
131. Geyer H, Krauss AG, Klein W, Richter E, Korte F (1980) Chemosphere 9:277
132. Geyer H, Scheunert I, Korte F (1987) Chemosphere 16:239
133. Geyer H, Scheunert I, Korte F (1986) Regul Toxicol Pharmacol 6:313
134. Muir DCG, Fairchild WL, Yarechewski AL, Whittle DM (1992) In: Gobas FAPC, McCorquodale JA (eds) Chemical Dynamics in Fresh Water Ecosystems. Lewis, Boca Raton Ann Arbor London Tokyo, pp 187–210
135. Soto AM, Chung KL, Sonnenschein C (1994) Environ Health Perspect 102:380
136. Soto AM, Sonnenschein C, Chung KL, Fernandez MK, Olea N, Serrano FO (1995) Environ Health Perspect 103[Suppl. 7]:113
137. Klotz DM, Beckman BS, Hill SM, McLachlan JA, Walters MR, Arnold SF (1996) Environ Health Perspect 104:1084
138. a. Körner W, Hanf V, Schuller W, Bartsch H, Kreienberg R, Hagenmaier H (1996) In: Olie K, Louw R, Denison M, Hutzinger O, Rappe C, van Zorgl J (eds) Organohalogen Compounds, Formation & Sources, Analysis, Quality Control, vol 27. Dioxin 96, University Amsterdam, ISBN 3-928379-49-6, pp 297–302. b. Coldham NG, Dave M, Sivapathasundaram S, McDonnell DP, Connor C, Sauer MJ (1997) Environ Health Perspect 105(7):734
139. Gimeno S, Gerritsen A, Bowmer T, Komen H (1996) Nature 384:221
140. Korach KS, Sarver P, Chae K, McLachlan JA, McKinney JD (1988) Mol Pharmacol 33: 130
141. McLachlan JA, Newbold RR, Korach KS, Hogan M (1987) In: Roloff MV, Wilson AW (eds) Human risk assessment: the roles of animal selection and extrapolation. Taylor and Francis, London, pp 187–193
142. McLachlan JA, Korach KS, Newbold RR, Degen GH (1984) Fundamental Appl Toxicol 4:686
143. Schneider SL, Alks V, Morreal CE, Sinha DK, Dao TL (1976) J Natl Cancer Inst 57:1351
144. Arnold SF, Klotz DM, Collins BM, Vonier PM, Guilette LJ, McLachlan JA (1996) Science 272:1489
145. a. McLachlan JA (1997) Science 277:459. b. Debrow C, Routledge E, Sheehan D, Waldock M, Sumpter J (1996) The Identification and Assessment of Oestrogenic Substances in Sewage Treatment Works Effluents. United Kingdom Environment Agency, MAFF Fisheries, Almondsbury, Bristol, UK. c. Harries JE, Sheehan DA, Jobling S, Matthiessen P, Neall P, Sumpter J, Tylor T, Zaman N (1997) Environ Toxicol Chem 16(3):534. d. Renner

R (1998) Environ Sci Technol 32(1):8A. e. Safe S (1998): Environ Health Perspect 106 [Suppl 4]:1051
146. Ramamoorthy K, Wang F, Chen I-C, Norris JD, McDonnell DP, Leonard LS, Gaido KW, Bocchinfuso WP, Korach KS, Safe S (1998) Endocrinology 138:1520
147. Ramamoorthy K, Wang F, Chen I-C, Safe S, Norris J, McDonnell DP, Gaido KW, Korach KS (1997) Science 275:405
148. Ashby J, Lefevre PA, Odum J, Harris CA, Routledge EJ, Sumpter JP (1997) Nature 385: 494
149. Shelby MD, Newbold RR, Tully DB, Chae K, David VL (1996) Environ Health Perspect 104(12):1296
150. Geyer H, Scheunert I, Korte F (1986) Regul Toxicol Pharmacol 6:313
151. Geyer H, Scheunert I, Korte F (1987) Chemosphere 16(1):239
152. Budavari S, O'Neil MJ, Smith A, Heckleman PE, Kinneary JF (eds) (1996) The Merck Index. An Encyclopedia of Chemicals, Drugs, and Biologicals, 12th edn. Merck, Whitehouse Station, N.J.
153. Mackay D, Shiu WY, Ma KC (1992) Illustrated Handbook of Physical-Chemical Properties and Environmental Fate for Organic Chemicals. Vol I: Monoaromatic Hydrocarbons, Chlorobenzenes, and PCBs. Lewis, Boca Raton, FL
154. Mackay D, Shiu WY, Ma KC (1992) Illustrated Handbook of Physical-Chemical Properties and Environmental Fate for Organic Chemicals. Vol II: Polynuclear Aromatic Hydrocarbons, Polychlorinated Dioxins, and Dibenzofurans. Lewis, Boca Raton, FL
155. Mackay D, Shiu WY, Ma KC (1992) Illustrated Handbook of Physical-Chemical Properties and Environmental Fate for Organic Chemicals. Vol V: Pesticides. Lewis, Boca Raton, FL
156. Rippen G (1997) Handbook of Environmental Chemicals (in German) Vol 1-6. Ecomed, Landsberg/Lech, ISBN 3-609-73235-0
157. Keith L, Johnson T (1997) Environmental Endocrine Disruptors. A Handbook of Property Data. Wiley, Chichester, UK
158. a. Gülden M, Turan A, Seibert H (1997) Endocrinically Active Substances in Surface Waters (in German). Umweltforschungsplan des Bundesministeriums für Umwelt, Naturschutz und Reaktorsicherheit - Wasserwirtschaft. Forschungsbericht 102 04 279 UBA-FB 97-068. Texte 46/97 Umweltbundesamt, Berlin, Germany. b. Hansen LG (1998): Environ Health Perspect 106[Suppl 1]:171. c. Li M-H, Hansen LG (1997): Rev Toxicol 1:71
159. Wania F, Mackay D (1996) Environ Sci Technol 30 (9):390A
160. Simonich SL, Hites RA (1995) Science 269:1851
161. Ernst W (1977) Chemosphere 6 (11):731
162. Ernst W, Weigelt S, Weigelt V, Rosenthal H, Hansen P-D, Chutsch M (1987) Development of a Bioaccumulation Test Using the Mussel *Mytilus edulis* Sect. 1-3 (in German). Umweltforschungsplan des Bundesministeriums des Inneren. Umweltchemikalien. Report No. UBA-FB 106-020-24/05. Umweltbundesamt, Berlin, Germany
163. Zaroogian GE, Heltsche JF, Johnson M (1985) Environ Toxicol Chem 4:3
164. Bruijn JD, Busser F, Seinen W, Hermens J (1989) Environ Toxicol Chem 8:499
165. a. Hawker DW, Connell DW (1988) Environ Sci Technol 22:382. b. Brodsky J, Ballschmiter K (1988) Fresenius Z Anal Chem 331:295
166. Simpson CD, Wilcock RJ, Smith TJ, Wilkins AL, Langdon AG (1995) Bull Environ Contam Toxicol 58:364
167. Sijm DTHM, Wever H, Vries PJ de, Opperhuizen A (1989) Chemosphere 19 (1-6):263
168. a. Fox K, Zaunke, G-P, Butte W (1994) Ecotoxicol Environ Saf 28:99. b. Schweinfurth H, Länge R, Miklautz H, Schauer G (1997) Umweltverhalten und aquatische Toxizität von Ethinylestradiol. In: Münchner Beiträge zur Abwasser-, Fischerei- und Flußbiologie, vol 50. Oldenbourg, München Wien, pp 39-54
169. Ahel M, Giger W (1993) Chemosphere 26(8):1471
170. Eklund R, Bergman A, Granmo A, Berggren M (1990) Environ Pollut 64:107
171. Butte W, Fox K, Schmidt A, Zaunke G-P (1993) Bioaccumulation at the Borderline of Lipophilicity and Molecular Structures (in German). Umweltforschungsplan des

Bundesministers für Umwelt, Naturschutz und Reaktorsicherheit. Umweltchemikalien. UFO PLAN Ref. No. 106 03 091. Umweltbundesamt, Berlin, Germany
172. Butte W, Paul C, Willig A, Zaunke G-P (1988) Relationship between the Structure of Phenols and Their Bioaccumulation, Evaluated with a Flow-Through Fish-Test (OECD No. 305E) (in German). Umweltforschungsplan des Bundesministers für Umwelt, Naturschutz und Reaktorsicherheit. Umweltchemikalien. UFO PLAN Ref. No. FKZ 106 02 053. Umweltbundesamt, Berlin, Germany
173. a. Geyer HJ, Scheunert I, Brüggemann R, Langer D, Oxynos K, Kettrup A, Steinberg, CEW, Nyholm N (1995) Paper presented at the 15th International Symposium on Chlorinated Dioxins and Related Compounds. August 21–25, 1995, Edmonton, Canada. Organohalogen Compounds 24:429. b. Jekat FW, Meisel ML, Eckard, R, Winterhoff H (1994) Toxicol Lett 71:9. c. Jobling S, Sumpter JP (1993) Aquat Toxicol 27:361
174. Metcalf RL (1980) In: McLachlan JA (ed) Estrogens in the Environment. Proceedings of the Symposium on Estrogens in the Environment, Raleigh, North Carolina, USA, September 10–12, 1979. Elsevier, New York Amsterdam Oxford, p 203
175. Neumann F, Schenk B, Schleuser H, Schweikert HU, Eckert K-G, Haen E (1996) In: Forth W, Henschler D, Rummel W, Starke K (eds) Allgemeine und spezielle Pharmakologie und Toxikologie. Spektrum Akademischer Verlag, Heidelberg Berlin Oxford, pp 581–637
176. Schlumpf M, Lichtensteiger W (1996) In: Schlumpf M, Lichtensteiger W (eds) Sinkt die Fertilität? Daten zur Wirkung von Umweltchemikalien auf Fortpflanzungsprozesse bei Mensch, Wirbeltieren und wirbellosen Tieren. Verlag Kind und Umwelt. Universität Zürich, Schweiz. ISBN 3-9520483-2-1
177. Turan A (1996) In: Gies A, Wenzel A (eds) Endocrinically Active Chemicals in the Environment. Expert Round. Berlin, 9–10 March 1995. Umweltbundesamt, Berlin, p 15
178. Bradlow L, Davis DL, Lin G, Sepkovic D, Tivari R (1995) Environ Health Perspect 103[Suppl 7]:147
179. Lustenhouver JWA, Olie K, Hutzinger O (1980) Chemosphere 9:501
180. Lahaniatis E, Clausen E, Bieniek D, Korte F (1985) Chemosphere 14(2):233
181. Geyer HJ, Scheunert I, Rapp K, Kettrup A, Korte F, Greim H, Rozman K (1990) Toxicology 65:97
182. Geyer HJ, Scheunert I, Rapp K, Gebefügi I, Steinberg C, Kettrup A (1993) Ecotoxicol Environ Saf 26:45
183. Safe S (1990) Crit Rev Toxicol 21(1):51
184. Geyer HJ, Schramm K-W, Scheunert I, Schughart K, Buters J, Wurst W, Greim H, Kluge R, Steinberg CEW, Kettrup A, Madhukar B, Olson JR, Gallo MA (1997) Ecotoxicol Environ Saf 36:213
185. Cook PM, Kuehl DW, Walker MK, Peterson RE (1991) In: Gallo MA, Scheuplin RJ, Van der Heijden KA (eds) Biological Basis for Risk Assessment of Dioxins and Related Compounds. Banbury Report 35. Cold Spring Harbor Laboratory Press, Cold Spring Harbor, N.Y., pp 143–167
186. Peterson RE, Theobald HM, Kimmel GL (1993) Crit Rev Toxicol 23:283
187. Geyer HJ, Steinberg CE, Scheunert I, Brüggemann R, Schütz W, Kettrup A, Rozman K (1993) Toxicology 83:169
188. Geyer HJ, Scheunert I, Brüggemann R, Matthies M, Steinberg C, Zitko V, Kettrup A, Garrison W (1994) Ecotoxicol Environ Saf 28:53
189. Lassiter RR, Hallam TG (1990) Environ Toxicol Chem 9:585
190. Geyer HJ, Scheunert I, Filser JG, Korte F (1986) Chemosphere 15(9–12):1495
191. Geyer HJ, Scheunert I, Korte F (1987) Chemosphere 16(1):239
192. Geyer HJ, Scheunert I, Korte F (1986) Regul Toxicol Pharmacol 6:313
193. Schmieder P, Lothenbach D, Tietge J, Erickson R, Johnson R (1995) Environ Toxicol Chem 14(10):1735
194. Karickhoff SW (1981) Chemosphere 10:833
195. Gellert RJ (1978) Environ Res 16:131
196. Yalkowsky SH, Banerjee S (1992) Aqueous Solubility, Methods of Estimation for Organic Compounds. Dekker, New York, New York

197. a. Smith JH, Mabey WR, Bohonos N, Holt BR, Lee SS, Chou TW, Bomberger DC, Mill T (1978) Environmental Pathways of Selected Chemicals in Freshwater Systems, Part II Laboratory Studies, Report No. EPA-600/7-78-074, US Environmental Protection Agency, Athens, GA. b. Gallo MA, Scheuplein RJ, van der Heijden KA (1991) Biological Basis for Risk Assessment of Dioxins and Related Compounds. Banbury Report 35. Cold Spring Harbor Laboratory, Cold Spring Harbor, N.Y.
198. Broman D, Näf C, Rolff C, Zebühr Y (1991) Environ Sci Technol 25(11):1850
199. Kulka M (1965) Can Pat. 702, 144, Appl May 3, 1961, granted J 19, 1965, 13 pp Chemical Abstract 62, 16261 a, b (1965)
200. Ballschmiter K, Bacher R (1996) Dioxine: Chemie, Analytik, Vorkommen, Umweltverhalten und Toxikologie der halogenierten Dibenzo-*p*-dioxine und -dibenzofurane, VCH, Weinheim, ISBN 3-527-28692-6
201. Geyer HJ, Muir DCG, Scheunert I, Steinberg C, Kettrup A (1992) Chemosphere 25(8-10):1257
202. Geyer HJ, Muir DCG, Scheunert I, Steinberg C, Kettrup A (1994) ESPR-Environ Sci Pollut Res 1(2):75
203. Bruggeman WA, Opperhuizen A, Wijbenga A, Hutzinger O (1984) Environ Toxicol Chem 7:173
204. Muir DCG, Marschall WK, Webster GRB (1985) Chemosphere 14:829
205. Muir DCG, Yarechewski AL, Knoll A, Webster GRB (1986) Environ Toxicol Chem 5:261
206. Gobas FAP, Schrap SM (1990) Chemosphere 20:495
207. Loonen H, Tonkes M, Parsons JR, Govers HAJ (1996) Aquat. Toxicol 30:153
208. Gellert RJ (1978) Environ Res 16:131
209. a. Bell MA, Ewing RA, Lutz GA, Alley EG (1978) Reviews of the Environmental Effects of Pollutants: I Mirex and Kepone. US EPA Health Effects Research Laboratory, Cincinnati OH 45268, EPA-600/1-78-013 July 1978. b. Huckins JN, Stalling DL, Petty JD, Buckler DR, Johnson BT (1982) J Agric Food Chem 30:1020. c. Comba ME, Norstrom RJ, Macdonald CR, Kaiser KLE (1993) Environ Sci Technol 27(10):2198
210. a. Ivie GW, Gibson JR, Bryant HE, Begin JJ, Barnett JR, Dorough HW (1974) J Agric Food Chem 22(4):646. b. Kaiser KLE (1974) Science 185:523. c. Muir DCG, Ford CA, Stewart REA, Smith TG, Addison RF, Zink ME, Bèland P (1990) Can Bull Fish Aquat Sci 224:165
211. a. Veith GD, DeFoe DL, Bergstedt BV (1979) J Fish Res Board Can 36:1040. b. Skaar DR, Johnson BT, Jones JR, Huckins JN (1981) Can J Fish Aquat. Sci 38:931. c. Gobas FAPC, Clark KE, Shiu WY, Mackay D (1989) Environ Toxicol Chem 8:231. d. Skea JC, Simonin HJ, Jackling S, Symula J (1981) Bull Environ Contam Toxicol 27:79
212. a. Swackhammer DL, Charles MJ, Hites, RA (1987) Anal Chem 59:913. b. Jansson B, Widequist U (1983) Int J Environ Anal Chem 13:309. c. Fingerling G, Hertkorn N, Parlar H (1996) Environ Sci Technol 30:2984. d. Saleh MA (1983) J Agric Food Chem 31:748
213. Voldner EC, Li YF (1995) Sci Total Environ 160/161:201
214. Saleh MA (1991) Toxaphene: Chemistry, Biochemistry, and Environmental Fate. In: Ware GW (ed) Reviews Environ Contam Toxicol, vol 118. Springer, Berlin Heidelberg New York, pp 1-85
215. Rice CP, Samson PJ, Noguchi GE (1986) Environ Sci Technol 20:1109
216. Bidleman TF, Walla MD, Muir DCG, Stern GA (1993) Environ Toxicol Chem 12:701
217. Muir DCG, de Boer J (1995) Trends Anal Chem 14:56
218. a. Parlar H, Fingerling G, Angerhöfer D, Christ G, Coelhan M (1997) In: Eganhouse RP (ed) Molecular Markers in Environmental Geochemistry. ACS Symposium Series 671, American Chemical Society, Washington, DC, Chapter 23, pp 346-364. b. Fingerling G, Coelhan M, Angerhöfer D, Parlar H (1997) Organohalogen Compounds 33:17. c. Coelhan M, Fingerling G, Angerhöfer D, Parlar H (1998) UWSF - Z Umweltchem Ökotoxicol 10:37. d. Fingerling G (1995) Thesis, University Kassel, Germany
219. Parlar H (1991) Nachr Chem Tech Lab 39:26
220. Musial CJ, Uthe HF (1983) Int J Environ Anal Chem 14:117

221. Stern GA, Muir DCG, Ford CA, Grift NP, Dewailly E, Bidleman TF, Walla MD (1992) Environ Sci Technol 26:1838
222. Buser H-R, Müller MD (1994) Environ Sci Technol 28:119
223. Luckas B, Vetter W, Fischer P, Heidemann G, Plötz J (1990) Chemosphere 21:13
224. a. Burhenne J, Hainzl D, Xu L, Vieth B, Alder L, Parlar H (1993) Fresenius J Anal Chem 346:779. b. Parlar H, Angerhöfer D, Coelhan M, Kimmel L (1995) Organohalogen Compounds 26:357
225. Vetter W, Krock B, Luckas B (1997) Chromatographia 44:65
226. Kidd K, Muir DCG (1997) Unpublished Results
227. Alder L, Beck H, Khandker S, Karl H, Lehmann J (1997) Chemosphere 34:1389
228. Vetter W, Scherer G, Schlabach M, Luckas B, Oehme M (1993) Fresenius J Anal Chem 349:552
229. Krock B, Vetter W, Luckas B, Scherer G (1996) Chemosphere 33:1005
230. a. Andrews P, Vetter W (1995) Chemosphere 31:3879. b. Oehme M, Kallenborn R (1995) Chemosphere 30:1739
231. Spehar RL, Veith GD, DeFoe DL, Bergstedt BV (1979) Bull Environ Contam Toxicol 21:576
232. a. Meylan WM, Howard PH (1995) LOGKOW Octanol-Water Partition Coefficient Program, Version 1.37. Syracuse Research Corporation, Chemical Hazard Assessment Division, Environmental Chemistry Center, Syracuse, NY 13210, USA. b. Meylan WM, Howard PH (1995) J Pharmacol Sci 84:83. c. Morris JJ, Hughes LR, Glen AT, Taylor PJ (1991) J Med Chem 34(1):447. d. Nakagawa Y, Izumi K, Oikawa N, Sotomatsu T, Shigemura M, Fujita T (1992) Environ Toxicol Chem 11:901. e. Allner B, Nagel R, Bauer ERS, Stahlschmidt-Allner P (1998): Environ Toxicol Chem, submitted
233. Kuehl DW (1981) Chemosphere 10:231
234. Poppe A, Alberti J, Friege H, Rönnefahrt B (1988) Vom Wasser 70:33
235. Ehmann J, Ballschmiter K (1989) Fresenius Z Anal Chem 332:904
236. Friege H, Stock W, Alberti J, Poppe A, Juhnke I, Knic J, Schiller W (1989) Chemosphere 18:1367
237. Rönnefahrt B (1987) Dtsch Lebensm Rundsch 83:214
238. Fürst P, Krüger C, Meemken H-J, Groebel W (1987) J Chromatography 405:311
239. Wester PG, van der Valk F (1990) Bull Environ Contam Toxicol 45:69
240. Bouraly M, Millischer RJ (1989) Chemosphere 18:2051
241. Van Haelst AG, Zhao Q, van der Wielen F, Govers HAJ, De Voogt P (1996b) Ecotoxicol Environ Saf 34:35
242. Van Haelst AG, Loonen H, van der Wielen FWM, Govers HAJ (1996a) Chemosphere 32:1117
243. Gobas FAPC, Zhang X (1992) Chemosphere 25:1961
244. Van Haelst AG, Zhao Q, van der Wielen WM, Govers HAJ (1996c) Chemosphere 33: 257
245. Van Haelst AG, Heesen PF, van der Wielen FWM, Govers HAJ (1994) Chemosphere 29:1651
246. Körner W, Hauf V, Schuller W, Bartsch H, Kreienberg R, Hagenmaier H-P (1996) In: Olie K et al. (eds) Organohalogen Compounds, vol 27. Dioxin 96. Formation & Sources, Analysis and Quality Control, pp 297
247. Watanabe I, Tatsukawa R (1990) In: Freij L (ed) Proceedings on the Workshop on Brominated Aromatic Flame Retardants. Skoloster, Sweden, 24–26 October 1989. Solna, Sweden, National Chemicals Inspectorate, pp 63–71
248. IPCS (1994) International Programme on Chemical Safety. Polybrominated Biphenyls. Environmental Health Criteria 152, World Health Organization, Geneva Switzerland
249. Pijnenburg AMCM, Everts JW, de Boer J, Boon JP (1995) Rev Environ Contam Toxicol 141:1
250. Zitko V, Hutzinger O (1976) Bull Environ Contam Toxicol 16:665
251. Oliver BG, Niimi AJ (1985) Environ Sci Technol 19(9):842
252. Gobas FAPC, Clark KE, Shiu WY, MacKay D (1989) Environ Toxicol Chem 8:231

253. Butte W, Fox K, Schmidt A, Zauke G-P (1993) Bioakkumulation im Grenzbereich von Lipophilie und molekularen Strukturen. Forschungsbericht Nr. 10603091
254. Sundström G, Hutzinger O, Safe S (1976) Chemosphere 5(1):11
255. Buser HR, Bosshardt HP, Rappe C (1978) Chemosphere 7:109
256. Korach KS, Sarver P, Chae K, McLachlan JA, McKinney JD (1988) Mol Pharmacol 33:130
257. McKinney JD, Waller CL (1994) Environ Health Perspect 102:290
258. Waller CL, Minor DL, McKinney JD (1995) Environ Health Perspect 103:702
259. a. Sellström U, Kierkegaard A, de Witt C, Jansson B (1998): Environ Toxicol Chem 17(6):1065. b. WHO (1994) Environmental Health Criteria No. 162: Brominated Diphenyl Ethers. World Health Organization, Geneva, Switzerland
260. Boer J de, Dao QJ (1993) Overview of bromodiphenylether data in aquatic biota and sediments. RIVO Report CO20/93. DLO-Rijksinsituut voor Visserijon der zoek, Netherlands Institute for Fisheries Research, Ijmuiden, Netherlands
261. OECD (1994) Selected Brominated Flame Retardants. Risk reduction monograph 3, Organisation for Economic Cooperation and Development, Environment Directorate, Paris
262. Kok JJ, Kok A, Brinkman UA (1979) J Chromatography 171:269
263. Sundström G, Hutzinger O (1976) Chemosphere 5:187
264. Andersson Ö, Blomkvist G (1981) Chemosphere 10:1051
265. Boer J de (1989) Chemosphere 18:2131
266. Rimkus G, Wolf M (1991) Proceedings of Euro Food Chemistry VI 22–26. September in Hamburg, pp 732–738
267. Watanabe I, Kashimoto T, Tatsukawa R (1987) Chemosphere 16:2389
268. Watanabe I, Tatsukawa R (1989) Proceedings Workshop on Brominated Aromatic Flame Retardants, 24–26 Oktober in Skokloster, Swedish National Chemical Inspectorate, Solna Sweden, pp 63–71
269. Jansson B, Asplund L, Olsson M (1987) Chemosphere 16:2343
270. Sellström U (1996) Polybrominated Diphenyl Ethers in the Swedish Environment. ITM-Report 1996:45, Stockholm University, Stockholm
271. Hagenmaier H, She J, Benz T, Dawidowsky N, Düsterhöft L, Lindig C (1992) Chemosphere 25, 1457
272. a. Loganathan BG, Kannan K, Watanabe I, Kawano M, Irvine K, Kumar S, Sikka HC (1995) Environ Sci Technol 29:1832. b. Boer J de, Wester PG, Klamer HJC, Lewis WW, Boon JP (1998) Nature 394:28
273. Cramer PH, Ayling RE, Thornburg KR, Stanley JS, Remmers JC, Breen JJ, Schwemberger J (1990) Chemosphere 20:821
274. Stanley JS, Cramer PH, Thornburg KR, Remmers JC, Breen JJ, Schwemberger J (1991) Chemosphere 23:1185
275. Kierkegaard A, Balk L, Sellström U, Tjärnlund U, Örn U, Wit C de, Jansson B (1995) Abstract book of the fifth SETAC Europe Congress, Copenhagen, June 25–28, 1995, p 221
276. UBA (1992) Weiterführende Untersuchung zur Bildung von polybromierten Dioxinen und Furanen bei der thermischen Belastung flammgeschützter Kunststoffe und Textilien, Teilvorhaben 1 und 2. Umweltbundesamt (German Environmental Protection Agency), Berlin, Germany, Texte 45/92
277. Buser H-R (1986) Environ Sci Technol 20:404
278. Thoma H, Hutzinger O (1987) Chemosphere 16:1353
279. Thoma H, Hauschulz G, Knorr E, Hutzinger O (1987) Chemosphere 16:277
280. Dumler R, Lenoir D, Thoma H, Hutzinger O (1989) J Anal Appl Pyrolysis 16:153
281. Dumler R (1989) Brandversuche zur Bildung von bromierten Dibenzofuranen und -dioxinen aus flammgeschützten Kunststoffen. Doctoral Thesis, University Bayreuth
282. Clausen E, Lahaniatis ES, Bahadir M, Bieniek D (1987) Fresenius Z Anal Chem 327:297
283. Lahaniatis ES, Bergheim W, Bieniek D (1991) Toxicol Environ Chem 31/32:521
284. Lenoir D, Kampke-Thiel K (1995) In: Fire and Polymers II, Materials and Tests for Hazard Prevention. ACS Symposium Series 599, Ed. GL Nelson, American Chemical Society, Washington, D.C., pp 377–392

285. Pohle H (1990) UWSF-Z Umweltchem Ökotoxicol 2:148
286. a. Darnerud PO, Sinjari T (1996) In: van den Berg M, Brouwer A, Bergman A, Birnbaum L (eds) Organohalogen Compounds. Short papers, Ecotoxicology, Toxicology, Metabolism, Toxicokinetics. University Amsterdam, The Netherlands vol 29, pp 316–319. b. Haglund PS, Zook DR, Buser H-R, Hu J (1997): Environ Sci Technol 31:3281. c. Asplund L, Athanasiadou M, Eriksson U, Sjödin A, Börjeson H, Bergman Å (1997) Organohalogen Compounds 33:355. d. Marsh G, Bergman Å., Bladh L-G, Gillner M, Jakobsson E (1998) Organohalogen Compounds 37:305. e. Meerts, IATM, Marsh G, van Leeuwen BJ, Luijks EAC, Jakobsson E, Bergman Å, Brower A (1998) Organohalogen Compounds 37:309
287. Mason G, Denomme MA, Safe L, Safe S (1987) Chemosphere 16:1729
288. Mason G, Farrell K, Keys B, Piskorska-Pliszczynska J, Safe L, Safe S (1986) Toxicology 41:21
289. Moore JA, McConnell EE, Dalgard DW, Harris MW (1979) Ann NY Acad Sci 320:151
290. Zacharewski T, Harris M, Safe S, Thoma H, Hutzinger O (1988) Toxicology 51:177
291. Löser E, Ivans I (1989) Chemosphere 19:759
292. Safe S (1990) CRC Crit Rev Toxicol 21:51
293. CBC (1982) Chemical Biotesting Center, Chemicals Inspection and Testing Institute, Tokyo, Japan (unpublished Report) cited in WHO (1994) Environmental Health Criteria No. 162. Brominated Diphenylethers
294. Jordan VC (1995) Annu Rev Pharmacol Toxicol 35:194
295. Kurz J, Ballschmiter K (1995) Fresenius Z Anal Chem 351:98
296. Kurz J (1994) Doctoral Thesis, University Ulm, Germany
297. Sundström G, Hutzinger O (1976) Chemosphere 5:305
298. Paasivirta J, Koistinen J (1994) In: Kiceniuk JW, Ray S (eds) Analysis of Contaminants in Edible Aquatic Resources, chapter 6. VCH, Weinheim New York
299. Nevalainen T, Koistinen J, Nurmela P (1994) Environ Sci Technol 28:1341
300. Matsunaka S (1975) In: Kearney PC and Kaufman DD (eds) Herbicides, vol 2. Marcel Dekker, New York, N.Y., pp 709–739
301. Rappe C, Buser HR, Bosshardt HP (1979) Ann NY Acad Sci 320:1
302. Nilson C-A, Renburg L (1974) J Chromatograph 89:325
303. Koistinen J, Paasivirta J, Suonperä M, Hyvärinen H (1995) Environ Sci Technol 29:2541
304. Paasivirta J, Tarhanen J, Soikkeli J (1986) Chemosphere 15:1429
305. Gara A, Andersson K, Nilsson CA, Norström A (1981) Chemosphere 10:365
306. Lake JL, Rogerson PF, Norwood CB (1981) Environ Sci Technol 15:549
307. Koistinen J, Vuorinen PJ, Paasivirta J (1993) Chemosphere 27:2365
308. Niimi AJ, Metcalfe CD, Huestis SY (1994) Environ Toxicol Chem 13:1133
309. Williams DT, Kennedy B, LeBel GL (1991) Chemosphere 23:601
310. Stanley JS, Cramer PH, Thornberg KR, Remmers JC, Breen JJ, Schwemberger J (1991) Chemosphere 23:1185
311. Koistinen J, Mussalo-Rauhamaa H, Paasivirta J (1995) Chemosphere 31:4259
312. Neely WB, Branson DR, Blau GE (1974) Environ Sci Technol 8:1113
313. Zitko V, Carson WG (1977) Chemosphere 6:293
314. Niimi AJ (1986) Aquat Toxicol 9:105
315. Chui YC, Addison RF, Law FCP (1990) Xenobiotica 20:489
316. Becker M, Phillips T, Safe S (1991) Toxicol Environ Chem 33:189
317. Kodavanti PRS, Ward TR, McKinney JD, Waller CL, Tilson HA (1996) Toxicol Appl Pharmacol 138:251
318. Nilsson CA, Norström Å, Anderson K, Rappe C (1978) In: Rao KR (ed) Pentachlorophenol: Chemistry, Pharmacology, and Environmental Toxicology. Plenum Press, New York, pp 313–324
319. Tulp MT, Sundström G, Martron LB, Hutzinger O (1979) Xenobiotica 9:65
320. Komsta E, Chu I, Villeneuva DC, Benoit FM, Murdoch D (1988) Arch Toxicol 62:258
321. Lans MC, Klasson-Wehler, Brouwer A (1994) Organohalogen Compounds 20:481
322. McKinney JD (1989) Environ Health Perspect 82:323

323. Lans MC, Spiertz C, Brouwer A, Koeman JH (1994) Eur J Pharmacol 270:129
324. Waller CL, McKinney JD (1995) Chem Res Toxicol 8:847
325. a. Opperhuizen A, Voors PL (1987) Chemosphere 16:2379. b. Oliver BG, Niimi AJ (1984) Environ Toxicol Chem 3:271
326. Barbetta L, Trowbridge T, Eldib IA (1988) Perfumer & Flavorist 13:60
327. Schlatter J, Hunyady G (1993) Bulletin des BAG 30:546
328. Ippen H (1994) Bundesgesundhbl 37:255
329. Qinghua Z (1993) Perfumer & Flavorist 18:47
330. Srinivasan SK, Reddy GO (1993) Indian Perfumer 37:344
331. Geyer HJ, Rimkus G, Wolf M, Attar A, Steinberg C, Kettrup A (1994) UWSF-Z Umweltchem Ökotoxicol 6:9
332. Rimkus G, Brunn H (1996) Ernährungs-Umschau 43:442
333. Schramm K-W, Kaune A, Beck B, Thumm W, Behechti A, Kettrup A, Nickolova P (1996) Water Res 30(10):2247
334. Yamagishi T, Miyazaki T, Horii S, Kaneko S (1981) Bull Environ Contam Toxicol 26:656
335. Yamagishi T, Miyazaki T, Horii S, Akiyama K (1983) Arch Environ Contam Toxicol 12:83
336. Rimkus G, Wolf M (1993) Dtsch Lebensm-Rundsch 89:171
337. Rimkus GG, Wolf M (1995) Chemosphere 30:641
338. Rimkus GG, Wolf M (1993) Lebensmittelchemie 47:31
339. Eschke H, Traud J, Dibowski H-J (1994) Vom Wasser 83:373
340. Chemicals Inspection and Testing Institute Japan (1992) Data of existing chemicals based on the CSCL Japan. Japan Chemicals Industry Ecology-Toxicology & Information Center
341. Kuhlmann H, Rimkus GG, Butte W (1999) Chemosphere (in preparation)
342. Butte W (1991) In: Nagel R, Loskill R (eds) Bioaccumulation in Aquatic Systems. VCH, Weinheim, pp 29–42
343. Butte W, Kellermann H-J, Kuhlmann H, Rimkus GG, Geyer HJ (1998) Chemosphere (in preparation)
344. Rimkus GG, Nausch I, Geyer HJ (1997) Chemosphere (in preparation)
345. McKay D (1982) Environ Sci Technol 16:274
346. Rimkus GG, Butte W, Geyer HJ (1997) Chemosphere 35:1497
347. Tarazona JV, Ortis JA, Carballo M, Munoz MJ (1993) Fresenius Environ Bull 2:84
348. Boleas S, Fernandez C, Tarazona JV (1996) Bull Environ Contam Toxicol 57:217
349. Kokot-Helbig K, Schmid P, Schlatter C (1995) Mitt Geb Lebensmitt Hyg 86:1
350. Ford RA, Api AM, Newberne PM (1990) Food Chem Toxicol 28:55
351. a. Rimkus GG (1998) In: Frosh PJ, Johansen JD, White R (eds) Fragrances. Beneficial and Adverse Effects. Springer, Berlin Heidelberg New York, pp 136–149. b. Allner B (1997) Toxikokinetik und -dynamik von 3,4-Dichloranilin beim dreistachlichen Stichling unter besonderer Berücksichtigung der Fortpflanzungsbiologie. Dissertation, Universität Mainz. c. Sauerwein H, Allner B (1998) Personal communication. d. Stahlschmidt-Allner P, Allner B, Römbke J, Knacker T (1997) ESPR - Environ Sci & Pollut Res 4(3):155. e. Schramm K-W (1998) Personal communication
352. Barbetta L, Trowbridge T, Eldib IA (1988) Perfumer & Flavorist 13:60
353. Ohloff G (ed) (1990) Riechstoffe und Geruchssinn. Springer, Berlin Heidelberg New York
354. Eschke H-D, Traud J, Dibowski H-J (1994) UWSF-Z Umweltchem Ökotoxicol 6:183
355. Eschke H-D, Dibowski H-J, Traud J (1995) UWSF-Z Umweltchem Ökotoxicol 7:131
356. Rimkus GG, Wolf M (1999) Chemosphere (in preparation)
357. Eschke H-D et al. (1999) Chemosphere (in preparation)
358. Eschke H-D, Dibowski H-J, Traud J (1995) Dtsch Lebensm-Rundsch 91:375
359. Rimkus GG, Wolf M (1996) Chemosphere 33(10):2033
360. Müller S, Schmid P, Schlatter C (1996) Chemosphere 33(1):17
361. a. Rimkus GG (1999): Toxicol Lett Special Issue (in press). b. Gebauer H, Bouter T (1997) Euro Cosmetics 1:30
362. Rimkus G, Brunn H (1996) Ernährungs-Umschau 43(12):442
363. Brunn H, Rimkus G (1997) Ernährungs-Umschau 44(1):4

364. Kramer VJ (1996) Rohm Haas Company, Spring House, PA, USA, personal communication to GG Rimkus
365. Lala DS, Mukherjee R, Schumman IG, Koch SS, Dardashti LJ, Nadzan AM, Croston GE, Evans RM, Heyman RA (1996) Nature 383:450
366. Mangelsdorf DJ, Umesono K, Evans R (1994) In: Spom MB, Roberts AB, Goodman DS (ed) The Retinoids. Academic Press, New York, pp 319–349
367. a. Balk F, Blok J, Okkerman PC, Ford RA (1997) Paper presented at the Seventh Annual Meeting of SETAC-Europe, Amsterdam, The Netherlands, April 6–10, 1997, Abstract book, p 111. b. Plassche EJ, Balk F (1997) Environmental Risk Assessment of the Polycyclic Musks AHTN and HHCB according to the EU-TGD Report N 601503 008. National Instiute of the Public Health and the Environment (RIVM), Bilthoven, The Netherlands. c. Ewald F (1998) Diploma thesis, University Oldenburg, Germany
368. Schauder S, Schrader A, Ippen H (1994) Göttinger Liste, Sonnenschutzkosmetik in Deutschland, 3 Aufl. Blackwell, Berlin
369. Statistisches Bundesamt (1993) Monatliche Produktion ausgewählter Körperpflegemittel in Deutschland, Wiesbaden
370. Ternes T (1993) Qualifizierung und Quantifizierung von Fremdstoffen in Wasser und Fischen aus unterschiedlich belasteten Gewässern mittels Gaschromatographie/Massenspektrometrie. Ph. Thesis, Johannes Gutenberg-University, Mainz
371. Nagtegaal M, Ternes TA, Baumann W, Nagel R (1997) UWSF - Z Umweltchem Ökotoxicol 9:79
372. Ternes T (1996) ESWE-Institut für Wasserforschung und Wassertechnologie GmbH, Wiesbaden, Germany. Personal Communication to H Geyer
373. Hany J, Nagel R (1995) Dtsch Lebensmitt Rdsch 91(11):341
374. Ternes TA, Hany J, Baumann W, Nagel R (1995) Fresenius J Anal Chem 351:790
375. Kazuo I, Sukeji S, Yohya W (1990) J Chrom 513:321
376. Ro KW, Choi JB, Lee MH, Kim JW (1994) J Chrom 688:375
377. Schwack W, Rudolph T (1996) GIT Fachz Lab 4:373
378. a. Van den Heuvel WJA, Forbis AD, Halley BA, Ku CC, Jacob TA, Wislocki PG (1996) Environ Toxicol Chem 15(12):2263. b. Davies IM, McHenery JG, Rae GH (1997): Aquaculture 158:263
379. Springer JP, Arison BH, Hirshfeld JM, Hoogsteen K (1981) J Am Chem Soc 103:4221
380. Uyeda N, Kobayashi T, Ishizuka K, Fujiyoshi Y (1978-79) Chimica Scripta 14:47
381. Belfroid A, Meiling J, Dreuth H-J, Hermens J, Seinen W, Van Gestel K (1995) Ecotoxicol Environ Saf 31:185
382. Nicolaou KC (1996) Angew Chem 108:645
383. Murata M, Naoki H, Matsunaga S, Satake M, Yasumoto T (1994) J Am Chem Soc 116:7098
384. Yasumoto T, Murata M (1993) Chem Rev 93:1897
385. Gusovsky F, Daly JW (1990) Biochem Pharmacol 39(11):1633
386. Zeeman M (1996) Comparative immunotoxicology and risk assessment, chapter 27. In: Stolen JS, Fletcher TC, Bayne CJ, Secombes CJ, Zelikoff JT, Twerdok LE, Anderson DP (eds) Modulators of Immune Responses. SOS Publications, Fair Haven, N.J., pp 317–329
387. Zitko V (1981) Uptake and excretion of chemicals by aquatic fauna. In: Stokes PM (ed) Ecotoxicology and the Aquatic Environment. Pergamon Press, New York, pp 67–78
388. Bräse S (1997) The Scripps Research Institute La Jolla, California, personal communication to HJ Geyer
389. Mebs D (1992) Gifttiere. Ein Handbuch für Biologen, Toxikologen, Ärzte, Apotheker. Wissensch. Verlagsges. mbH, Stuttgart
390. Jansson B, Anderson R, Asplund L, Litzen K, Nylund K, Sellström U, Widequist U, Olsson M (1993) Environ Toxicol Chem 12:1163
391. US EPA (1985) Risk Assessment for chlorinated paraffins: Effect on fish and wildlife. Environmental Protection Agency, Health and Environmental Review Division, Toxicology Section, Washington DC, Dec. 18, 1985

392. Department of the Environment (1994) Environmental Hazard Assessment: Chlorinated Paraffins. Toxic Substances Division. Directorate for Air, Climate and Toxic Substances. Building Research Establishment. Garston, Watford WD2 7JR, UK, ISBN 0-85-125-6279
393. Department of Indian Affairs and Northern Development Canada (1994) State of Knowledge Report of the UN ECE Task Force on Persistent Organic Pollutants. For the Economic Commission for Europe (ECE) Convention on Long-Range Transboundary Air Pollution. Environmental Service and Research Division. Ottawa, Ontario, Canada
394. Sijm DTHM, Sinnige TL (1995) Chemosphere 31 (11/12):4427
395. Brooke LT (1993) Accumulation and lethality for two freshwater fishes (fathead minnow and bluegill) to nonylphenol. Report to the U.S. EPA for Work Assignment No. 1-12 of Contract No. 68-C1-0034. Lake Superior Research Institute, University of Wisconsin-Superior, Superior, WI. September 30, 1993, and personal communication
396. Call DJ, Brooke LT, Lu P-Y (1980) Arch Environ Contam Toxicol 9:699
397. Veith G (1981) Personal communication to H. Geyer
398. Lien G (1999) Personal communication to H. Geyer
399. Granmo A, Eklund R, Berggren M, Magnusson K (1991) Toxicity of 4-nonylphenol to aquatic organisms and potential for bioaccumulation. Proceedings, Swedish Environmental Protection Agency Seminar on Nonylphenol Ethoxylates/Nonylphenol, Saltsjobaden, Sweden, February 6-8, pp 53-75
400. Staples CA, Weeks J, Hall JF, Naylor CG (1998) Environ Toxicol Chem 17:2470
401. Gustafsson K, Björk M, Bureau S, Gilek M (1999) Environ Toxicol Chem 18(6):1218
402. Scheunert I (1992) In: Ebing W (ed) Terrestrial Behavior of Pesticides. ISBN 3-540-54238-8 Springer-Verlag Berlin Heidelberg New York, pp 25-75
403. Gattermann R, Rimkus G, Hecker M, Biselli S, Hühnerfuss H (1999) Posterpresentation on the 9th Annual Meeting of SETAC-Europe, 25-29 May 1999, Leipzig, Germany
404a. Panter GH, Thompson RS, Beresford N, Sumpter JP (1999) Chemosphere 38(15):3579
404b. Ternes T, Wilken R-D (eds) (1999) Drugs and Hormones as Pollutants of the Aquatic Environment: Determination and Ecotoxicological Impacts, Special Issue Sci Total Environ 225 (1,2)
405. Cheek AO, Kow K, Chen J, McLachlan JA (1999) Environ Health Perspect 107:273

Internal Effect Concentration: Link Between Bioaccumulation and Ecotoxicity for Organic Chemicals

Dick T.H.M. Sijm* and Joop L.M. Hermens

Environmental Toxicology Section, Research Institute of Toxicology (RITOX), Utrecht University, P.O. Box 80.158, NL-3508 TD Utrecht, The Netherlands.

This paper reviews the concept and the use of *internal* effect concentrations. Bioaccumulation plays a very important role in this concept, and is part of the process which results in that chemicals attain body burdens and eventually *internal* effect concentrations in an organism which cause adverse effects. Hydrophobic compounds elicit their toxicity at low *external* concentrations because their high bioaccumulation properties allow critical or lethal body burdens in organisms to be reached already at those low environmental ambient concentrations. First, a concise overview is provided of bioaccumulation models, bioaccumulation parameters, and factors which influence bioaccumulation of organic chemicals for aquatic, benthic and terrestrial organisms. Second, a brief overview is given on *external* and *internal* effect concentrations. The concept and assumptions related to the *internal* effect concentrations are dealt with in more detail. Third, bioaccumulation and effects are linked through the concept of *internal* effect concentration. Bioaccumulation kinetics can be used to describe and predict concentrations of organic compounds in an organism. Established relationships can be used for this purpose, which include physical-chemical and physiological parameters, in addition to ambient concentrations in the environment, such as in water, sediment and soil, and in food. The use of predicted concentrations and *internal* effect concentrations of organic compounds enables one to evaluate ecotoxicological risk for these compounds. Since the *internal* concentration adds all molar concentrations of individual chemicals as one molar concentration, the *internal* concentration thus deals with additivity of a mixture. Biomimetic extraction and molar detection techniques are discussed and suggested to offer a useful tool to assess the total amount of bioaccumulatable organic compounds.

Keywords: Bioaccumulation; Body Burden; Mixtures; Hydrophobicity; LC50.

* Present address: National Institute of Public Health and the Environment P.O. Box 1, NL-3720 BA, Bilthoven, The Netherlands. E-mail: Dick.Sijm@rium.nl

The Handbook of Environmental Chemistry, Vol. 2 Part J
Bioaccumulation (ed. by B. Beek)

1 Introduction

Ever since the presence of an organochlorine pesticide (DDT) in gull eggs and singing birds was related to egg shell thinning and death, respectively, the link between bioaccumulation and ecotoxicological effects was identified. DDT accumulated through the foodchain in gull eggs, was found to be the causative agent for egg shell thinning [1], and caused the death of many singing birds, after they had consumed worms that had been exposed to DDT [2].

Even at low ambient concentrations, some organic compounds such as DDT resulted in toxic effects, due to their high bioaccumulation properties. In early ecological toxicity studies, the aqueous concentration of organic chemicals necessary to cause lethality in fish (LC50) was found to decrease with increasing hydrophobicity, expressed as the octanol/water partition coefficient (K_{ow}) [3].

In general, hydrophobic organic compounds have a strong tendency to bioaccumulate in aquatic organisms [4]. Therefore, it is not surprising that there is an inverse relationship between LC50 and hydrophobicity, since the more hydrophobic a chemical is, the more it will accumulate in an organism. Bioaccumulation is thus linked and usually needs to precede effects. This is even more clear when we refer to studies performed in the late 19th and the early 20th century when it was recognized that it is the *internal* critical dose in an organism that leads to the effects [5–7]. However, current risk assessment is still based on *external* concentrations [8]. At present, since more knowledge is available on bioaccumulation and on *internal* and *external* effect concentrations, it may be worthwhile to reconsider the earlier thoughts on relating ecotoxicological effects to *internal* concentrations, and to involve the role of bioaccumulation of hydrophobic organic chemicals, following the thoughts and approaches of some earlier studies [7, 9–16]. The present study does not include the toxicity of metals, since there are great species differences in toxicity because many organisms react differently with regard to detoxification of metals.

A few of the major drawbacks of relating ecotoxicological effects to *external* concentrations are that i) organisms in the field are exposed to mixtures of many compounds, ii) some chemicals do not show (acute) toxic effects at aqueous concentrations below their aqueous solubility but do show effects as a result of biomagnification through the food-chain or show effects in a mixture, and iii) the bioavailable fraction of the compound is sometimes difficult to determine, thus giving rise to problems in the interpretation of *external* concentrations. Most of the drawbacks are thus related to the exposure concentration of a compound in the environment. The *internal* effect is more directly related to the concentration at the target of an organism, although it is not always clear what this target is. In addition, the *internal* effect concentration would be able to deal with mixtures of compounds, and with different exposure regimes that affect the bioavailability of a compound in the environment.

For a broad applicability applied to either lethal or sublethal effects, the *internal* effect concentration (expressed as mol kg^{-1} or mol kg_{lipid}^{-1}) approach should meet a couple of conditions. The following examples refer to lethality. The first condition may be that an organism dies when a distinct *internal* effect concentration, the lethal body burden, of a specific chemical has been reached. The second condition is that any individual dies when it has attained this lethal body burden. The third condition is that the lethal body burden is independent of time of death or exposure concentration. In the latter case it may take longer to die at a lower exposure concentration and shorter to die at a higher concentration, but in either case, when the lethal body burden has been reached, it should be the same for both conditions. The fourth condition is that all chemicals which have the same mechanism of action have the same lethal body burden. The latter thus enables one to deal with additivity, since the individual chemicals of a mixture, all of which have the same mechanism of action, will contribute equally to the body burden on a molar basis.

The aim of this review is first to describe bioaccumulation in different types of aquatic, benthic and terrestrial organisms, second, to describe some ecotoxicological effects, and third, to link bioaccumulation and ecotoxicological effects.

2 Bioaccumulation

Organisms need to take up chemicals before toxic effects are elicited. The rate at, and the route by, which the toxicants are taken up depends on both the organism and the compound, such as on the habitat and physiology of the organism and on the physical-chemical properties of the compound. To understand how *internal* concentrations are built up in an organism resulting from exposure to the chemical in either the ambient environment or in food, a short overview on bioaccumulation is given.

Aquatic, benthic or terrestrial organisms will be exposed to a variety of chemicals in water, food, sediment or soil. This exposure may lead to uptake and to adverse effects, including death under specific conditions. In most cases it is the ambient water which is the prime route through which xenobiotics accumulate for most aquatic and for some benthic and terrestrial organisms. In other cases it is the food, sediment or soil which is the prime route through which xenobiotics accumulate. Hydrophobic organic chemicals tend to bioaccumulate in almost any species. Knowledge on bioaccumulation and the role of the physical-chemical properties of the compound and of the characteristics of the organism and its environment is therefore of paramount importance.

Bioaccumulation can simply be viewed as the process of a chemical moving from an organism's medium (sediment, soil or water) or diet into the organism. Uptake by respiratory organs (gills and skin) exposed to water can be an important route for aquatic, benthic and terrestrial animals. Uptake by the gastro-intestinal tract is the major uptake route for sediment and soil ingesting organisms, and for animals higher in the foodchain, such as mammals or fish-eating birds.

2.1 Bioaccumulation Models

Bioaccumulation results in higher concentrations of compounds in organisms than in their ambient environment (sediment, soil or water) or in their food. When uptake occurs from water, bioaccumulation is called bioconcentration. When these higher concentrations in organisms results from food only, bioaccumulation is called biomagnification. When different routes are important, such as (additional) uptake from sediment or soil, it is called bioaccumulation in a general way. In the following sections a brief description will be given for the different models which describe bioconcentration, biomagnification and bioaccumulation.

2.1.1 *Bioconcentration*

Bioconcentration models are used when the exchange of the chemicals is via water. Since most of the theoretical models have been developed for aquatic organisms, we will first discuss a bioconcentration model for those organisms. The exchange of chemicals between water and organism is usually described by a first-order one-compartment model, relating the concentration in the organism to that in water [17, 18]:

$$\frac{dC_b}{dt} = k_{u,w}C_w - k_e C_b \tag{1}$$

where C_b is the concentration in the organism (mol kg^{-1}), t is time (d), $k_{u,w}$ is the uptake rate constant (l kg^{-1} day^{-1}), C_w is the concentration in water (mol l^{-1}) and k_e is the elimination rate constant (day^{-1}). The first-order one-compartment model assumes that either biota and the ambient environment of concern, such as water, soil, sediment or food, is one homogeneous compartment, and that the exchange rate constants are independent of concentration.

In the present study, bioconcentration, biomagnification and bioaccumulation models are presented using models which describe the concentrations of chemicals in the organisms and environment and food. Other models use fugacities to describe the bioaccumulation processes [e.g. 19, 20]. For the sake of simplicity, however, we continue with describing the models based on concentrations.

Elimination, or the reduction of the concentration, may be the result of several processes, such as passive excretion (physical-chemical elimination), growth dilution, biotransformation of the chemical, and reproductive transfer [21].

At steady-state, the concentrations of the chemical in the aquatic organism and that in water do not change any longer. In that case, the ratio of these concentrations in organism and water is reflected by the bioconcentration factor (BCF), which is equal to the ratio of the uptake and elimination rate constant:

$$BCF = \frac{C_b}{C_w} = \frac{k_{u,w}}{k_e} \tag{2}$$

Organic hydrophobic chemicals tend to be stored in the lipid parts of an organism. Differences in lipid content between organisms thus result in differences in bioconcentration factors. Therefore, the BCF in fish is usually normalized for the lipid content of the studied organism [22], resulting in

$$BCF_L = \frac{BCF}{L_w} \tag{3}$$

where BCF_L is the lipid-normalized bioconcentration factor (l kg_{lipid}^{-1}), and L is the lipid content of the organism (kg_{lipid} $kg_{wet\ weight}^{-1}$).

2.1.2
Biomagnification

When organisms are predominantly exposed to the chemicals via ingestion, Eq. (1) can be rewritten as

$$\frac{dC_b}{dt} = k_{u,fd}\, C_{fd} - k_e C_b \tag{4}$$

where $k_{u,fd}$ is the uptake rate constant for food (kg kg^{-1} day^{-1}) and C_{fd} is the concentration of the chemical in food (mol kg^{-1}). A further refinement for the food uptake rate constant is often used to distinguish better between the uptake efficiency of chemical from food after uptake in the gastro-intestinal tract (E_{fd},) and the rate of food uptake or the amount of food consumed per day (V_{fd} in $kg_{food}\,kg^{-1}\,day^{-1}$) [23]:

$$k_{u,fd} = E_{fd} \cdot V_{fd} \tag{5}$$

The biomagnification factor (BMF) is equal to the ratio of the uptake and elimination rate constant at steady state, similar to the bioconcentration factor:

$$BMF = \frac{C_b}{C_w} = \frac{k_{u,fd}}{k_e} = \frac{E_{fd} V_{fd}}{k_e} \tag{6}$$

2.1.3
Bioaccumulation

The term bioaccumulation is used when the exchange of the chemicals is via water, sediment and/or soil. For benthic and terrestrial species, the equations describing bioaccumulation from the ambient environment are analogous to those for bioconcentration in aquatic organisms (Table 1).

While many studies report on relationships between physical-chemical and physiological properties on the one hand and bioconcentration on the other for aquatic organisms [18, 24–38], very few data are available for benthic and soil organisms [e.g. 20, 39].

Analogous to the steady-state bioconcentration factor (BCF) and the biomagnification factor (BMF), the biota-to-sediment-accumulation factor ($BSAF_{sed}$) and the biota-to-soil-accumulation factor ($BSAF_{soil}$) are defined as:

$$BSAF_{sed} = \frac{C_b}{C_{sed}} = \frac{k_{u,sed}}{k_e} \tag{7}$$

$$BSAF_{soil} = \frac{C_b}{C_{soil}} = \frac{k_{u,soil}}{k_e} \tag{8}$$

Either for soil or for sediment, the BSAF is usually expressed as the ratio of the lipid-normalized concentration in the organism and the organic carbon normalized concentration in the sediment or soil:

$$BSAF_L = \frac{C_b/L}{C_{soil/sed}/f_{oc}} \tag{9}$$

where L is the lipid content of the organism and f_{oc} the organic carbon fraction of the sediment or soil.

It is often assumed that, particularly in the aqueous environment, there is a steady-state situation, i. e. that the concentrations of pollutants in the water and the suspended solids is in equilibrium. Hendriks [40] verified this assumption. He found that the ratios of concentrations in different organisms and those in suspended solids of a series of organic compounds were not significantly different from the calculated ratios that were based on existing bioaccumulation and sorption relationships. The organisms that were studied were chironomidae, mollusca, crustacea and a number of fish species.

Table 1. Bioaccumulation models for different organisms

Bioconcentration is described by: $dC_b/dt = k_{u,i} \cdot C_{ambient} - k_e \cdot C_b$

species	ambient environment	uptake rate constant	elimination rate constant
aquatic	water (C_w in mol l^{-1})	$k_{u,w}$ (l kg^{-1} day^{-1})	k_e (day^{-1})
benthic	sediment ($C_{sediment}$ in mol kg^{-1})	$k_{u,sed}$ (kg kg^{-1} day^{-1})	k_e (day^{-1})
	water (C_w in mol l^{-1})	$k_{u,w}$ (l kg^{-1} day^{-1})	k_e (day^{-1})
terrestrial	soil (C_{soil} in mol kg^{-1})	$k_{u,soil}$ (kg kg^{-1} day^{-1})	k_e (day^{-1})
	water (C_w in mol l^{-1})	$k_{u,w}$ (L kg^{-1} day^{-1})	k_e (day^{-1})
all species	food (C_{fd} in mol kg^{-1})	$k_{u,fd}$ (kg kg^{-1} day^{-1})	k_e (day^{-1})

2.2 Bioaccumulation Parameters

For a number of organic compounds, such as DDT, polychlorinated benzenes (PCBzs), biphenyls (PCBs), dibenzo-*p*-dioxins (PCDDs), dibenzofurans (PCDFs), and naphthalenes (PCNs), relationships between physical-chemical, physiological and bioaccumulation parameters have been established [18, 37, 41–43], which will be evaluated in the following sections.

2.2.1 *Uptake from Water*

Uptake rate constants for aquatic organisms have been found to increase with increasing hydrophobicity for chemicals with a log K_{ow} up to approximately 3, are relatively constant for chemicals with a log K_{ow} higher than 3, and decrease slightly for chemicals with a log K_{ow} higher than 6 [18, 37, 42, 44]. In addition, uptake rate constants are related to organism weight. For fish, an empirical allometric relationship between the uptake rate constant and weight (W, g) has been derived for chemicals with a log K_{ow} between 3 and 6 [43]:

$$k_{u,w} = (550 \pm 16)\, W^{-(0.27 \pm 0.05)} \tag{10}$$

Typical values for the uptake rate constants of hydrophobic chemicals range between approximately 1000 l kg^{-1} day^{-1} for small fish, such as guppy of 0.1 g, and 130 l kg^{-1} d^{-1} for larger fish, such as rainbow trout of 750 g. It must be noted that a theoretically better founded relationship for the uptake rate constant does not exclusively rely on weight and K_{ow}, but also includes ventilation rate of the organism, molecular weight of the chemical, ambient temperature and others [23, 37, 45, 46].

Uptake rate constants for other animals are much less documented, but can be related to the organisms' ventilation rate (respiration rate), since uptake rate constants of the chemicals can be described as the product of the uptake efficiency from the ambient environment (E_w) and of the ventilation rate (V_w, in l kg^{-1} day^{-1}) for uptake via the aqueous phase:

$$k_{u,w} = E_w \cdot V_w \tag{11}$$

It is assumed that the ventilation rate is an organism specific property, which will usually increase with decreasing weight, and which will be higher for homeothermic organisms than for poikilothermic organisms. The uptake efficiency from the exposure medium, however, is related to a more complex suite of factors. For water the efficiency will depend on, e.g., the bioavailable fraction, the presence of dissolved organic carbon in the water [47–50], on temperature [43], and on the hydrophobicity of the chemical [18, 37, 44].

In some studies, the relationship between uptake efficiency and the ventilation rate has been demonstrated [51]. Under hypoxic conditions, rainbow trout increased their ventilation volume which resulted in a higher uptake rate constant for a hydrophobic chemical, since the uptake efficiency remained constant. At very high ventilation rates, however, the uptake efficiency decreased with increasing ventilation flow, which resulted in a constant uptake rate constant at high ventilation rate [51]. It was shown later that ventilation rate related uptake rate constants were only found for relatively large fish of 5–10 g or more, while uptake rate constants were independent of flow rate for smaller fish [52–54].

2.2.2
Elimination to Water

Elimination rate constants for aquatic organisms have been found to be inversely related to K_{ow} within one organism. They further decrease with increasing weight and increasing lipid content of the organism [18, 37, 42]. Elimination rate constants for small fish such as guppy range between 10 day^{-1} for less hydrophobic chemicals (log $K_{ow} < 3$) to less than 0.001 d^{-1} for very hydrophobic chemicals (log $K_{ow} > 6$).

Less information is available for other organisms, but in general, for organisms that are in direct contact with the aqueous environment, which include aquatic, and many benthic and terrestrial organisms, elimination rate constants decrease with increasing hydrophobicity for very hydrophobic chemicals and are fairly constant for less hydrophobic chemicals [37, 55]. For extremely hydrophobic chemicals, it is likely that not elimination to water, but elimination via the faeces, is the predominant route of excretion for aquatic organisms [56]. For a terrestrial species as the earthworm, it has been shown that there are distinct differences between excretion of chlorinated benzenes to water or to soil, which indicated that water is not the predominant route of excretion either [57].

Elimination is sometimes expressed as a half-life ($t_{1/2}$), i.e.

$$t_{1/2} = \frac{\ln 2}{k_e} \tag{12}$$

The lower the elimination rate constant of a chemical is, the higher will be its half-life, and therefore the longer it takes to reduce the concentration of that chemical in an organism. The half-life thus more clearly shows the persistence of a chemical in an organism. In general, smaller organisms will show shorter half-lives for the same chemical. The half-life of a given chemical will thus increase with the size of an organism [37].

2.2.3
Bioconcentration Factors

Several correlations between bioconcentration factors in fish and K_{ow} have been published [4, 18, 25, 27, 32, 37, 42, 58–60]. BCF-values for aquatic organisms may reach numbers up to a million or more for very hydrophobic chemicals. For example, the BCF of 1,4-difluorobenzene in guppy is 32 l kg^{-1} [61] and the BCF of 2,2',5,5'-tetrachlorobiphenyl in goldfish is 1.6 10^6 l kg^{-1} [23].

2.2.4
Uptake from Food

Uptake efficiencies of hydrophobic chemicals have been reported and vary between 10 and 90% [42, 62, 63]. Several factors make it difficult to draw general conclusions or establish a relationship between physico-chemical parameters and uptake efficiency, such as that the uptake efficiency will depend on food composition [64–67], feeding rate [68], and on the developmental stage or age of the fish [21]. Given the large variety in experiments with PCBs, however, the average absorption efficiency of 50 ± 25% (n = 101) and the average uptake rate constant of 0.0082 ± 0.0049 kg kg^{-1} day^{-1} (n = 64) for PCBs in aquatic and benthic species, are relatively constant [63].

2.2.5
Biomagnification Factors

Biomagnification factors of organic compounds for aquatic organisms and aquatic mammals have been reported only for the very hydrophobic chemicals and may reach values as high as 70 [42, 69, 70]. To estimate the concentration of a xenobiotic in a predator, biomagnification factors are multiplied by the concentration in the prey and thus result in high concentrations in the predator.

Much less data are available for BMF than for BCF values. Even more than bioconcentration, biomagnification is highly dependent on the persistence and the hydrophobicity of the chemical.

2.2.6
Uptake from Sediment and Soil

For sediment and soil, the uptake efficiency will depend on the exposure regime and on the organism [20, 71–73]. While there is a three to four orders of magnitude variation in uptake rate constants of PCBs from sediment [63], the average equilibrium $BSAF_{sed}$ values of PCBs, PAHs and some pesticides showed less variability for several benthic organisms, which included infaunal deposit feeders, filter feeders and benthically-coupled fish [74], although Parkerton [66] showed that individual $BSAF_{sed}$ values may differ four orders of magnitude.

2.2.7
Bioaccumulation Factors for Sediment and Soil

Much less data are available for BSAF than for BCF values. Tracey and Hansen [74] collected several sediment organic carbon (f_{oc}) and organism lipid (lipid) normalized BSAF-values that were found to be relatively constant: approximately 1 for PCBs, 0.3 for PAHs, and 1.4 $kg_{oc}\ kg_{lipid}^{-1}$ for organo-pesticides in several types of benthic species [74]. However, the BSAFs for PCBs were lower for PCBs, having a log K_{ow} either smaller than 5.99 or larger than 7.27. Parkerton [66], however, found a more than four orders of magnitude difference in BSAFs for hydrophobic chemicals in benthic invertebrates. In addition, in a study which reported BSAFs in eel, large BSAF values of up to 70 $kg_{oc}\ kg_{lipid}^{-1}$ were also found for organo-pesticides [75].

For soil, organic carbon and lipid normalized $BSAF_{soil}$ in earthworms for a series of polychlorinated benzenes and biphenyls were between 0.4 and 6 $kg_{oc}\ kg_{lipid}^{-1}$ [76].

Both for soil and for sediment, BSAFs have been reported which seem to vary more than four orders of magnitudes for hydrophobic organic compounds. Location specific factors, such as disequilibrium between overlying water, diet and sediment, biomagnification, and feeding preferences and strategies [20], however, significantly modulate BSAFs, and further studies are required to assess the influence of these specific factors.

2.3
Factors Which Influence Bioaccumulation

Many studies have focussed on the uptake and bioaccumulation from water, and have resulted in models. Most of these existing models for the steady-state BCF are valid only for non-ionised organic chemicals and less for ionised chemicals or organometallic compounds. For practical purposes, a kind of worst-case BCF can be estimated for non-ionised organic chemicals based on the published BCF-K_{ow} correlations.

Bioaccumulation can be influenced by several factors, which results in most cases apparently, and in some cases actually, in low bioaccumulation factors.

2.3.1
Absence of Steady State

The elimination rate constants for the more hydrophobic chemicals are low and therefore it will take a long period of time to reach steady state. The time needed to reach a steady state for very hydrophobic chemicals can be in the order of months or even longer [77], which can be much larger than the lifespan of the organism, as in the case of fish [21] or phytoplankton [78]. Absence of steady state will thus lead to an apparently lower BCF.

2.3.2
Limited Uptake by Steric Hindrance

A lack of gill membrane permeation for uptake from water has been proposed for large molecules which have an effective cross section larger than 0.95 nm [79] or which exceed a length of 4.3 nm [60], although this limiting value may be species-dependent [43, 80]. Hydrophobic organic chemicals which are "big" molecules may thus show a very low bioaccumulation behaviour. Limited uptake will thus lead to an actual low BCF.

2.3.3
Differences Between Properties of Octanol and Membrane Lipids

Based on thermodynamic arguments, it has been suggested that octanol does not accurately represents fish lipids [81]. Lipid solubility has been proposed as an additional parameter based on the argument that lipids are more structured than octanol [82]. Partition coefficients with other solvents (triolein-water) and membrane vesicles are measured and related to K_{ow} as well as to BCF [32, 83, 84]. In general, these latter partition coefficients fairly correlate with K_{ow} with systematic lower values at high K_{ow}. These differences may lead to an apparently lower BCF. The consequences are that K_{ow} overestimates bioaccumulation for very hydrophobic compounds.

2.3.4
Bioavailability

A low BCF of hydrophobic compounds might also be related to a reduced bioavailability. In that case, however, the lower BCF is related to an experimental problem [49, 50], and the apparently low bioaccumulation factor is a result of an overestimated concentration in the ambient environment. Usually the aqueous concentration is determined after liquid-liquid extraction of a water sample. The overestimation of the concentration in water results from analytical difficulties which fails to differentiate between available compounds and non-available compounds that are, for example, associated to particles.

In water several types of materials may affect the bioavailability, such as dissolved organic carbon (DOC), particulate organic carbon (POC), etc. In sediment and soil, other factors affect the bioavailability of organic compounds, such as the hydrophobicity of the contaminant, the contact time between contaminant and soil/sediment, the nature and amount of organic carbon and other soil/sediment characteristics, the behaviour of soil/sediment organisms, etc. [85]. Bioavailability may thus lead to an apparently lower BCF.

2.3.5
Biotransformation

Biotransformation increases the elimination rate of the parent compound, which does not necessarily mean that the biotransformation product, the metabolite(s), will be eliminated from the organism, but chemicals which are biotransformed relatively fast will have a low BCF [58, 59, 80, 86, 87].

The former four arguments influence the bioaccumulation of the more hydrophobic chemicals with log K_{ow} above 5 to 6, while an influence of biotransformation is possible for all kinds of chemicals. It must be emphasised that the development of the arguments is implicitly based on the assumption that K_{ow} should be a good descriptor.

2.4
Concluding Remarks on Bioaccumulation

To describe the uptake of an organic compound by an organism which enables one either to determine or to predict the *internal* concentration, bioaccumulation models have been developed. Aquatic, benthic and terrestrial organisms will take up contaminants from their ambient environment and their food. Most bioaccumulation models include one physical-chemical parameter, the K_{ow}, as a single descriptor to predict uptake, elimination and bioaccumulation of these organic contaminants in the organisms. However, it is clear that, in addition to the physical-chemical properties of the contaminant, properties of the surrounding environment and the behavior of the organism are also very important.

Many studies have focussed on the uptake and bioaccumulation from water, and have resulted in models. Most of these existing models for the steady-state BCF are valid only for non-ionised organic chemicals and less for ionised chemicals or organometallic compounds. For practical purposes, a kind of worst-case BCF can be estimated for non-ionised organic chemicals based on the published BCF-K_{ow} correlations.

To describe bioaccumulation, physiological properties of the organism need to be included in addition to a chemical property, such as K_{ow}. Furthermore, many chemicals are known to bioconcentrate to a lesser extent. There is some evidence that this reduced bioaccumulation is due to a size or shape cut-off effect in membrane permeation but an exact value is difficult to set. Other reasons for lower bioconcentration factors are related to biotransformation. It is not possible yet to apply discrete equations for these kind of "deviating" compounds. Other descriptors will have to be developed and applied that describe the underlying processes for the deviating behaviour. Parameters which relate the size of the molecule, and also parameters that represent differences in potency for biotransformation, will be important.

The studies which have focussed on the uptake and bioaccumulation from food, sediment or soil show that many factors significantly influence bioaccumulation, such as food composition, feeding rate, developmental stage or age, the hydrophobicity of the contaminant, the contact time between contaminant and soil/sediment, the nature and amount of organic carbon and other soil/sediment characteristics, the behavior of soil/sediment organisms, etc.

3
Ecotoxicological Effects

Ecotoxicological effects of organic chemicals can be related to *external* and *internal* effect concentrations. Earlier studies already showed that a lot of information is available on *external* effect concentrations for different classes of compounds and different organisms. The main focus of this section therefore is on *internal* effect concentrations.

3.1
External Effect Concentrations

Many compounds exert adverse effects in organisms, dependent on various factors, such as their concentration, their mechanism of action and the type of organism. A relationship between an ecotoxicological effect, which is a biological activity, of a compound and its chemical structure or its physical-chemical properties is not arbitrary. While the biological activity may either be an acute or a chronic effect, most of the present studies relate to acute effects, i.e. lethality. Hansch, who is the pioneer of quantitative structure-activity relationships (QSARs), has given a rationalisation of such a relation in several of his publications [88–90]. The reason for summarising the theoretical background of such relationships is that an understanding of the assumptions behind estimation models for toxicity will enable one to evaluate QSAR studies in a more detailed manner.

The biological activity of a toxicant is dependent on:

- the probability that a chemical reaches its site of action (Pr_1), which is called the toxicokinetic phase;
- the probability that a chemical interacts with a receptor or target molecule (Pr_2), which is called the toxicodynamic phase; and
- the *external* concentration (C) or dose to which the organism is exposed.

For a particular effect, the number of molecular events or the concentration of the target molecules (C_t) that has interacted with a toxicant is constant. So, C_t can be written as

$$C_t = a \cdot Pr_1 \cdot Pr_2 \cdot C = \text{constant} \tag{13}$$

Logarithmic transformation of the latter equation yields

$$\log 1/C = b + \log Pr_1 + \log Pr_2 \tag{14}$$

where a and b are constants.

The rate or equilibrium constants of each of these aforementioned processes will depend on structural characteristics or physical-chemical properties. Because of the variety of these processes, relationships between effect parameters and physical-chemical properties are usually more complex than those for chemical processes.

Effect parameters in hazard or risk assessment of chemicals for the aquatic environment are usually based on *external* effect concentrations for a few types of organisms. In general, simple overall criteria such as survival or inhibition of growth and reproduction are measured. Common types of test species are algae, crustaceans (for instance *Daphnia*) and fish. Effect concentrations are usually expressed as 50% effect concentrations (LC50 or EC50) or as no-observed effect concentrations (NOECs).

The class of relatively unreactive chemicals which act, at least in acute toxicity tests, as narcotics [3] is the best known class of compounds for which several QSARs are established. Those chemicals exert the so-called base-line toxicity. Studies from Könemann [3] and Veith et al. [91] have shown that *external* effect concentrations such as LC50s or NOECs for these chemicals depend on the octanol-water partition coefficient according to the following equation:

$$\log C = A \log K_{ow} + B \tag{15}$$

where A and B are constants.

Two examples, one for LC50 to guppy [3] and one for NOECs to *Daphnia magna* [92] are given in the next two equations:

$$\text{guppy: } \log LC50\ (\text{mol l}^{-1}) = -0.87 \log K_{ow} - 1.1 \quad (n = 50, r^2 = 0.97) \tag{16}$$

$$\textit{Daphnia magna}\text{: } \log NOEC\ (\text{mol l}^{-1}) = -0.95 \log K_{ow} - 2.0 \quad (n = 10, r^2 = 0.95) \tag{17}$$

The lower value for the intercept (the "B" constant) in the latter equation is due to the more sensitive endpoint (growth reduction vs survival), whereas in both cases the "A" constant is close to unity.

For a number of ecotoxicological endpoints, such as survival and growth reduction, relationships between LC50 or EC50 and one or more physical-chemical properties are available for many aquatic, and in a lesser extent for benthic and terrestrial species for different mechanisms of action [93–104].

While it is in general more clearly defined what the exposure concentration is in the aqueous environment, it is more difficult to measure the actual exposure concentration in soil and sediment. In the latter case it is also more difficult to show a clear relationship between effects and exposure. For example, the influence of sorption on bioavailability and thus on toxicity is very important for soil toxicity testing [e.g. 105].

3.2
Internal Effect Concentrations

As stated earlier, it is in most cases the *internal* critical concentration which can be more closely related to an (ecotoxicological) effect. Exceptions may be strong acids or other toxicants which act on the outer surface of an organism. The concept of the *internal* critical concentration is illustrated in Fig. 1 which shows that an organism which takes up a chemical from its environment may accumulate this chemical until a certain critical body burden has been attained, which then kills the organism. Recently, McCarty et al. [10–16], van Hoogen and Opperhuizen [9] and others [61, 106] have indeed shown that *internal* con-

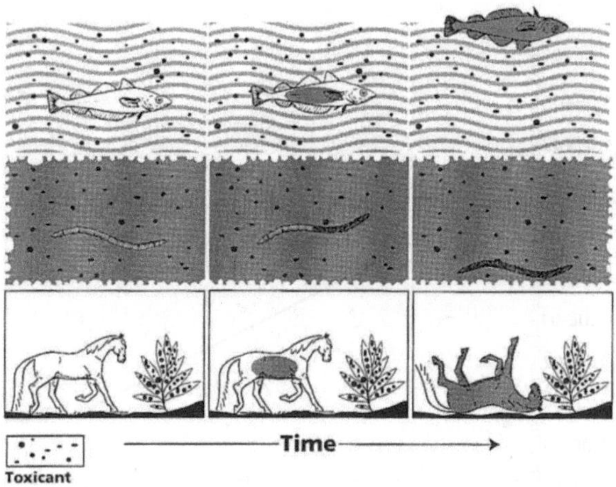

Fig. 1. The concept of attaining an *internal* effect concentration in time as the result of bioaccumulation. An organism is exposed to a contaminant from the ambient environment, which can be water *(top)* or soil *(middle)*, or from food *(bottom)*. The more it has taken up the higher its *internal* concentration will be until a critical *internal* concentration is reached, e.g. the lethal body burden, and the associated effect, e.g. death, is elicited

centrations of halogenated organic chemicals in fish causing death are fairly constant: about 2–8 mmol kg^{-1}. McCarty et al. [11, 13, 14] have mathematically explained this as follows. The fairly constant *internal* effect concentration or lethal body burden (LBB) is the result of the bioconcentration factor (BCF), which increases with K_{ow}, and the *external* effect concentration (LC50), which decreases with K_{ow} (Fig. 2):

$$LBB = LC50 \cdot BCF \quad (18)$$

or

$$\begin{aligned} \log LBB &\approx \log(LC50) + \log(BCF) \\ &\approx (-\log K_{ow} + b_1) + (\log K_{ow} + b_2) \\ &\approx b_1 + b_2 \\ &\approx \text{constant} \end{aligned} \quad (19)$$

where b_1 and b_2 are constants.

In the following sections lethal body burden associated with some mechanisms of actions will be discussed first, which will then be followed by a critical discussion of the assumptions behind the *internal* effect concentration.

3.2.1 *Mechanisms of Actions*

While lethal body burdens of narcotic chemicals are in the range 2–8 mmol kg^{-1}, LBBs of chemicals with other mechanisms of actions in fish are usually lower. McKim and Schmieder [107] and McCarty and Mackay [16] have collected toxi-

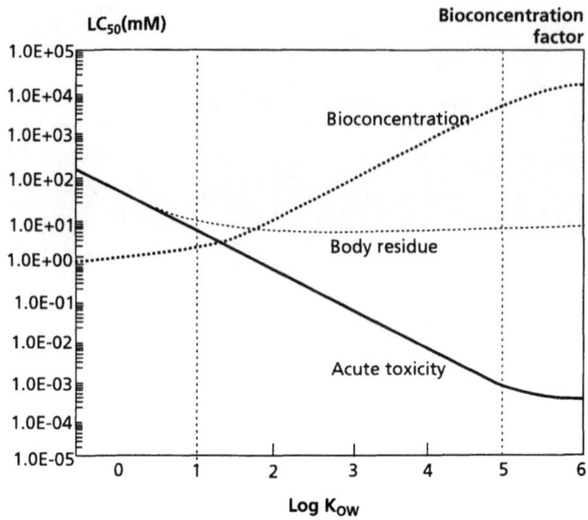

Fig. 2. A simplified example of the general relationships between on the one hand the octanol/water partition coefficient (K_{ow}) and on the other hand *internal* effect concentrations (body residues), bioconcentration and acute toxicity for narcotic organic chemicals and small freshwater fish [16], reproduced with permission

city data and bioconcentration factors for six classes of chemicals, each with a specific mechanism of toxic action for fish, which had been described earlier [97, 101]. The calculated lethal body burdens responsible for these different mechanisms of actions according to Eq. (18) are provided in Table 2 and Fig. 3.

3.2.2
Variation in Internal Effect Concentrations

Table 2 and Fig. 3 show that each mechanism of action has one, but in some cases a rather broad range of, *internal* effect concentrations for aquatic organisms. Therefore there is not one distinct value of the lethal body burden associated with one mechanism of action, but rather a range of *internal* concentrations that is related to an ecotoxicological effect. Some other questions which can be asked to validate the use of the *internal* effect are: how large is the interspecies variation in *internal* effect concentration (for two types of mechanisms of action), how large is the intraspecies variation in *internal* effect concentration (for one type of mechanism of action), and what is the time and concentration dependent influence on the *internal* effect concentration (for one type of mechanism of action).

3.2.2.1
Interspecies Variation for One Mechanism of Action

The first condition in working with the *internal* concentration concept is that once any organism has reached the lethal body burden it will die. Also, for each

Table 2. Lethal body burdens (LBB) in fish associated to different mechanisms of action, according to McKim and Schmieder [107], and extended with data for polychlorinated dibenzo-*p*-dioxins (PCDDs) [86, 108], and organotin compounds [109, 110]

Mechanism of action	Examples	LBB (mmol kg^{-1})
non-polar narcosis	MS-222, octanol	2.8–10
polar narcosis	phenols, anilines	0.17–4.6
uncouplers	pentachlorophenol, 2,4-dinitrophenol	0.06–0.33
AChE inhibitors	malathion, carbaryl	0.009–0.76
Respiratory blocker	rotenone	0.0028
Respiratory irritant	acrolein, benzaldehyde	0.0014–2.1
Ah-mediated toxicity	TCDD	0.00004–2.0
Organotin induced toxicity	organotin	0.0014–0.026

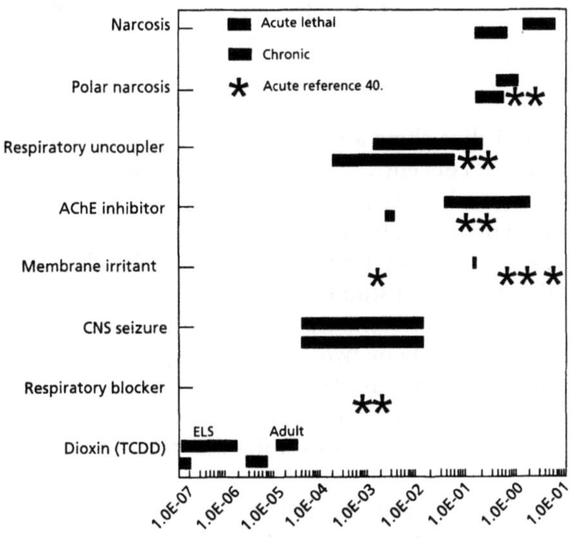

Fig. 3. Calculated body burdens (in mmol l^{-1}) associated with different acute and chronic toxicity endpoints for fish exposed to eight categories of organic chemicals. From McCarty and Mackay [16], reproduced with permission

sublethal effect a distinct *internal* effect concentration is assumed. Hitherto, for both narcotic, e.g. polychlorinated benzenes and biphenyls (Table 3), and polar narcotic compounds, e.g. chlorinated phenols and anilines (Table 4), sufficient information is available to study this assumption.

Table 3 shows that, for different aquatic, benthic and terrestrial organisms, the lethal body burdens vary approximately by two orders of magnitude, but most of the values are in the range as predicted by McCarty [15], i.e.

Table 3. Interspecies variation in experimentally determined LBBs for narcotic chemicals (polychlorinated benzenes and polychlorinated biphenyls)

Species	LBB (mmol kg^{-1})	Reference
Amphipod (*Hyalella azteca*)	0.1–0.6	[111]
Brook trout	0.4	[112]
Crab	1.4–4.8	[113]
Chironomus riparius	0.14–1.7	[114]
Chinook salmon fry	0.012–0.013	[115]
Coho salmon	2.2	[116]
Daphnia magna	3.1	[117]
Earthworm	0.19–2.5	[118]
Fathead minnow	2–8	[61]
Fathead minnow	2.7–3.0	[119]
Guppy	0.7	[62]
Guppy	2.1–2.7	[9]
Guppy	2–8	[61]
Lake trout fry	0.0072–0.03	[115]
Mosquitofish	2.3–8.3	[120]
Rainbow trout	0.29–2.4	[119]

2–8 mmol kg^{-1}, and thus show a significant reduction in the variation of the ecotoxicological effect concentrations compared to the more than five orders of magnitude differences that are found in *external* effect concentrations for this type of mechanism of action. However, one distinct lethal body burden cannot be used and Table 3 shows that there is variation in the LBB data for the different organisms that have been studied.

Table 4 shows that, for different organisms, the lethal body burdens for polar narcotics vary approximately by two orders of magnitude, and thus again show a significant reduction in the variation of the ecotoxicological effect concentrations compared to the more than five orders of magnitude differences that are found in *external* effect concentrations for this type of mechanism of action [e.g. 121].

One distinct lethal body burden cannot thus be used for either the polar or the nonpolar narcotic compounds, since there is again a significant variation in the data for the different organisms that have been studied.

3.2.2.2
Intraspecies Variation for One Mechanism of Action

A second condition in working with the *internal* concentration concept is that, once any individual within a population has reached the lethal body burden, it will die. This condition was recently studied by a few groups which showed that, although there is a small range of LBBs within one population of fish, there is still not a single value that will cause death (Table 5). It has thus been shown that intraspecies variation does occur. One of the explanations for the intraspecies variation is differences in lipid content: the survival-of-the-fattest concept

Table 4. Interspecies variation in experimentally determined lethal body burdens for polar narcotic chemicals (chlorinated phenols and anilines)

Species	LBB (mmol kg^{-1})	Reference
Brown trout	0.03 – 0.91	[122]
Earthworm (*Eisenia fetida*)	0.08 – 1.1	[121]
Fathead minnow	1.1 – 1.7	[123]
Goldfish	0.32 – 1.64	[124]
Goldfish	0.19 – 1.84	[125]
Guppy	0.7 – 1.8	[126]
Rainbow trout	0.23 – 0.93	[107]

[127]. It can be argued that, on a wet weight basis, fatter individuals may accumulate higher body burdens of toxicants before being affected. Lipid normalisation should, in this case, diminish intraspecies variation. However, lipid content only explains approximately 50% of the variation (Table 5).

An additional explanation for the observation that lipid only explains approximately 50% of the variation in *internal* effect concentration may be that the different lipids of an organism do not evenly contribute to storage in target tissues [128], and that lipid normalisation may thus not be appropriate. The assumption, however, that the *internal* concentration is a distinct value is not valid. Intraspecies differences do exist and cannot be explained by intraspecies differences in lipid content alone, although the variation in LBB within a population is less than an order of magnitude.

3.2.2.3
Time and Concentration Dependency

A third condition in working with the *internal* concentration concept is the following. It may take a long time when exposed to a relatively low concentration or a small time when exposed to a relatively high concentration to reach the lethal body burden, but once the organism has reached this lethal body burden it will die (Fig. 4).

Table 5. Intraspecies variation in wet weight lethal body burden (LBB) and the contribution of lipid content (lipid) to explain intraspecies variation in fish

Compound	Fish	LBB (mmol kg^{-1})	Influence of lipid on variability (%)	Reference
1,2,4-CBz	fathead minnow	2.2	59	[129]
1,1,2,2-TCE	fathead minnow	2.5	43	[129]
1,2-CBz + 1,4-CBz	fathead minnow	3.5	53	[129]
1,2-CBz + 1,4-CBz	fathead minnow	4.4	60	[129]
naphthalene	fathead minnow	8 ± 3.1	82	[130]
1,2,4-CBz	fathead minnow	14 ± 4.5	41	[130]

CBz = chlorobenzene; TCE = tetrachloroethane.

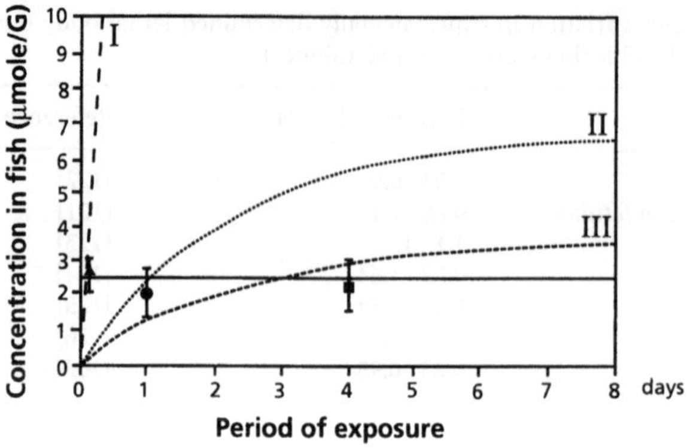

Fig. 4. Time and exposure concentration dependent concentrations in fish in addition to the lethal body burden *(horizontal solid line)* for 1,2,3-trichlorobenzene. The *dotted lines* are theoretical curves calculated with a bioaccumulation model. Exposure concentrations are: 55.9 μmol l^{-1} (I), 3.78 μmol l^{-1} (II), and 1.92 μmol l^{-1} (III). The symbols represent the mean of the *internal* effect concentrations of ten fish [9], reproduced with permission

Van Hoogen and Opperhuizen [9] indeed showed that, irrespective of the time required to kill the fish, which ranged from 0.1 to 8 days, the LBB for three chlorobenzenes were very similar, i.e. the range of the LBBs was 2.02–2.71 mmol kg^{-1} (Fig. 4).

However, Kleiner et al. [131] found lower LBBs for fish that died after a few hours than for the fish that died after 12 h of aqueous exposure to pentachloroethane. Also van Wezel et al. [129] found that fish died shortly after exposure (< 50 h) to an aqueous solution of commercial PCB mixtures had a lower LBB than the fish that died after longer times (> 50 h). Furthermore, de Maagd [130] showed that increasing exposure time increased the LBBs of naphthalene and of 1,2,4-trichlorobenzene in fathead minnow. Somewhat contradictory to this was that de Bruijn et al. [132] found that fish which were killed shortly after exposure had a higher LBB than the fish which died after longer exposure to waterborne organophosphorus pesticides. This same phenomenon was found by de Wolf et al. [126] who clearly showed that fish exposed to 2,3,4,5-tetrachloroaniline and died shortly after exposure to a relatively high aqueous concentration had a significantly higher LBB than fish which were exposed to the same compound but to a lower aqueous concentration. The "high" LBB was 1.8 ± 1.0 mmol kg^{-1} and the "low" LBB was 0.7 ± 0.5 mmol kg^{-1}. In addition, Mortimer and Connell [113] showed a decrease in LBB in time for a series of chlorinated benzenes in the crab *Portunus pelagicus* (L) with increasing exposure time. Also Chaisuksant et al. [120] showed a decrease in LBB for two chlorinated benzenes and two brominated benzenes in mosquitofish in time. Furthermore, Ohayo-Mitoko and Deneer [133] showed a clear correlation between concentration (and thus time) and LBB for two organophosphorus pesticides, for which higher LBBs were found at the higher exposure concentration and the shorter time-to-death, but for two other organophosphorus pesticides, similar LBBs were found at "low" and "high" exposure concentrations.

No clear indication is thus obtained as to whether or not exposure time affects the LBB and more studies are required to elucidate this. Arguments for a time dependent LBB may be that intraspecies differences exist which results in longer survival of the more tolerant fish, that physiological adaptation could make fish more tolerant, and that slow *internal* distribution could favour high concentrations at target sites relative to non-target sites or vice versa, that *internal* distribution could favour high concentrations at 'narcotic' target sites relative to target sites for more specific toxicity.

3.3 Concluding Remarks on Ecotoxicological Effects

Ecotoxicological effects, such as acute or sublethal responses, can be related to both *external* and *internal* concentrations. The former is still used in risk assessment procedures, while the latter has recently been investigated for its potential use in risk assessment. *External* concentrations may vary by many orders of magnitude for different chemicals, even when they exert the same mechanism of action. The variability in *internal* concentrations is much smaller. The assumptions which form the basis for a broad applicability of the *internal* concentration, namely that for a given mechanism of action, i) there would be no intraspecies variation, ii) there would be no interspecies variation, and iii) there would be no time or concentration dependency, have been studied. It was found that no assumption was completely valid. However, given the magnitude of variability found, these variations are much less than those which are found for *external* concentrations, while some of the reasons for the variations in the *internal* effect concentrations may be similar for the variation in *external* effect concentrations.

4 Bioaccumulation and Ecotoxicity

Overviews of QSAR studies for aquatic toxicity of chemicals which show narcosis are extensively discussed in several publications [93, 94]. At first sight, it is quite remarkable that QSAR equations for all kinds of different species are so similar. On the other hand, the explanation is rather simple. It is generally accepted that the mechanism of narcosis is not a very specific process and each compound has the same intrinsic activity. In other words: the *external* concentration of a compound at a fixed effect (e.g. narcosis or death) is only a function of the probability of a compound to reach its site of action. For many chemicals for which bioaccumulation is not influenced by biotransformation reactions, this probability is correlated to the octanol-water partition coefficient (K_{ow}) and this explains directly the correlation between K_{ow} and the *external* effect concentrations.

4.1 Predicting Ecotoxicological Effects

It has been shown that most ecotoxicological effects can be related to *internal* effect concentrations in the organism, critical for that effect, such as death. It has also been shown that for many chemicals, attaining high concentrations in organisms is the result of bioaccumulation from the ambient environment or food, which can be described by bioaccumulation and biomagnification kinetics. Depending on i) the mechanism of action of the chemical, ii) the organism, and iii) the physical-chemical properties of the chemical, the time to elicit an ecotoxicological effect may thus be predicted. This assumes that each mechanism of action can be associated with a distinct *internal* effect concentration or lethal body burden for acute effects.

4.2
Bioaccumulation and Lethal Body Burden

When the concentration in an aquatic organism which causes an ecotoxicological effect is replaced by the lethal body burden, when Eqs. (1), (2) and (18) are combined and resolved, and when a constant exposure concentration is assumed, then ecotoxicological effects can be related to aqueous exposure of chemicals:

$$LBB = C_b(t = t_{LBB}) = \frac{k_{u,w}}{k_e} C_w(1 - e^{-k_e t_{LBB}}) = BCF \cdot C_w \cdot (1 - e^{-k_e t_{LBB}}) \quad (20)$$

where t_{LBB} is the time (days) when the organism dies, and $C_b(t = t_{LBB})$ the lethal concentration in the organism (mol kg^{-1}) at the time of death.

It must be noted that in order to predict when the concentration in the organism is high enough to reach the LBB, the bioaccumulation factor, the ambient concentration and the elimination rate constants should be available. Also, for other exposure routes than water, information on those factors is required.

Equation (20) can thus be used to estimate if or when an organism will die at a given exposure concentration. If the exposure concentration is too low, the LBB will not be attained in the organism. If the exposure concentration is high enough the LBB will be attained at time $t = t_{LBB}$.

Analogously, when Eqs. (4), (6) and (18) are combined, ecotoxicological effects can be related to uptake from food:

$$LBB = C_b(t = t_{LBB}) = \frac{k_{u,fd}}{k_e} \cdot C_{fd}(1 - e^{-k_e t_{LBB}}) = \quad (21)$$

$$= \frac{E_{fd} \cdot V_{fd}}{k_e} \cdot C_{fd}(1 - e^{-k_e t_{LBB}}) = BMF \cdot C_{fd} \cdot (1 - e^{-k_e t_{LBB}})$$

Similarly, ecotoxicological effects can be related to uptake from sediment or soil:

$$LBB = C_b(t = t_{LBB}) = \frac{k_{u,sed}}{k_e} \cdot C_{sed} \cdot (1 - e^{-k_e t_{LBB}}) = \quad (22)$$

$$= BSAF_{sed} \cdot C_{sed} (1 - e^{-k_e t_{LBB}})$$

$$LBB = C_b(t = t_{LBB}) = \frac{k_{u,soil}}{k_e} \cdot C_{soil}\,(1 - e^{-k_e t_{LBB}}) = \tag{23}$$

$$= BSAF_{soil} \cdot C_{soil}\,(1 - e^{-k_e t_{LBB}})$$

Equations (20)–(23) include bioaccumulation kinetics, and thus enable us to predict when organisms will attain lethal body burdens. The most important bioaccumulation parameters, and the relationships between the bioaccumulation parameters and physical-chemical and physiological factors, which are required can either be found in the literature or need to be studied. The equations can thus be used to predict if organisms are at risk and will experience adverse effect at a given *external* exposure concentration. Time will thus be a variable, whereas the *external* exposure concentration in either water or food will be the given input parameters in this exercise. The equations can also be used to estimate the *external* concentration which will lead to adverse effects at a given exposure time. Then, *external* exposure concentration will be a variable, whereas the time required for eliciting effects will be a constant.

In all the equations, the elimination rate constant, k_e, is an important parameter. It is the elimination rate constant which determines whether or not the concentration in an organism is in steady-state with that in the environment or the food. For chemicals which are not extremely hydrophobic and for small organisms, elimination rate constants are, in general, relatively high, and therefore steady-state will be reached in a few hours or days. In that case, provided the ambient exposure concentration is high enough, adverse effects will be shown in a short period of time. For the more hydrophobic chemicals and for bigger organisms, however, steady-state may be reached only after several days or weeks, if at all. For those situations it may thus also take some time to attain a body burden high enough to elicit effects.

For exposure to water, food, sediment or soil, some general relationships exist which enable us to predict the concentration in many organisms. However, in particular for the latter three types of exposure, little data are available. In addition, the present knowledge for derivation and application of the relationships is based on only a few classes of organic compounds, such as polycyclic aromatic hydrocarbons and chlorinated benzenes and biphenyls.

Prediction of ecotoxicological effects for other types of chemical classes as well as for foodchain transfer is less founded, and should be studied further to extend our knowledge and the applicability in using the *internal* effect concentrations.

4.3
Biomimetic Monitoring of *Internal* Concentrations

It is well known that the effects of narcosis type chemicals are completely concentration additive [92, 134–136]. Intrinsically, these chemicals are equally toxic. In other words: body burdens at a certain effect are the same for all compounds within this "toxicological class". The differences in aqueous effect concentrations of chemicals with base-line toxicity are only due to differences in

Table 6. No-Effect Body Residues for narcotic chemicals at three different effect endpoints

Endpoint	No-effect Body Residues (mmol kg_{lipid}^{-1})[a]
1. mortality (fish)	25
2. sublethal effects (fish)	5.0
3. ecosystem level effects (HC_5)	0.25

[a] Data from [138]. Please note that the no-effect body residue for mortality is about a factor of 2 lower than the lethal body residue (ca. 50 mmol/kg).

bioaccumulation factors, e.g. [16]. Lethal body burdens or critical body residues for base-line toxicity at a few well known endpoints or effects are given in Table 6.

One would like to know the total *internal* molar concentration of these compounds in organisms in the environment instead of the *external* aqueous concentrations of individual compounds. A parameter which measures this total, accumulated, body residues (TBRs) of organic chemicals will be a useful tool in risk assessment of effluents and surface waters.

The parameter TBR gives information on the total bioaccumulation of mixtures of chemicals in the aqueous phase. Information on TBR is useful in order to get an impression of the total load of organic chemicals in aquatic organisms in a toxicologically relevant manner. The fact that it includes chemicals with a high potential for bioaccumulation is an important advantage. Because total concentrations instead of individual compounds are determined, the outcome also includes those chemicals which are usually not measured because they cannot be identified or because their concentrations are below the individual detection limits.

Besides being a parameter for the bioaccumulation of mixtures, it is also a measure for the total residues of chemicals with base-line toxicity, including the contributions of chemicals with specific modes of action to this overall base-line toxicity. If the total residues exceed a certain effect level, there is reason for concern. If the residues are below the critical effect levels, however, effects cannot be ruled out because chemicals with more specific modes of action may be present.

The advantage of working with body residues is that, e.g. for chemicals (and chemical mixtures) with only base-line toxicity, the No-Effect Body Residue is relatively constant for a certain endpoint. Because of that, the evaluation of the effects of mixtures can be performed by using the equation: TBR/No-Effect Body Residue < 1.0. The current evaluation of mixture effects based on *external* aqueous concentrations is based on the equation Σ {PEC/PNEC} < 1.0 and this equation can only be used if the concentrations of all individual chemicals are known.

This new parameter, TBR, also has its limitations. The total body residues are usually measured in or on a biomimetic hydrophobic phase as a surrogate for biota. Other properties than hydrophobicity alone can influence the bioaccumulation in biota. For example, the molecular size of a chemical may decrease the uptake [79, 137], and also chemicals that are biotransformed relatively fast will have lower bioconcentration factors than predicted by their hydrophobicity alone [87, 139]. Uptake of very hydrophobic compounds may also take place via routes other than simply via diffusion; in those cases uptake via food or sediment particles may become the predominant routes for uptake [20, 21, 65, 140]. Bearing in mind these limitations, the results from this procedure can be interpreted properly.

In the early 1980s the use of the semi-permeable membrane devices (SPMD) method was proposed to simulate bioaccumulation by Södergren [141] and Huckins and co-workers [142, 143]. The principle of the SPMD is that a semi-permeable membrane containing pores similar to those assumed in fish membranes is filled with a lipid surrogate, such as triolein, and exposed in water for some days or weeks. Organic compounds in the water will diffuse through the membrane and accumulate in the lipid. SPMD is thus a surrogate for measuring bioaccumulation in aquatic organisms. The advantages of SPMD are that it is a simple method and that it showed fairly good agreement with uptake in aquatic organisms. The disadvantages are that fouling of the membrane and loss of the solutes or lipids can occur. The former problem is difficult to overcome, the latter more easy by adding a standard compound with a known concentration in the lipid. The final concentration of this standard after exposure will indicate this loss.

Recently, another simple method for measuring TBRs has been developed [138, 144]. The experimental procedure to measure the surrogate parameter TBR is based on two important features:

a. a biomimetic extraction procedure, and
b. the determination of total molar concentrations.

With regard to the biomimetic extraction procedure, measurement of TBRs can in principle be carried out in biota, but this will need a very extensive clean-up in order to purify the samples from compounds such as proteins and lipids. Instead of working with biota, the use of a solid phase extraction on a hydrophobic phase is chosen in order to mimic the uptake by organisms. This biomimetic extraction has been described by Verhaar et al. [138]. A “biomimetic extraction” is an extraction technique in which a chemical is extracted from the aqueous phase in a hydrophobicity-dependent manner. In other words, the method does not select chemicals but accumulates the more hydrophobic compounds more efficient than less hydrophobic compounds, similar to the bioconcentration process in biota. This condition can only be met by keeping the aqueous concentration practically constant during the extraction or concentration process (see Fig. 5), which may take days or weeks. The aqueous concentration will remain constant only if the amount of hydrophobic material, into which a compound is partitioning, is extremely small compared to the volume of the aqueous phase. A solid phase extraction disk (Empore disk), which is chemically bound C_{18} embedded in a Teflon matrix, was selected. The rationale for this choice was the fact that bioconcentration in biota is related to the hydrophobicity of organic chemicals and that partitioning onto C_{18} is a good

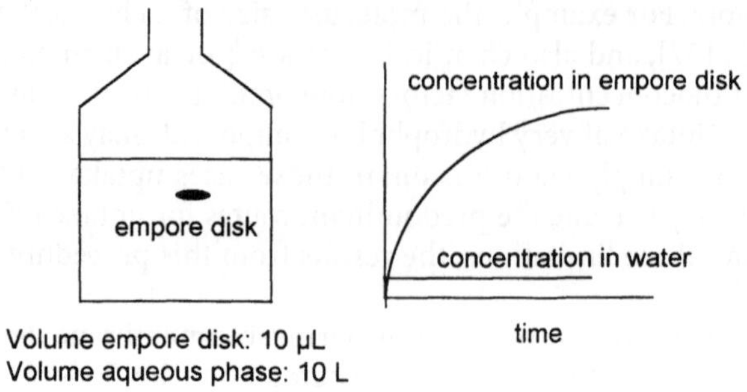

Fig. 5. The principle of a biomimetic extraction

measure for hydrophobicity. Total Body Residues in biota can be estimated from measured total concentrations on the Empore disk.

With regard to the determination of total molar concentrations, procedures for measuring total molar concentrations should, in principle, fulfill the following two conditions: (i) the response of individual chemicals must be equal, and (ii) the response of individual chemicals in a mixture must be additive. Two analytical techniques, namely vapour pressure osmometry and GC-MS (total ion current), were tested. Both techniques are, with some limitations, able to quantify total molar concentrations of organic compounds [138, 144]. Results of the application of this procedure to effluents and surface water were recently published [144]. TBR gives information on the total bioaccumulated compounds from water samples. Moreover, using these total body burdens, base-line toxicity effects can be predicted, including the contributions of chemicals with specific modes of action to the overall base-line toxicity. The advantage of the parameter is that it determines total molar concentrations of organic chemicals, including those chemicals which are usually not measured because they cannot be identified or because their concentrations are below the detection limits of individual compounds.

4.4 Gaps of Knowledge

Uptake of contaminants is very likely to precede effects, since first the contaminant has to reach the receptor, which can be very specific or non-specific, to exert its adverse biological effect. Since uptake is an important part of the bioaccumulation process, the relationship between bioaccumulation and ecotoxicity is shown. However, there are a number of gaps in knowledge which prevent a broad use of bioaccumulation models to be incorporated in predicting ecotoxicological effects.

First of all, a clear classification of contaminants with respect to their ecotoxicological effects is a prerequisite [101, 104]. This should provide a better insight into the most important physical-chemical properties that are related to a specific ecotoxicological effect. Second, the effects should be clearly described.

It must be distinguished whether the effect of concern is acute or subacute. Then, for each class of chemicals and/or for each ecotoxicological effect, information is required on both bioaccumulation and ecotoxicological effects. Furthermore, it has also to be taken into account that, in addition to physical-chemical properties, physiological properties of the organism of interest and environmental conditions will also determine bioaccumulation and possibly effects.

The use of the *internal* effect concentration may be of great help in classifying chemicals and their effects. However, for that purpose, more data on *internal* effect concentrations associated with different mechanisms of action in different organisms are a prerequisite.

A few examples will be given to indicate that for some processes or effects insufficient knowledge is available to use both information from bioaccumulation and ecotoxicology.

Sex related differences in bioaccumulation will occur for species higher in the foodchain or for very hydrophobic chemicals, where biomagnification is the predominant process. Biomagnification factors between trophic levels are in the order of 3–20 on a lipid weight basis, and hydrophobicity and persistence play a significant role in the uptake from the gastro-intestinal tract for the polychlorinated aromatics, such as PCBs, PCDDs and PCDFs [69]. If organochlorines have low or zero rates of metabolism, excretion may be so slow that the contaminant builds up with age in the animal. This has been shown to be true, especially for male marine mammals. Female marine mammals have as an additional route of excretion the elimination of the more hydrophobic chemicals via lactation or by giving birth, and are more likely to come in steady-state with their diet, and dispose of some high hydrophobic chemicals [70]. These sex related differences are difficult to model, since they are highly dependent on the size of litter, the lactation period, etc.

Another example is to relate subacute effects to *internal* concentrations. Two examples will be given, one for a well-known class of chemicals, and one for a well-known subacute effect.

The well-known class of chemicals is the dioxin-type chemicals, such as PCBs, PCDDs and PCDFs. These chemicals have caught the attention because many of those congeners bioaccumulate to potentially toxic concentrations, especially high in the foodchain [145–147]. Sijm and Opperhuizen [108] critically reviewed both environmental concentrations of PCDDs and PCDFs in fish, and body burdens which elicited acute and subacute effects. They concluded that, in some environmental regions, concentrations in fish are close to those eliciting ecotoxicological effects, indicating the high potential risk of these classes of organochlorine chemicals, the same conclusion which was earlier suggested by Cook et al. [145]. However, large species differences were observed for both the acute and subacute effects. This shows that, in addition to a sometimes broad range of *internal* effect concentrations for different chemicals with one mechanism of action as is shown for one organism (Table 2, Fig. 3), *internal* effect concentrations related to one chemical in different species also show a broad range. A very distinct *internal* effect concentration is thus difficult to determine for different species.

A well-known subacute effect is the growth reduction in algae. Hitherto, only *external* effect concentrations have been reported for this type of subacute effect, since experimental problems make it difficult to determine those *internal* effect concentrations, and existing bioaccumulation models for, e.g., fish, do not apply to algae, e.g. [78]. It must be noted that algae and other small organisms are prone to diffusive uptake for contaminants from the ambient environment for which the link between bioconcentration and the *internal* effect concentration concept would be very promising.

In addition to gaps in knowledge for currently existing classes of chemicals and ecotoxicological effects, other mechanisms of actions that are currently not yet studied, or other processes, may require further studies. For example, recently it has become clear that phototoxic effects may be a realistic problem for polycylic aromatic hydrocarbons (PAHs) in aquatic and benthic organisms. The amount of UV-light which is required for phototoxicity, is thus an example of a parameter which was not introduced earlier as an important environmental parameter to describe or predict toxicity [148]. Other examples are if the *internal* effect concept can be used for metals and organometals in risk assessment [149, 150].

Furthermore, most existing risk assessment and ecotoxicological effects are related to (physico-chemical properties of) the parent compound. Chemicals, however, may be biotransformed by organisms. This may be very species-specific, and, in addition, may result in the formation of lesser or more toxic metabolites. Neither the *internal* nor the *external* concentration is then a good representative measure for toxicity.

4.5 Concluding Remarks on Bioaccumulation and Ecotoxicity

Many structure-activity relationships can be used to deal with mixture toxicity. Bioaccumulation models in combination with *internal* effect concentration may provide a good means to better predict when organisms are at risk. It must be noted, however, that in many cases there is significant variation in these *internal* effect concentrations, although even larger variation is found for *external* effect concentrations. The variation in the *external* effect concentrations is partly related to the variation in bioaccumulation and partly to interspecies and intraspecies variation.

When more knowledge is available on *internal* effect concentrations, biomimetic monitoring may be a useful tool to estimate the environmental risk of organisms in the field, and at present can already be used for narcotic effects. Already mixed-function oxygenase system components and antioxidant enzymes are related to contaminant body burdens in marine bivalves in the field [151], which indicates the potential of the use of *internal* concentrations as parameters for ecotoxicological effects.

Most of the *internal* effect concentrations that are described in this chapter are related to the in vivo situation. However, this approach may also be of value for in vitro studies. Recently, examples of relatively constant *internal* concentrations have been given for the inhibition of yeast H^+-ATPase, chinese hamster ovary cell

Na^+/K^+-ATPase and for human skin fibroblast Na^+/K^+-ATPase [152]. *External* effect concentrations were combined with tissue/water partition coefficients to estimate the *internal* effect concentrations. For these latter studies, *external* effect concentrations showed a much greater variation than the *internal* effect concentrations, as is found for in vivo *external* and *internal* effect concentrations.

5 Conclusions

Ecotoxicological effects due to organic chemicals are usually the result of uptake and bioaccumulation of the chemical from the ambient environment or the food, followed by a toxicodynamic process which actually results in eliciting the final effect. It is recognized that it is an *internal* concentration which should be related to the ecotoxicological effect. Bioaccumulation is thus a very important process which results in attaining relatively high body burdens of hydrophobic chemicals in organisms at relatively low ambient concentrations. Bioaccumulation kinetics can be used to describe and predict the concentrations of compounds in an aquatic, benthic or terrestrial organism, for which size of the organism, its lipid content, and the hydrophobicity (K_{ow}) of the chemical are the key parameters. In particular for aquatic organisms, and to a lesser extent for benthic and terrestrial, bioaccumulation is fairly described by existing relationships, whereas biomagnification is only poorly described and predicted for many (organochlorine) chemicals. Important bioaccumulation parameters, such as absorption efficiency from food and biotransformation in organisms, are poorly understood. In addition, limited knowledge on bioavailability hinders the predictability of bioaccumulation.

For different ecotoxicological effects and different mechanisms of actions, critical or lethal body burdens (*internal* effect concentrations) have been determined. It is shown that these *internal* effect concentrations show much less variability than the *external* effect concentrations. The assumptions that each mechanism of action is connected to a distinct *internal* effect concentration, that there are no inter- and intraspecies variations in the *internal* effect concentrations, and that there are no time- or concentration-dependent variations in the *internal* effect concentrations, are not completely justified. However, the variation in the *internal* effect concentrations are much less than those for the *external* effect concentrations. The comparison of a predicted concentration and critical body burden of a compound in an organism may enable one to evaluate the ecotoxicological risk for that compound. One of the major advantages of the *internal* effect concentration approach is that it more easily deals with additivity. Chemicals for which no individual *external* effect concentration can be determined, e.g. very hydrophobic chemicals, may contribute to toxicity when present in large mixtures. Since the *internal* concentration is the sum of all concentrations of the individual chemicals expressed as a molar concentration in the organism, the *internal* concentration thus deals with additivity of a mixture. With respect to additivity, biomimetic extraction and molar detection techniques offer a very useful tool to assess the total amount of bioaccumulatable organic compounds.

References

1. Hickey JJ, Anderson DW (1968) Science (Washington, D.C.) 162:271
2. Mitchell RT, Blaggrough HP, van Etten RC (1953) J Wildl Manage 17:45
3. Könemann H (1981) Toxicology 19:209
4. Mackay D (1982) Environ Sci Technol 16:274
5. Meyer H (1899) Naunyn-Schmiederbergs Arch Exp Pathol Pharmakol 42:109
6. Overton E (1901) Studien über die Narkose zugleich ein Beitrag zur allgemeinen Pharmakologie. Jena, Gustaf Fisher
7. Ferguson J (1939) Proc Roy Soc London B 127:387
8. van Leeuwen CJ, Hermens JLM (1995) Risk assessment of chemicals: an introduction. Kluwer Academic Publishers, Dordrecht, The Netherlands
9. van Hoogen G, Opperhuizen A (1988) Environ Toxicol Chem 7:213
10. McCarty LS, Hodson PV, Craig GR, Kaiser KLE (1985) The use of quantitative structure-activity relationships to predict the acute and chronic toxicities of organic chemicals to fish. Environ Toxicol Chem 4:595
11. McCarty LS, Mackay D, Smith AD, Ozburn GW, Dixon DG (1991) Sci Total Environ 109/110:515
12. McCarty LS (1986) Environ Toxicol Chem 5:1071
13. McCarty LS (1987) Relationship between toxicity and bioconcentration for some organic chemicals I: examination of the relationship. In: Kaiser KLE (ed) QSAR in environmental toxicology-II. Reidel, Dordrecht, The Netherlands, p 207
14. McCarty LS (1987) Relationship between toxicity and bioconcentration for some organic chemicals II: application of the relationship. In: Kaiser KLE (ed) QSAR in environmental toxicology-II. Reidel, Dordrecht, The Netherlands, p 221
15. McCarty LS (1990) A kinetics-based analysis of quantitative structure-activity relationships in aquatic toxicity and bioconcentration bioassays with organic chemicals. Ph.D. thesis. University of Waterloo, Waterloo, Ontario, Canada
16. McCarty LS, Mackay D (1993) Environ Sci Technol 27:1719
17. Branson DR, Blau GE, Alexander, HC, Neely WB (1975) Trans Am Fish Soc 104:785
18. Gobas FAPC, Opperhuizen A, Hutzinger O (1986) Environ Toxicol Chem 5:637
19. Gobas FAPC (1993c) Ecolog Modelling 69:1
20. Morrison HA, Gobas FAPC, Lazar R, Haffner GD (1996) Environ Sci Technol 30:3377
21. Sijm DTHM, Seinen W, Opperhuizen A (1992) Environ Sci Technol 26:2162
22. Geyer H, Scheunert I, Korte F (1985) Chemosphere 14:545
23. Bruggeman WA, Martron LBJM, Kooiman D, Hutzinger O (1981) Chemosphere 10:811
24. Neely WB, Branson DR, Blau GE (1974) Environ Sci Technol 8:1113
25. Kenaga EE (1980) Environ Sci Technol 14:553
26. Kenaga EE, Goring CA (1980) Relationship between water solubility, soil sorption, octanol-water partitioning and bioconcentration of chemicals in biota. In: Eaton JG, Parrish PR, Hendricks AC (eds) Aquatic toxicology, vol. 707. American Society for Testing and Materials, ASTM, Philadelphia, PA, p 78
27. Könemann H, van Leeuwen K (1980) Chemosphere 9:3
28. Geyer H, Sheehan D, Kotzias D, Freitag D, Korte F (1982) Chemosphere 11:1121
29. Oliver BG, Niimi AJ (1983) Environ Sci Technol 17:287
30. Veith GD, Kosian P (1983) Estimating bioconcentration potential from octanol-water partition coefficients. In: Mackay D, Paterson S, Eisenreich SJ, Simmons MS (eds) Physical behaviour of PCBs in the Great Lakes. Ann Arbor Science Publishers, Ann Arbor, MI, p 269
31. Geyer H, Politzki G, Freitag D (1984) Chemosphere 13:269
32. Chiou CT (1985) Environ Sci Technol 19:57
33. Hawker DW, Connell DW (1986) Ecotox Environ Safe 11:184
34. Hawker DW, Connell DW (1989) Environ Sci Technol 23:961
35. Connell DW, Hawker DW (1988) Ecotox Environ Safe 16:242

36. Nendza M (1991) QSARs of bioconcentration: validity assessment of log P_{ow}/log BCF correlations. In: Nagel R, Loskill R (eds) Bioaccumulation in aquatic systems: contributions to the assessment. VCH Publishers, Weinheim, Germany, p 43
37. Sijm DTHM, van der Linde A (1995) Environ Sci Technol 29:2769
38. Hendriks AJ (1995) Chemosphere 30:265
39. Belfroid A, van Wezel A, Sikkenk M, van Gestel K, Seinen W, Hermens J (1993) Ecotox Environ Safety 25:154
40. Hendriks AJ (1995) Aquat Toxicol 31:1
41. Veith GD, Defoe DL, Bergstedt BV (1979) J Fish Res Board Can 36:1040
42. Opperhuizen A, Sijm DTHM (1990) Environ Toxicol Chem 9:175
43. Sijm DTHM, Pärt P, Opperhuizen A (1993) Aquat Toxicol 25:1
44. McKim J, Schmieder P, Veith G (1985) Toxicol Appl Pharmacol 77:1
45. Norstrom RJ, McKinnon AE, DeFreitas ASW (1976) J Fish Res Board Can 33:248
46. Neely WB (1979) Environ Sci Technol 13:1506
47. Black MC, McCarthy JF (1988) Environ Toxicol Chem 7:593
48. Schrap SM, Opperhuizen A (1989) Hydrobiologia 188/189:573
49. Schrap SM (1991) Comp Biochem Physiol 100:13
50. Hamelink JL, Landrum PF, Bergman HL, Benson WH (1994) Bioavailability: physical, chemical, and biological interactions. SETAC Special Publications Series, Lewis Publishers, CRC Press, Boca Raton, Fl, USA
51. McKim J, Goeden HM (1982) Comp Biochem Physiol 72:65
52. Opperhuizen A, Schrap S M (1987) Environ Toxicol Chem 6:335
53. Streit B, Siré EO, Kohlmaier GH, Badeck FW, Winter S (1991) Ecol Modelling 57:237
54. Sijm DTHM, Verberne ME, Pärt P, Opperhuizen A (1994) Aquat Toxicol 30:325
55. Tolls J, Sijm DTHM (1995) Environ Toxicol Chem 14:1675
56. Gobas FAPC, Clark KE, Shiu WY, Mackay D (1989) Environ Toxicol Chem 8:231
57. Belfroid A, Sikkenk M, Seinen W, van Gestel K, Hermens J (1994) Environ Toxicol Chem 13:93
58. de Bruijn J, Hermens J (1991) Environ Toxicol Chem 10:791
59. Deneer JW, Sinnige TL, Seinen W, Hermens JLM (1987) Aquat Toxicol 10:115
60. Opperhuizen A (1986) Bioconcentration of hydrophobic chemicals in fish. In: Poston, TM, Purdy R (eds) Aquatic toxicology and environmental fate, vol 9. American Society for Testing and Materials, Philadelphia, p 304
61. Sijm DTHM, Schipper M, Opperhuizen A (1993) Environ Toxicol Chem 12:1117
62. Opperhuizen A, Schrap S M (1988) Chemosphere 17:253
63. van den Berg M, Sijm DTHM (submitted) Crit Rev Toxicol
64. Gobas FAPC, McCorquodale JR, Haffner GD (1993) Environ Toxicol Chem 12:567
65. Gobas FAPC, Zhang X, Wells R (1993) Environ Sci Technol 27:2855
66. Parkerton TF (1993) Estimating toxicokinetic parameters for modeling the bioaccumulation of non-ionic organic chemicals in aquatic organisms. Ph.D. Thesis, The State University of New Jersey, U.M.I., Ann Arbor, Michigan
67. Sijm DTHM, Bol J, Seinen W, Opperhuizen A (1993) Arch Environ Contam Toxicol 25: 102
68. Clark CE, Mackay D (1991) Environ Toxicol Chem 10:1205
69. Muir DCG, Norstrom RJ, Simon M (1988) Environ Sci Technol 22:1071
70. Kim GB, Tanabe S, Iwakiri R, Tatsukawa R, Anamo M, Miyazaka N, Tanaka H (1996) Environ Sci Technol 30:2620
71. Harkey GA, Lydy MJ, Kukkonen J, Landrum PF (1994) Environ Toxicol Chem 13:1445
72. Kukkonen J, Landrum P F (1995) Aquat Toxicol 32:75
73. Landrum PF, Faust WR (1991) Effect of variation in sediment composition on the uptake rate coefficient for selected PCB and PAH congeners by the amphipod, Diporeia sp., In: Mayes M A, Barron MG (eds) Aquatic toxicology and risk assessment, vol 14. American Society for Testing and Materials, ASTM STP 1124, Philadelphia, PA, p 263
74. Tracey GA, Hansen DJ (1996) Arch Environ Contam Toxicol 30:467
75. van der Oost R, Opperhuizen A, Satumalay K, Heida H, Vermeulen NPE (1996) Aquat Toxicol 35:21

76. Belfroid A, Seinen W, van den Berg M, Hermens J, van Gestel K (1995) Environ Toxicol Chem 14:605
77. Hawker DW, Connel DW (1985) Chemosphere 14:1205
78. Swackhamer DL, Skoglund RS (1993) Environ Toxicol Chem 12:831
79. Opperhuizen A, van der Velde EW, Gobas FAPC, Liem AKD, van der Steen JMD, Hutzinger O (1985) Chemosphere 15:1871
80. Sijm DTHM, Wever H, Opperhuizen A (1993) Environ Toxicol Chem 12:1895
81. Opperhuizen A, Serné P, van der Steen, JMD (1988) Environ Sci Technol 22:286
82. Banerjee S, Baughman GL (1991) Environ Sci Technol 25:536
83. Gobas FAPC, Lahittete JM, Garofalo G, Shiu WY, Mackay D (1988) J Pharm Sci 77:265
84. Dulfer WJ, Govers HAJ (1995) Environ Sci Technol 29:2548
85. Belfroid A, Sijm DTHM, van Gestel CAM (1996) Environ Rev (In press)
86. Sijm DTHM, Opperhuizen A (1988) Chemosphere 17:83
87. de Wolf W, de Bruijn JHM, Seinen W, Hermens JLM (1992) Environ Sci Technol 26:1197
88. Hansch C (1971) In: Ariëns EJ (ed) Drug design, vol 1. Academic, New York, p 271
89. Hansch C (1973) In: Cavallito CJ (ed) Structure-activity relationships, vol 1. Pergamon, Oxford, p 75
90. Hansch C (1978) In: Chapman NB, Shorter J (eds) Correlation analysis in chemistry: recent advances. Plenum, New York, p 397
91. Veith GD, Call DJ, Brooke LT (1983) Can J Fish Aquat Sci 40: 743
92. Hermens J, Broekhuyzen E, Canton H, Wegman R (1985) Aquat Toxicol 6:209
93. Hermens J (1989) Quantitative structure activity relationships of environmental pollutants. In: Hutzinger O (ed) Handbook of environmental chemistry, vol. 2E. Springer, Berlin Heidelberg New York, p 111
94. Lipnick RL (1995) Structure-activity relationships. In: Rand GM (ed) Fundamentals of aquatic toxicology, second edition, effects, environmental fate, and risk assessment. Taylor and Francis, Washington, DC, p 609
95. Hermens J, Busser F, Leeuwangh P, Musch A (1985) Toxicol Environ Chem 9:219
96. Schultz TW, Holcombe GW, Phipps GL (1986) Ecotox Environ Safe 12:146
97. McKim JM, Bradbury SP, Niemi GJ (1987) Environ Health Persp 71:171
98. Veith GD, Broderius SJ (1987) Structure-toxicity relationships for industrial chemicals causing type (II) narcosis syndrome. In: Kaiser KLE (ed) QSAR in environmental toxicology-II.Reidel, Dordrecht, The Netherlands, p 385
99. Deneer JW, Seinen W, Hermens JLM (1988) Aquat Toxicol 12:185
100. Deneer JW, Sinnige TL, Seinen W, Hermens JLM (1988) Aquat Toxicol 195
101. Bradbury SP, Henry TR, Carlson RW (1990) Fish acute toxicity syndromes in the development of mechanism-specific QSARs. In: Karcher W, Devillers J (eds) Practical applications of quantitative structure-activity relationships (QSAR) in environmental chemistry and toxicology. Kluwer, Dordrecht, p 295
102. van Gestel CAM, Ma W-C (1990) Chemosphere 21:1023
103. van Leeuwen CJ, van der Zandt PTJ, Aldenberg T, Verhaar HJM, Hermens JLM (1992) Environ Toxicol Chem 11:267
104. Verhaar HJM, van Leeuwen CJ, Bol J, Hermens JLM (1992) Chemosphere 25:471
105. van Gestel CAM, Ma W-C (1988) Ecotox Environ Safe 15:289
106. van Wezel, AP (1995) Residue-based effects of narcotic chemicals in fish and in lipid bilayers. Ph.D. Thesis, Utrecht University, Utrecht
107. McKim JM, Schmieder PK (1991) Bioaccumulation: does it reflect toxicity? In: Nagel R, Loskill R (eds) Bioaccumulation in aquatic systems. Contributions to the assessment. VCH Publishers, Weinheim, Germany, p 161
108. Sijm DTHM, Opperhuizen A (1996) Dioxins: an environmental risk for fish? In: Beyer WN, Heinz GH, Redmond-Norwood AW (eds) Environmental contaminants in wildlife. Interpreting tissue concentrations. A Special SETAC Publication, p 209
109. Tas JW, Seinen W, Opperhuizen A (1991) Comp Biochem Physiol 100:59
110. Tas JW, Keizer A, Opperhuizen A (1996) Bull Environ Contam Toxicol 57:146
111. Borgmann U, Norwood WP, Ralph KM (1990) Arch Environ Contam Toxicol 19:558

112. Mauck WL, Mehrle PM, Mayer FL (1978) J Fish Res Board Can 35:1084
113. Mortimer MR, Connell DW (1994) Ecotox Environ Safe 28:298
114. Roghair, personal communication
115. Broyles RH, Noveck MI (1979) Toxicol Appl Pharmacol 50:299
116. Mayer FL, Mehrle PM, Sanders HO (1977) Arch Environ Contam Toxicol 5:501
117. Pawlisz AV, Peters RH (1993) Environ Sci Technol 27: 2801
118. Belfroid A, Seinen W, van Gestel K, Hermens J (1993) Chemosphere 12: 2265
119. van Wezel AP, Sijm DTHM, Seinen W, Opperhuizen A (1995) Chemosphere 31: 3201
120. Chaisuksant Y, Yu Q, Connell D. Ecotoxicol Environ Safe (submitted)
121. Lanno RP, McCarty LS (submitted) Worm bioassays: what knowledge can be applied from aquatic toxicity testing? Presentation at the SETAC conference in Vancouver, 1995
122. Hattula ML, Wasenius VM, Reunanen H, Arstila AU (1981) Bull Environ Contam Toxicol 26:295
123. van Wezel AP, Punte SS, Opperhuizen A (1995c) Environ Toxicol Chem 14:1579
124. Kobayashi K, Akitake A, Manabe K (1979) Bull Jap Soc Fish 45:173
125. Kishino T, Kobayashi K (1995) Water Res 29:431
126. de Wolf W, Seinen W, Opperhuizen A, Hermens JLM (1992) Chemosphere 25:853
127. Lassiter RR, Hallam TG (1990) Environ Toxicol Chem 9:585
128. Ewald G, Larsson P (1994) Environ Toxicol Chem 13:1577
129. van Wezel AP, de Vries DAM, Kostense S, Sijm DTHM, Opperhuizen A (1995) Aquat Toxicol 33:325
130. de Maagd PGJ (1996) Polycylic aromatic hydrocarbons: fate and effects in the aquatic environment. Ph.D. Thesis, Utrecht University, Utrecht, The Netherlands.
131. Kleiner C, Erickson R, Flandt J, Highland T (1993) Relationship of toxicokinetics to effects of pentachloroethane to fathead minnows. Poster presented at the 14th annual meeting of the Society of Environmental Toxicology and Chemistry (SETAC), Houston, TX
132. de Bruijn J, Yedema E, Seinen, W, Hermens J (1991) Aquat Toxicol 20:111
133. Ohayo-Mitoko GJA, Deneer JW (1993) Sci Total Environ, Suppl. 559
134. Broderius SJ, Kahl MD, Hoglund MD (1995) Environ Toxicol Chem 14:1591
135. Deneer JW, Sinnige TL, Seinen W, Hermens JLM (1988) Aquat Toxicol 12:33
136. Hermens J, Leeuwangh P, Musch A (1985) Ecotox Environ Safe 9:321
137. Anliker R, Moser P, Poppinger D (1988) Chemosphere 17:1631
138. Verhaar HJM, Busser FJM, Hermens JLM (1995) Environ Sci Technol 29:726
139. Sijm DTHM, Wever H, Opperhuizen A (1989) Chemosphere 19:475
140. Thomann RV, Connolly JP, Parkerton JF (1992) Environ Toxicol Chem 11:615
141. Södergren A (1987) Environ Sci Technol 21:855
142. Huckins JN, Tubergen MW, Manuweera GK (1990) Chemosphere 20:533
143. Ellis GS, Huckins JN, Rostad CN, Smitt CE, Petty DD, MacCarthy P (1995) Environ Toxicol Chem 14:1875
144. van Loon WMGM, Wijnker FG, Verwoerd ME, Hermens JLM (1996) Anal Chem 68:2916
145. Cook PM, Kuehl DW, Walker MK, Peterson RE (1991) Bioaccumulation and toxicity of TCDD and related compounds in aquatic ecosystems. In: Banbury report 35. Biological basis for risk assessment of dioxins and related compounds. Cold Spring Harbor Laboratory Press, p 143
146. van der Weiden MEJ, van der Kolk J, Bleumink R, Seinen W, van den Berg M (1992) Aquat Toxicol 24:123
147. van den Berg M, de Jongh J, Poiger H, Olson J R (1994) Crit Rev Toxicol 24:1
148. Ankley GT, Erickson RJ, Phipps GL, Mattson VR, Kosian PA, Sheedy BR, Cox, JS (1995) Environ Sci Technol 29:2828
149. Crommentuijn T, Doodeman CJAM, Doornekamp A, van der Pol JJC, Bedaux JM, van Gestel CAM (1994) Environ Toxicol Chem 13:1781
150. Niimi AJ, Kissoon GP (1994) Arch Environ Contam Toxicol 26:169
151. Solé M, Porte C, Albaigés J (1994) Aquat Toxicol 30:271
152. van Wezel AP, Opperhuizen A (1995) Crit Rev Toxicol 25:255

Long-term Effects of Bioaccumulation in Ecosystems

Jaakko Paasivirta

Department of Chemistry, University of Jyväskylä, Finland, P.O. Box 35, SF-40351 Jyväskylä, Finland

Extensive damage to organisms and declines in wildlife populations have been observed together with long-term bioaccumulation and biomagnification of persistent xenobiotic chemicals. Heavy metals, especially organic or biomethylated mercury, lead, cadmium and organic tin compounds have caused environmental damage through bioaccumulation on a local scale. Effects on wildife caused by bioaccumulation of persistent organochlorine compounds are more widespread. However, the causal relationship between a biomagnified compound and the long-term effects have been established in only a few cases. Metabolic transformations, and occurrence of several toxic contaminants together in many cases, complicate evaluations of the sources of long-term effects. Environmental fate, exposure of biota and biomagnification of a chemical can be predicted by modelling from its properties and from ecological, geological and climatic conditions of the recipient environment. Model predictions can be refined by experimental factors obtained from results of the field studies. Empirical estimates of hazardous bioaccumulation or biomagnification are obtained from field analyses of different trophic levels. Trend analyses of biomagnified contaminants and their effects can be utilized in prognosis of future development and in evaluation of the need for further action to protect the environment and human health.

The Handbook of Environmental Chemistry, Vol. 2 Part J
Bioaccumulation (ed. by B. Beek)

List of Symbols and Abbreviations

ACE	aromatic chloroethers
AHH	aryl hydrocarbon hydroxylase
ALAD	aminolevulinic acid dehydratase
BCF	bioconcentration factor
Bp	boiling point
CHL	chlordanes
CYMD	chlorocymenenes
CYMS	chlorocymenes
DBT	dibutyltin
DDD	tetrachlorodiphenylethane (1,1-dichloro-2,2-bis(4-chlorophenyl) ethane)
DDE	dichloro-diphenyl-dichloroethene (1,1-dichloro-2,2-bis (4-chlorophenyl)ethene)
DDT	dichloro-diphenyl-trichloroethane (1,1,1-trichloro-2,2-bis (4-chlorophenyl)ethane)
dw	dry weight
EI	eggshell (thickness) index
EROD	etoxyresorufin-O-deethylase
ESM	eyed stage mortality (of hatching fish juveniles)
FERM	fertilizing mortality (of fish eggs)
GCOL	fish egg colour
fw	fresh weight (wet tissue)
HCBz	hexachlorobenzene
HCH	hexachlorocyclohexanes
k_B	biodegradation rate,
k_H	hydrolysis rate,
k_P	photodegradation rate
LIND	γ-HCH
lw	lipid weight (in fat)
MBT	monobutyltin
MFO	mixed fuction oxydases
$\delta^{15}N$	nitrogen isotope ratio
OCC	organochlorine compound
OCS	octachlorostyrene
P	vapour pressure
PBA	polybromoanisoles
PBB	polybromobiphenyls
PBDE	polybromodiphenyl ethers
PCA	polychloroanisoles (compounds) or principal component analysis (statistical treatment)
PCB	polychlorobiphenyls
PCBA	polychlorobiphenyl anisoles
PCBOH	polychlorobiphenylols
PCDD	polychlorodibenzo-*p*-dioxins
PCDE	polychlorodiphenyl ethers

PCDF	polychloridibenzofurans
PCC	toxaphene (polychlorinated camphene, TOX)
PCN	polychloronaphthalenes
PCPA	polychlorophenoxyanisoles
PCT	polychloroterphenyls
PCV	polychloroveratroles
PeCP	pentachlorophenol
POP	persistent organic pollutant
QSAR	quantitative structure-activity relationship
RPCBB	alkyl polychlorobibenzyls
RPCFL	alkyl polychlorofluorenes
RPCN	alkyl polychloronaphthalenes
RPCPH	alkyl polychlorophenanthrenes
S	solubility in water
SCHL	sum of chlordane residues (CHL)
SDDT	DDE + DDD + DDD
SPCB	total PCB content
TBT	tributyltin
TBTO	bis(tributyltin) oxide
TCDD	2,3,7,8-tetrachlorodibenzo-*p*-dioxin
345TCG	3,4,5-trichloroguaiacol
246TCP	2,4,6-trichlorophenol
TeCG	tetrachloroguaiacol
TeCC	tetrachlorocatechol
TeCP	2,3,4,6-tetrachlorophenol
TEF	toxic equivalency factor (potency compared to TCDD)
TEQ	toxic equivalent (concentration or load compared to TCDD)
TML	tetramethyl lead
TotHg	total mercury content
TOX	toxaphene (PCC)
TPT	triphenyltin
YSM	yolk sac mortality (of juvenile fish)

1 Introduction

Man-made chemicals have caused deaths of wildlife populations due to serious dumpings, industrial discharges and accidental spills. In many cases the causal linkage between certain toxic chemical and damage has been obvious. In many other cases epidemic deaths of animal populations or vegetation has been suspected to be caused by an acute exposure to chemicals but not verified. Ecological damage from chronic exposure is even more difficult to explain. Long-term studies on the occurrence of anthropogenic chemicals in the environment, their effect potencies and their monitoring in connection with specific episodes have, however given some specific answers. These results justify the conclusion that some persistent anthropogenic chemicals are causing ecological damage such as the decline of populations through breeding losses or developmental damage as a result of bioaccumulation and biomagnification in the environment.

2 Observed Effects from Bioaccumulation

2.1 Mercury

Mercury is mobilized in the environment mainly from sources related to human activities [1, 2]. Its ecotoxic effects were first observed in Sweden during the period 1948–65, where excess mortality of seed-eating birds was found to be caused by methyl mercury used as seed dressing fungicide [3, 4]. Serious human mass poisonings from seafood in the early 1950s in Minamata and the early 1960s in Niigata were due to alkylmercury discharged from chemical manufacturing plants being bioaccumulated in fish and crustacea consumed by people [5]. Symptoms, e.g. visual field impairment, unsteadiness, frequent falls, circling movements, convulsions and death, were also observed in cats of Minamata and in swine fed with alkylmercury-dressed grain in the USA [6, 7].

A large human catastrophe (6530 hospitalized, 459 died) took place in the winter 1971–1972 in Iraq, where people ate homemade bread prepared from wheat seed that had been treated with methylmercurial fungicide [6, 7]. A food chain transfer of alkylmercury occurred in Mexico in 1969, where a farmer and five of his neighbours fed treated grain to hogs and then ate the contaminated pork. The pigs died or developed blindness, and several family members suffered methylmercury poisoning [6, 7].

In the environment mercury circulates as vaporized element Hg^0, as inorganic mercury salts (mainly $HgCl_2$), as dimethyl mercury (H_3C-Hg-CH_3) and as monomethyl mercury (ClHg–CH_3) [2]. The latter organic mercury compounds are significantly more toxic than elemental or inorganic mercury. While inorganic mercury is methylated by microbia in terrestrial and aquatic solids to the persistent, lipophilic methylmercury, which is bioaccumulating at a high rate and being enriched in the food chain, serious risk of long term damage to humans and wildlife is expected from mercury releases. The investigations carried out during 1965–1975 showed that enriched methyl mercury concentrations in fish were widespread on a global scale [2, 4]. In addition to local industrial discharges, dredging and especially reservoir construction mobilized mercury which then bioaccumulated and biomagnified in aquatic species [6, 8]. In La Grande 2 reservoir, mercury levels as high as 3000 ng g^{-1} fw were measured in fish [9]. Further bioaccumulation of mercury in reservoir areas was considered to form a threat to loons, bald eagles, ospreys and other predator species [10].

Humans, cats and birds are sensitive to mercury poisoning but fish are not. The lifetime of methyl mercury in fish (400–1000 days) is significantly longer than in mammals (in humans 70–76 days, small mammals only few days) [6, 11]. Mercury in fish is more than 90% in methylated form. Accordingly, the most important environmental mercury hazard is the bioaccumulated methyl mercury in seafood eaten by humans and animals. There are numerous examples of elevated mercury levels in high fish consumers, but not many reported toxic symptoms in these populations [6].

Wild mammals collect mercury via their food: in southern Ontario in 1973–74 fischerotter, marten and mink had high (330–710 ng g^{-1} fw) but fox, raccoon and skunk low (50–99 ng g^{-1} fw) total mercury contents in muscles [12]. This corresponds well with the different feeding habits of species: more fish-eating animals get higher mercury contamination. Any toxic effects from mercury were not reported in mammals in rural areas. Near contamination sources, however, small mammals have shown developmental disturbances (genetic aberrations and asymmetry) thought to be due to elevated mercury levels [13]. Marine mammals at the top of the aquatic food chain could be expected to accumulate harmful levels of mercury. Like some terrestrial mammals, however, they seem to metabolize organic mercury to inorganic mercury. This adaptation effect prevents toxic consequences and is perhaps the result of evolution during long periods of geological time [14]. A review of Hg accumulation in organs of wild terrestrial mammals in relation to dietary habits, sex and age was published in 1986 [15].

2.2 Other Heavy Metals

Besides mercury, lead (Pb), cadmium (Cd) and tin (Sn) are the most hazardous heavy metals which can bioaccumulate to toxic levels [1, 6, 16]. Industrially manufactured organolead and organotin compounds are emitted from traffic and other technical uses, and can be serious bioaccumulating ecotoxicants. Both organic and inorganic lead compounds are bioaccumulating. Microbial biomethylation takes place in the environment with lead and tin, but not with cadmium [1, 16].

Lead is converted in nature to tetramethyl lead (TML) which is bioavailable to such a degree that 10–24% of the total lead content in fish muscle consists of TML [1, 6]. Accordingly, environmental hazard from TML is not as great as from methyl mercury [6]. Bioaccumulation of lead compounds has caused human sickness and ecologial damage. Aqueous emissions of alkyl lead has been a source of mass mortality for water birds [17–19], and ingestion of lead shot has been connected to increased avian mortality [20, 21]. Reduction of the enzyme delta-ALAD has been observed as a biomarker of lead intoxication [22]. Wood ducks near mining and smelting sites collected lead up to 8 µg g^{-1} levels in their blood and 14 µg g^{-1} in their livers. The Pb concentrations correlated negatively with ALAD and with nesting success, showing population damage as a result of lead bioaccumulation [23]. Lead shot has also contaminated soil in shooting ranges. Transformation of pellets to bioavailable forms of lead has been shown to cause high concentrations of lead and toxic effects in exposed small mammals [24, 25]. In the aquatic ecosystem, significant bioaccumulation of lead causing adverse effects seems to take place only near point sources of heavy lead pollution [26]. Restriction of the use of leaded gasoline and changing to other metals for shot has already greatly reduced the overall environmental hazard from lead pollution.

The toxic threat from bioaccumulating cadmium was demonstrated by human "itai-itai" disease in Japan 1947. Industrial discharge of cadmium in the Jintsu River area was exposing humans fatally (more than 100 deaths) via contaminated drinking water [1, 6, 20]. Cadmium has both acute and long-term toxicity to mammals because it is not eliminated but instead accumulates in the liver, kidneys and bones [20]. Inorganic cadmium accumulates in biota because it binds tightly to sulfur-containing proteins such as metallothionein [27]. Bioconcentration factors for cadmium from water to some insects, snails and amphipods are as high as 90 000 [28]. Cadmium-metallothionein is stored in hepatopancreas of crustaceans: extraordinarly high Cd concentrations were measured in hepatopancreas and green glands of lobsters near a lead smelter [29]. Record high levels measured for Cd were in scallops, being 200–500 $\mu g\ g^{-1}$ fw in whole organisms and 2000 $\mu g\ g^{-1}$ dw in hepatopancreas [20, 30]. According to present literature, ecotoxic effects of bioaccumulated cadmium are local incidences and not of global concern.

Organic tin compounds used in stabilizers, pesticides and marine antifouling paints cause local or regional ecological problems [6]. Inorganic tin is biomethylated similar to mercury and lead [16, 31]. Bioconcentration factors (BCF) from water to fish were 1800 for bis(tributyltin) oxide (TBTO) [32]. Tributyltin (TBT) and triphenyltin (TPT) compounds had BCF of 50–600 to fish muscle and up to 5000 to liver and kidney [33]. In a marina contaminated with organotin compounds, BCF values of 5000–60 000 from water to blue mussels were measured. In this field study, the half life time for depuration of organic and total Sn were 40 and 25 days, respectively [34]. Bioaccumulation of TBT from sediments to deposit-feeding clams up to toxic levels has been observed [35]. Bioaccumulated tin causes shell-thickening in oysters [36], and sterility in juvenile and imposex (the growth of a penis and vas deferens in females) in adult dog-whelks [37]. Neurotoxic influence of TBTO bioaccumulation in fish has been indicated [38]. Algae seem to be able to collect relatively high amounts of TBT, but also degradate it to less toxic dibutyltin (DBT) and monobutyltin (MBT) compounds [39]. Accumulation of TBT and TPT in red sea bream was observed to take place more by direct uptake from water and less by dietary intake. Compared to PCB and methyl mercury, assimilation efficiency and the percentage retention of organotin compounds were low [40].

2.3
Other Metals

Aluminium, iron, zinc, chromium and copper are common metallic elements that are strongly bioaccumulating in their salt form. Their increased release from soil due to acid precipitation has toxic effects on plants and aquatic organisms in lakes [6, 16]. However, biomagnification of these elements is, in general, not significant. Therefore, their long-term effects in the environment are less likely to be connected to bioaccumulation than those of persistent lipophilic pollutants. However, aluminium is of great public concern because of its implication in several human disorders such as Alzheimer's disease and senile dementia [41]. These chronic effects support the irreversible accumulation of aluminium in certain tissues.

2.4 Organochlorine Compounds

Very much higher amounts of organohalogen compounds are produced in nature than manufactured, used and discharged in human activities [41–43]. But the anthropogenic organohalogens, especially organochlorine compounds (OCC), are responsible for all wide scale ecological damage associated with bioaccumulation of organic halogen compounds. OCCs of human origin often occur locally at high concentrations, while natural OCCs are diluted in the terrestrial and aquatic compartments. Human originated OCCs enter the natural environment in accidents, but more commonly from industrial discharge, pesticide and preservative usage, urban waste and especially from chlorination and combustion processes [44]. In general, covalently bound chlorine increases persistency and lipophilicity of an organic molecule, and thus enhances bioaccumulation and biomagnification to toxic levels.

The harmful ecological effects of organochlorine compounds were first shown by the decline in certain bird populations in areas where organochlorine pesticides, especially DDT, were heavily used, as Rachel Carson revealed in her book *Silent Spring* in 1962. A scientific explanation for this decrease in reproduction was found in the eggshell thinning effect of DDE, which is the major persistent metabolite of DDT [45]. In addition to DDT and its metabolites, persistent residues of many organochlorine pesticides have been accumulated in food chains globally [44]. They include aldrin, chlordane, lindane, heptachlor, dieldrin (persistent metabolite of aldrin), toxaphene and mirex [46]. Their bioaccumulative potential and various observed acute and chronic toxic effects on animals including estrogenity and teratogenity has led to the banning or severe restriction of their use in both industrialized and developing countries [47].

Hexachlorobenzene (HCBz) occurs in the environment in amounts that are orders of magnitude higher than its production for fungicidal usage and technical fluids [6]. Therefore, discharges as unwanted by-product and combustion products are major sources of environmental HCBz [48, 49]. Acute toxicity of HCBz is small, but its chronic effect causing hepatic porphyria in mammals is severe. Consuming treated seeds caused an epidemic of HCBz-induced porphyria cutanea tarda in Turkey from 1955 to 1959 involving 3000–5000 people with a mortality of 10% [48]. Biomagnification rate of HCBz in the aquatic-terrestrial food chain is similar to that of DDE [6].

Polychlorinated biphenyls (PCB) as environmental contaminants were first found in Baltic seals and fish [50]. PCBs are industrial products, about one millon tons being manufactured from 1929 to 1987 for use in electrical equipment, in closed power and heat transfer systems, as plasticizers, binders, paint, copypaper additives, adhesives etc. [6]. Their extreme persistency and lipophilicity caused their biomagnification to high levels, which were associated with cases of damage to the reproduction of sea-lions in California [51], seals [52] and mink [53] in the Baltic sea, area and birds in the Lower Great Lakes of North America [54]. The effects on embryos and juveniles of wild birds were the same as chick oedema disease which killed millions of broilers in the USA in 1957 due to contamination of their food by leaked PCB used as a heat transfer liquid [55]. Later, the symptoms were associated with pyrolysis products of PCBs [56–58], as were the human catastrophes in Japan (1968) and Taiwan (1979), where people consumed rice-oil which was contaminated with heated PCB [44].

The extremely toxic TCDD (2,3,7,8-tetrachlorodibenzo-*p*-dioxin) and related compounds first became known as the source of chloracne in industrial workers exposed during chlorine and chlorophenol production [59, 60]. After several documented occupational mass poisonings in industry manufacturing chlorophenols, polychlorodibenzo-*p*-dioxins (PCDD) and polychlorodibenzofurans (PCDF) were also found in emissions of thermal processes, industrial discharges, pesticides and preservatives [61]. They are widespread in the environment and have a high bioaccumulation power [62]. Dioxins have been included in notorious incidents. Herbicide 245-T used extensively as a leaf-dropping agent in the Vietnam war contained tens of ppm of TCDD. Accidental release of TCDD from production of 2,4,5-trichlorophenol to the environment followed by mass deaths of animals and contamination of people in Seveso, Italy, 1976 triggered great public concern about dioxins [63]. Chick oedema disease [55], Yusho oil poisoning in Japan and Yu Cheng disease in Taiwan from heated PCB contamination was found to be due to toxic PCDFs formed in the pyrolysis of PCBs [56–58]. PCDFs were most probably formed from ortho-OH substituted PCBs formed as major products during pyrolysis in the presence of oxygen [58].

The toxic effects of PCDDs and PCDFs were associated with their structure-related metabolism. Compounds having chlorine in lateral positions 2,3,7,8 of the dioxin or furan molecule (seven PCDDs and ten PCDFs out of a total of 210 structures possible; see Fig. 1) were found to fit closely to cytosolic receptors (Ah) of mixed function oxydase (MFO) enzymes which oxidize xenobiotics such as PAHs [64]. Because the substrates of MFO also include steroid hormones, certain vitamins, fatty acids and bile acids [65], the induction of MFO enzymes by dioxins and related compounds is associated with their long term toxic impact on both wildlife and humans [66, 67]. Binding of PCDDs and PCDFs to Ah receptors leads to hepatic MFO induction (e.g. AHH or EROD) which parallels dermal toxicity, thymic atrophy, reproductive effects, teratogenity, hepatoxicity and carcinogenity [63, 68]. However, hormone-like behaviour of TCDD and other dioxins is also demonstrated by their anti-tumour-promoting activity. Consequently, they are studied as potential cancer drugs [69].

Use of MFO induction potency as a measure of dioxin-like toxicity [70] is confused by the fact that many planar aromatic and heteroaromatic naturally formed compounds are also strongly bound to Ah receptors and potent MFO inducers [71]. However, toxic potency as equivalency factors (TEF) related to TCDD is generally used in emission control and toxic load estimation of dioxins and related compounds. TEFs are based on MFO-inductions, immunotoxicity and other biological response measurements and evaluated by expert groups. For each chemical in sample (food, tissue or emission) is calculated a toxic load value TEQ = TEF × concentration. Total TEQ load, based on the assumption that the effects are additive, is calculated as the sum of the TEQs of each compound [72, 73].

Structural similarity with toxic PCDDs and PCDFs explains the same MFO-induction and toxic effects of certain PCBs and TCDD. The structures of dioxin-like toxic PCDDs, PCDFs and PCBs are illustrated in Fig. 1. When chlorine substitution in the PCB molecule is in meta (3,3′,5,5′) and para (4,4′) positions, the molecule seeks its lowest energy configuration in the plane [74, 75]. The non-ortho chlorine substituted "coplanar" PCBs, such as 3,3′,4,4′-tetrachlorobiphenyl (PCB77), 3,3′,4,4′,5-pentachlorobiphenyl (PCB126) and 3,3′,4,4′,5,5′-hexachlorobiphenyl (PCB169) are the most toxic of the PCB congeners [74–77]. Mono-ortho coplanar PCBs, such as 2,3′,3′,4,4′-pentachlorobiphenyl (PCB105), 2,3′,4,4′,5-pentachlorobiphenyl (PCB118) and 2,3,3′,4,4′,5-hexachlorobiphenyl (PCB156) are also MFO inducers and could be taken into TEQ evaluations. The TEF approach can been extended to bromo analogues of PCDDs, PCDFs and PCBs, and also to polychlorodiphenyl ethers (PCDE) [77, 78]. The TEFs for PCDDs, PCDFs and PCBs have been evaluated internationally by toxicology expert groups [79, 80]. A bioaccumulation estimate for PCDDs, PCDFs and PCBs from Baltic wildlife analysis results [81–83] as TEQ loads is illustrated in Fig. 2.

Further bioaccumulating OCCs which are suspected, but less frequently observed in the field, to have harmful effects in ecosystems are polychloronaphthalenes (PCN) [84–87], polychloroterphenyls (PCT) [87–92], octachlorostyrene (OCS) [87, 93–98] and hexachlorobutadiene (HCBD) [96–100]. The major persistent and bioaccumulating OCCs discharged from bleaching of pulp or from chlorodisinfection of water are alkylaromatic chlorohydrocarbons – chlorocymenes (CYMS), chlorocymenenes (CYMD), alkyl polychlorobibenzyls

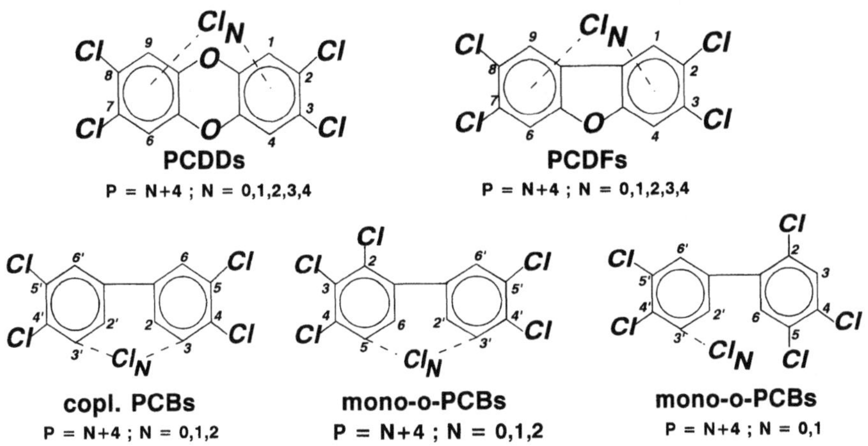

Fig. 1. Structures of the most toxic polychlorodibenzo-*p*-dioxins (PCDDs; seven compounds), polychlorodibenzofurans (PCDFs; ten compounds) and polychlorobiphenyls (three coplanar and and six mono-ortho-substituted PCBs)

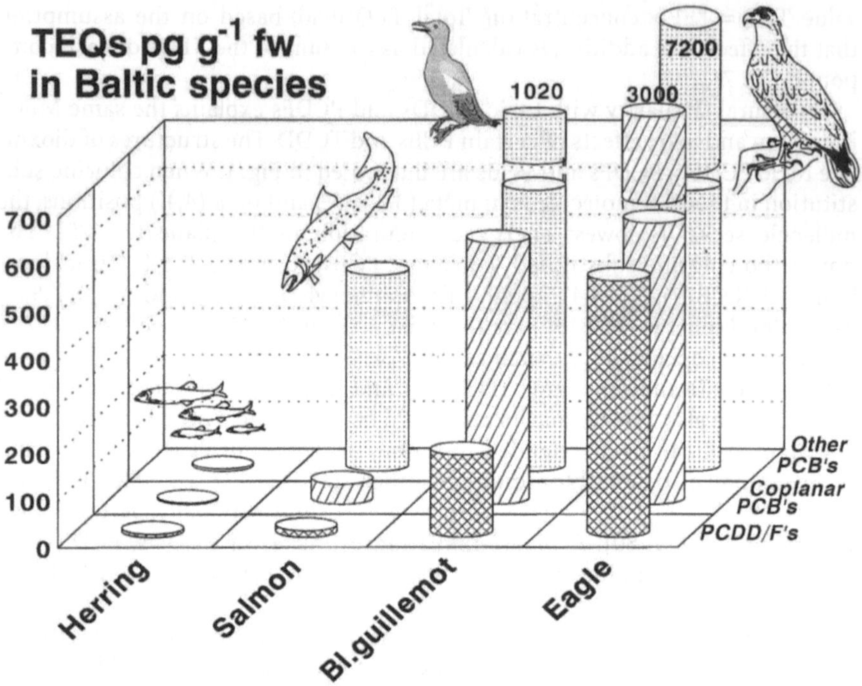

Fig. 2. Average contents of PCDDs, PCDFs and PCBs as toxic TCDD equivalents (TEQs) in Baltic sea animals in the 1980s [81–83]

(RPCBB), alkyl polychlorophenanthrenes (RPCPH), alkyl polychloronaphthalenes (RPCN) and alkyl polychlorofluorenes (RPCFL). Their bioaccumulation potency has been shown but their toxic effects have not been much studied to date [101–114].

In addition to PCDDs and PCDFs, some other groups of aromatic chloroethers (ACE) are of environmental concern [115, 116]. Biomethylation products of chlorophenols, chlorocatechols and chloroguaiacols – chloroanisoles (PCA) and chloroveratroles (PCV) – are well known as extremely potent off-flavours [117–119]. Although the observed tainting effects of PCAs and PCVs are local incidences from point source discharges, PCAs occur as globally distributed pollutants [120].

Major phenolic impurities of chlorophenol products, polychlorinated phenoxyphenols (PCPP) [121–125], transfer in the environment to their biomethylation products, polychlorophenoxyanisoles (PCPA) [126], which have been detected in Baltic wildlife and fish liver oil [115, 127, 128]. PCPAs also occur as minor neutral impurities in tetrachlorophenol preservative Ky-5 made by chlorination of phenol [129].

Other phenolic impurities in chlorophenol formulations are polychlorobiphenylols (PCBOH) [123] which are metabolites of PCBs [130] and also major products of the air-pyrolysis of PCBs [58]. Biomethylation products of PCBOHs and further metabolites of PCBs, polychlorobiphenyl anisoles (PCBA) [130], have been identified in fish liver oil and in Baltic whitetailed eagles [115, 127]. Polychlorodiphenyl ethers (PCDE) are the most abundant neutral impurities of technical chlorophenol formulations [121, 123]. The main PCDE congeners in Ky-5 were identified via model substance syntheses by Humppi [124, 129]. Additionally, a number of highly chlorinated PCDEs were synthesized and studied by Nevalainen [131–133] and Kurz [134] with their coworkers. PCDEs are bioaccumulating and enriched in biosphere as are PCBs, although their background levels are lower [82, 83, 115, 116, 134–137].

From the bromo-analogues of OCCs, bromoanisoles (PBA) occur and bioaccumulate in the marine environment [138, 139]. Bromoanisoles and the corresponding chloroanisoles, at least the 2,4,6-trisubstituted ones, can be, in great part, products of natural halogenation of anisoles or phenols [43, 140]. Fire retardant mixture of polybrominated biphenyls (PBB) seriously contaminated livestock and humans in an accidental poisoning of animal food in Michigan [141, 142]. Another fire retardant group of polybromodiphenyl ethers (PBDE) has caused widespread pollution of aquatic wildlife [143]. The concentrations of PBDEs in cod liver were highest in the southern and lowest in the northern North Sea and decreased over the time 1977–87, showing a point source pollution which had a decreasing trend [98].

3 Predicted and Observed Bioaccumulation in the Environment

3.1 Exposure Models

To assess the potential exposure of humans and the environment to chemical substances, mathematical modelling is needed to reduce the need of time-consuming and expensive analyses from field samples [144–146]. The models handle the major environmental processes in compartments of air, water, solids and plants. In soil/plant systems leaching, run-off and plant uptake, and in aquatic systems bioaccumulation are the major pathways of chemicals leading to contamination of food and drinking water of higher animals and humans [145, 147]. Steady state multimedia mass balance models are most popular for estimating exposure of biota in different compartments, starting from known emissions [146].

The environmental fate of the discharged chemical is predicted with modelling from the known properties of the compound and verified by analyses in the environment [146, 147]. A preliminary fate modelling only needs a limited number of properties of the chemical: molecular mass, water solubility (S), vapour pressure (P) in environmental temperatures, and bioconcentration factors (BCF) and rates of hydrolysis (k_H), photodegradation (k_P) and biodegradation (k_B) in the compartments (air, water, soil/sediment) of the model environment. Instead of expensive testing [148], some of these properties can be roughly estimated from readily available properties [149] such as octanol-water partition coefficient (S and BCF from Kow) or boiling point (P from Bp).

Degradation rates must be measured in the environment or under similar laboratory conditions. For the series of similar compounds, however, evaluation of the rate constants can also be done by semiempirical calculations from accumulated data using quantitative structure-activity relationships (QSAR) [149, 150].

The multimedia models include the main processes in the environment which influence the fate of the chemical. Transport processes determine the distribution, and transformation processes the persistence of the chemical in the environment. The main transport processes incorporated in models are 1) advection which transports the chemical in dissolved, gaseous, condensed or particulate phases, 2) dispersion as a result of turbulence and molecular diffusion, 3) volatilization determining air-water and soil-air transfers, 4) adsorption on soils and sediments, 5) bioaccumulation, 6) water phase heterogenous transport, particle settling, resuspension, sedimentation and sediment mixing, and 7) air phase heterogenous transport by wet and dry deposition. The main transformation phenomena modeled are 1) biodegradation, 2) hydrolysis, 3) phototransformation, and 4) speciation by dissociation to charged species and complex formation [151].

The multimedia models can be classified into four different levels [146, 147, 151]. In the level I model, equilibrium is assumed, and transformation of the chemical is excluded. Output consists of the relative concentrations (equilibrium distribution) of the substance in compartments of an environment. The compartments are air, water, soil, sediment, suspended sediment and fish (biota).

In the level II model, equilibrium is also assumed, but transformation and advection are taken into account. In addition to the steady-state concentrations, reaction and advection rates and residence times are obtained. The concentrations obtained are arbitrary, calculated from assumed total emission rate, but their ratios are characteristic to the environment and compound chosen.

The level III model gives similar output as level II, but with greater precision and in non-equilibrium conditions. Estimates of chemical quantities, concentrations and lifetimes in four compartments (air, water, soil and sediment) are obtained. Concentration in fish is given (as in Level II) only based on partition between biota (lipid) and water.

Level IV models assume non-steady state. They predict the time needed for the chemical to reach steady state when the releases are changed [146, 151].

In addition to multimedia models, a number of models for fate of chemicals with reduced numbers of compartments, like models for air, water/rivers/watercourses, soil/groundwater, air/plants, soil/plants etc., are widely used [145, 151]. Normally these models handle real environments as do multimedia models. The latter, however, also use hypothetical "generic" environments for preliminary estimation of the environmental hazard potential of a chemical [146]. Microcomputer "toolbox" CemoS [152, 153] is an integrated system used to simulate distribution of a chemical substance from continuous releases to air, water and soil from both diffuse and point sources in multimedia environment systems by transport to plants and by movement in the food chain. CemoS consists of the following nine models:

AIR	a one-dimensional box model for boundary surface releases to air
BUCKETS	a scoop chain model for transport in soil
CHAIN	a model for the food chain with three trophic levels
LEVEL1	a multimedia model (Mackay level 1) for equilibrium distribution assuming that the chemical is fully persistent
LEVEL2	a multimedia model (Mackay level 2) for steady state equilibrium distribution including advections and transformations
PLANT	a box model for uptake in plants
PLUME	a three-dimensional steady state model for point sources to air
SOIL	a one-dimensional model for vertical transport in soil
WATER	a one-dimensional stationary state box model for point sources in flowing waters

Bioaccumulation is estimated from the modeled concentration of chemical in air, water or solids depending on habitat of the exposed biota. Bioaccumulation models [145, 146, 154, 155] normally predict the concentration in the first trophic level, in producers (phytoplankton, plants), by the bioconcentration factor (BCF) which can be estimated as a function of Kow (bioconcentration by lipid/media partitioning). In aquatic systems, most simply: BCF = Lipid fraction × Kow; Concentration in producer = BCF × Concentration in water. Also, BCF for uptake of chemical by plants from air to leaves/needles and from soil to roots is linearly dependent on lipid content in plant tissue [145]. Bioaccumulation to higher trophic levels (herbivores, carnivores) takes place not only by partitioning, but also by biomagnification uptake via food and elimination by excretion and metabolism [154, 155]. Concentration of the chemical is also decreased due to the dilution effect from the growth of the animal [154]. All these processes can be successfully included in programs of the predictive environmental fate models [146].

3.2 Model Results Compared with Environmental Levels

A simple prediction of environmental fate can be made by a multimedia fugacity model for a hypothetical unit environment of 1 km^2 area [156]. An example of application is prediction of the levels of common organic pollutants HCH, 2,3,4,6-tetrachlorophenol (TeCP), DDT and chlordanes (CHL, sum of chlordane residues, *cis*-chlordane as representative molecule) in Bay of Bothnia fish [157]. The results are listed in Table 1. In level I distribution the importance of water and air media in distribution of HCH and dominance of solids for the others is clearly seen. Level II calculation, including bioconcentration, gave residence times and relative concentrations which could be compared with those observed.

The results give a rough approximation of the exposure of biota as a basis of environmental hazard from bioaccumulation. Deviation of the actual levels indicates that the very lipophilic substances DDT and CHL are accumulated in fish not only by lipid/water partitioning, but also via food. More accurate estimation of concentrations in fish can be obtained from predicted levels in water and suspended solids by biomagnification models including uptake by food

Table 1. Level I and II modelling results for organochlorines in Bay of Bothnia environment [157]

	HCH	TeCP	DDT	CHL
Level I				
% in Air	30.4	8.88	0.45	9.72
% in Water	47.9	18.1	0.46	3.06
% in Sediment	21.6	73.0	99.1	87.2
Level II				
Assumed daily input to km^2 (kg)	0.868	2.00	0.042	0.050
Residence time (days)	37	5	7233	395
Predicted concentration in fish (mg l^{-1})	0.203	0.232	6.08	0.341
Observed level in salmon 1982–85 (ng g^{-1} lw)	4.7	5.0	436	27.4
Relative predicted concentration	1	1.14	30	1.7
Normalized observed level	1	1.06	93	5.8

and elimination processes, e.g. by those of Thomann [154] or Clark et al. [155, 158].

A further modelling of the fate of lindane (LIND = γ-HCH), chlordanes (CHL as above) and toxaphene (TOX or PCC) was performed using the program FATEMOD [147]. It is a modification of the Mackay GENERIC program [146], which contains estimations at levels I, II and III.

The environments were the boreal Bay of Bothnia and a fictious Southern Sea which has sizes and fluxes the same but average temperature (25 instead of 2 C) and organic carbon fractions different (significantly lower) compared to the Bay of Bothnia. The values of emissions for level III estimation were derived by trial modelling to give approximately the same concentrations as observed in air, water or fish at the Bay of Bothnia [159–162]. Then, the same emissions were used in modelling the Southern Sea case. Some modelling results are shown in Table 2. According to the model, bioaccumulation of LIND and CHL was slightly higher and that of TOX about the same in the Bay of Bothnia compared to the "Southern Sea" [147].

Two specific models for estimation of the fate of discharged chemicals in watercourses were tried in a pulp mill recipient in Äänekoski, Central Finland [147]. The model EXWAT was developed for the characterization of the transport and fate of a chemical in surface water bodies at steady state [163, 164]. It is a box model with two compartments: fluid and sediment. The processes considered were 1) deposition and resuspension of suspended matter, 2) partitioning of chemicals between water and suspended matter in the fluid and between pore water and benthic sediment solids, 3) ionization equilibrium, 4) exchange between pore and fluid water as driven by dispersion, 5) sediment burial, 6) volatilization, 7) degradation, and 8) bioconcentration. PPEFF model is a three-segment version of the Quantitative Water-Air-Soil-Interaction (QWASI) fugacity model [146, 165].

Both EXWAT and PPEFF models could be readily applied to the Äänekoski watercourse. For EXWAT, the 18-km long region downstreams from the discharge point can be divided into five segments each containing 1-km long bo-

Table 2. FATEMOD results for Bay of Bothnia compared with similar more southern area and observed average levels in Bay of Bothnia. Ca, Cw, Csed and Cb are the concentrations in air, water, sediment and biota (fish), respectively [147]

Area Compound	Bay of Bothnia			Southern Sea		
	LIND	CHL	TOX	LIND	CHL	TOX
Level I						
% in Air	1.75	0.003	0.0005	1.75	8.64	1.60
% in Water	87.6	6.15	2.03	87.6	14.13	4.95
% in Sediment	10.4	91.6	95.7	10.4	84.3	93.4
Level II						
Res.time h	509	1274	4669	152	338	1264
Level III						
Emission kg h^{-1}						
to air	0.055	0.10	0.005	0.055	0.10	0.005
to water	4.00	0.36	1.355	4.00	0.36	1.335
Ca pg m^{-3}	29.4	64	3.2	66.7	93.0	12.5
Cw ng L^{-1}	2.3	0.25	0.65	0.84	0.15	0.63
Csed ng g^{-1} fw[a]	0.010	0.033	0.78	0.00043	0.015	0.073
Cb ng g^{-1} fw	0.72	3.4	21.6	0.27	2.2	26.7
Observed						
Ca [159]	30	3				
Cw [160]	1.5–2.3					
Cb Salmon [b]	1.72	34.3	134.8			
Cb Trout [b]	0.76	3.12	21.5			

[a] Conc. in fresh sediment; Csed dw is approximated by division by 0.37.
[b] From Paasivirta and Rantio [161] and Paasivirta et al. [162].

xes. The model gives concentrations of the chemical in each of the 18 boxes and in each compartment (water, suspended solids, sediments and biota). In the PPEFF run, a three-box version (Lake Kuhnamo, River Kapeenkoski and Lake Vatia) could be applied.

Three pulp-mill originated chlorophenolics, 2,4,6,-trichlorophenol (246TCP), 3,4,5-trichloroguaiacol (345TCG) and tetrachlorogiaiacol (TeCG) were modelled for two time periods – August 1986 and March 1987 – when their concentrations in discharge and in environmental samples had been intensively analyzed. Examples of the modelled and observed data are presented inTable 3. Assuming that other necessary environment and compound parameters for model were reasonably true independent data, degradation rates were to be fitted by the model to give the best agreement of measured and modelled concentration in water of all 18 boxes.

Model predictions of the concentrations in fish were in fair agreement with observed levels in pike (Table 3). In the case of EXWAT, this might be a coincidence, because lipid partition should cause lower levels (fat percent in pike muscles is only about 0.5), but this was perhaps compensated by food uptake by this predatory species. In the case of PPEFF, various uptake, growth and metabolism mechanisms are included in the model, and the fish concentration result was selected for the "large piscivores" class.

Table 3. Modelled (EXWAT and PPEFF) and observed concentrations in water and in fish (Pike, *Esox lucius*) at the Äänekoski watercourse. Sample places KUH, KAP, VAT and KUU are 2, 7, 15 and 18 km downstream of the discharge. Value of pH of water at all sampling places was 6.5 [147]

Compound	246 TCP		345 TCG		TeCG	
Time	Aug 86	Mar 87	Aug 86	Mar 87	Aug 86	Mar 87
Temp. °C	16	1	16	1	16 1	
k [a]	0.063	0.058	0.087	0.08	0.077	0.070
t1/2 d	11.0	12.0	7.97	8.66	9.00	9.90
Discharge g d^{-1}	227	271	600	718	584	699
Waterflow m^{-3} s^{-1}	60	38	60	38	60 38	
Conc. in water μg l^{-1}						
KUH EXWAT	.035	.062	.093	.162	.057	.090
KUH obs.	*.038*	*.100*	*.085*	*.362*	*.091*	*.104*
KUH PPEFF	.044	.084				
KAP EXWAT	.024	.036	.062	.093	.019	.021
KAP obs	*.014*	*.051*	*.026*	*.050*	*.010*	*.043*
VAT EXWAT	.016	.021	.039	.048	.007	.006
VAT obs	*.019*	*.012*	*.036*	*.057*	*.010*	*.013*
KUU EXWAT	.014	.019	.034	.043	.007	.006
KUU obs	*.017*	*.008*	*.036*	*.026*	*.010*	*.008*
KUU PPEFF	.035	.051				
Conc. in fish ng g^{-1}						
KUH EXWAT	3.75	5.61	6.9	10.2	9.2	9.7
KUH obs.	*6.98*	*4.70*	*8.0*	*9.9*	*2.0*	*3.2*
VAT EXWAT	2.51	3.25	4.3	5.3	3.6	3.0
VAT obs.	*6.99*	*2.52*	*8.0*	*4.0*	*3.0*	*1.5*
KUU EXWAT	2.22	2.94	3.8	4.8	3.3	2.8
KUU obs.	*2.52*	*3.42*	*11.9*	*7.1*	*3.2*	*1.9*
KUU PPEFF	1.48	11.0				

[a] Degradation rate constant (k d^{-1}) in water and solids fitted by the model.

More accurate model estimations of biomagnification are complicated by the dependence of the bioaccumulation process on the lipid/water distribution ratio expressed as log Kow. When log Kow < 5, only partitioning is important. Food chain biomagnification is well predictable for compounds having log Kow 5–7, as shown by comparison of the calculated concentrations in top predators and those observed in the field. When log Kow is > 7, food chain effects are sensitive to the chemical assimilation efficiency and phytoplankton BCF [154].

While the bioconcentration (BCF) factor from the primary producer is fairly well modelled from simple lipid partitioning, the models of biomagnification to higher levels must consider ingestion from food and elimination and dilution by growth mechanisms. One model of food ingestion mechanism assumes that the biomagnification occurs in the organism's tissue after the lipophilic xenobiotic has been transferred there from intestine coassimilated with lipid [166]. However, other laboratory and field studies support an alternative, the fugacity model of Gobas et al. [155, 158], where the intestinal absorption is controlled by the chemical diffusion of the xenobiotic molecule which is leaving the lipid before transfer through the intestinal wall. This model explains not only biomagnification in fish but the biomagnification process from mammalian mother to baby during breast feeding. For example, the PCB concentrations in mother's and embryo's blood lipid are equal at the birth, but then during lactation period (24 months for humans), PCB in infant blood lipid increases by a factor of two, but in mother's blood lipid decreases to one third [158, 167, 168]. Accordingly, baby is on the higher trophic level related to mother with biomagnification rate for PCB as high as 5.5 [158]. This lactation enrichment model could, in addition to diet and poor metabolism suggested, explain the very high bioaccumulation rates of dieldrin, PCB and DDT residues from low concentrations in water to marine cetaceans [169].

Environmentally hazardous chemicals are not only locally discharged (direct emissions) at the geographical region modelled; their long-range transport must also be considered. In particular, some persistent organic pollutants (POPs) occur at significant levels far away from their sources due to atmospheric transport. Arctic POP pollution, bioaccumulation and food chain enrichment is in great part due to global atmospheric transport [170–180]. The process of transport is successfully explained by the model of Wania and

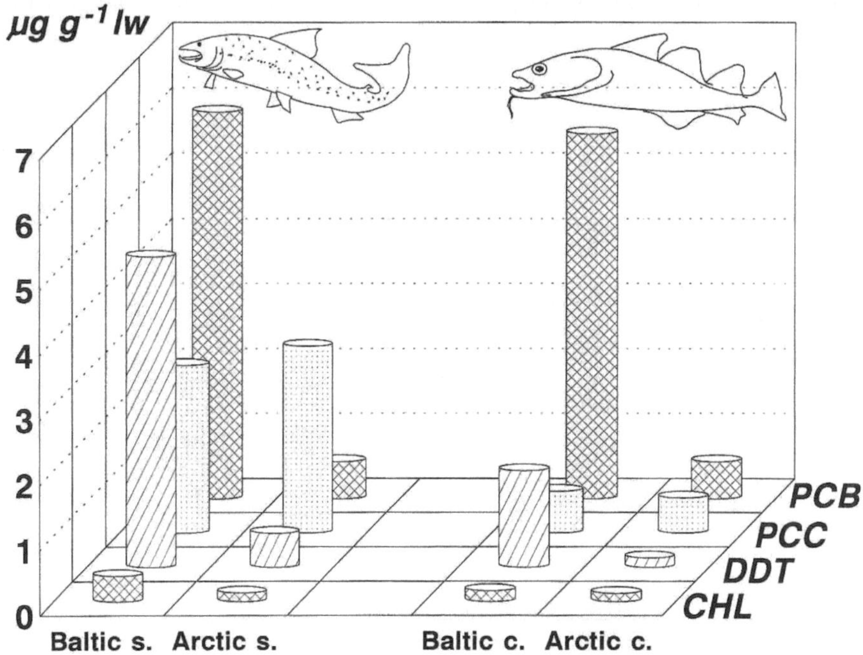

Fig. 3. Contents of polychlorobiphenyls (PCB), toxaphene (PCC), DDT residues and chlordane residues (CHL) in lipid of salmon muscle (s.) from Simojoki River (Baltic) and Teno River (Arctic), and in lipid of cod livers (c.) from Gulf of Finland (Baltic) and Vestertana fjord (Arctic) [175]

Mackay which incorporates the theory of global fractionation and cold condensation [181]. The model explains why toxaphene (PCC) residues in the Arctic ecosystem are at the same level as in more temperate areas [175, 182, 183]. Examples of toxaphene and other chlorohydrocarbon concentrations in Baltic and Arctic fish [183] are illustrated in Fig. 3.

4 Case Studies in the Field

4.1 Aquatic-Terrestrial Food Chain Bioaccumulation

Contents of xenobiotics in food chains of the freshwater lakes in Finland were intensively studied in the 1970s and their trends followed in the 1980s [184–190]. The levels of DDT residues and PCB were low, near background, but the mercury levels were elevated due to industrial discharges which had been stopped in 1968. These xenobiotics all biomagnify significantly (Fig. 4).

Concentrations of PCB, SDDT and mercury in adult fish-eating birds were orders of magnitude higher than in local fish [184, 187]. Study of eggs and juveniles of these bird species eliminated the influence of contaminants collected by adult birds [188]. Residues in eggs were an additional burden to the chick, which collected more biocides from food and diluted them by growth.

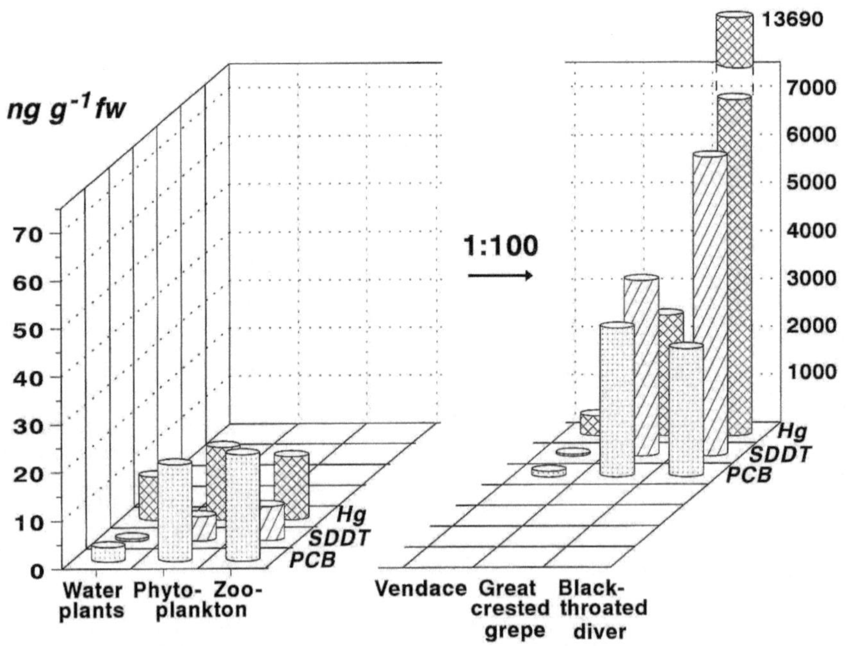

Fig. 4. Average concentrations in different trophic levels of the Lake Päijänne, Finland, in 1972–74 [184–187]

Therefore, instead of concentrations in eggs and juveniles, total amounts in specimens gave a better estimate of bioaccumulation of xenobiotics from food (fish) to the chicks (Fig. 5). During the nesting period the amount of pollutant was increasing in juveniles of two species for SDDT, five species for PCB, and for total mercury (TotHg) in all seven species studied. The relative biomagnification power of different compounds, however, could not be obtained from the amount ratios, but one must consider the different contents in food. A relative enrichment factor (E_{rf}) based on increase of the amount in juveniles was defined as

$$E_{rf} = \frac{A_{juv} - A_{egg}}{C_{ven}} \quad (1)$$

where A_{juv} = amount in juvenile, A_{egg} = amount in egg, and C_{ven} = concentration in vendace (main food of the chicks) [6, 157].

In cases where biomagnification was observed ($A_{juv} > A_{egg}$), variation in E_{rf} values between species were for mercury 0.1–1.14, for DDE 0.74–4.55 and for PCB 0.33–641. This great variation must be due to different food compositions (in addition to vendace) and metabolism of the species.

The method of comparing total amounts of egg and juvenile specimens has been used in a three-step terrestrial food chain study [191]. From 15 organochlorine compounds studied, concentrations of some PCB congeners, *p,p′*-DDE and hexachlorobenzene, indicated the highest biomagnification rates from oak

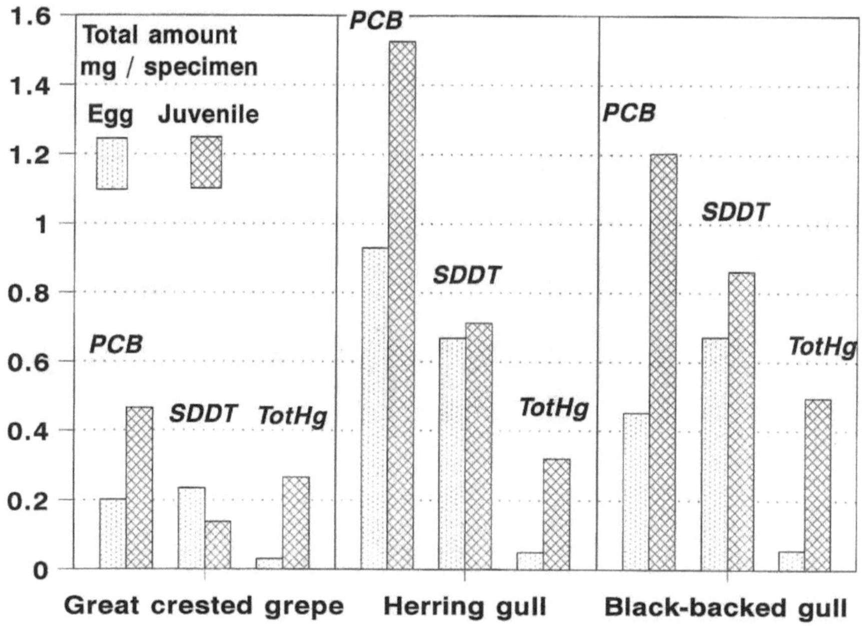

Fig. 5. Average total amounts in eggs and juveniles of three species of fish-eating bird at Lake Päijänne, Finland in 1972–74 [188]

leaves to caterpillars and further to birds (great tit specimens). Bird eggs had higher concentrations than juvenile birds. However, the amounts in eggs were lower than those in juveniles for PCB101, PCB138, PCB153 and PCB180 (numbering according to Ballschmiter et al. [192]), and for *p,p'*-DDE. The most abundant biomagnifying organochlorines in this study, PCB153 and PCB138, have been suggested to be estrogenic agents which might be responsible for impaired sperm mobility and, consequently, for the pollution originated nonfertility in the human population [193].

4.2 Empirical Estimates of Biomagnification

Comparison of contents at different trophic levels on a fresh weight basis is the simplest empirical estimate of biomagnification of a xenobiotic. One way to describe the estimate is to show the portions of the compounds as percentages of the combined contents at each trophic level [190, 194].

Comparison of the content distributions (Fig. 6) indicates the most significant biomagnification to tetrachloroguaiacol (TeCG), total mercury (TotHg, mainly methyl mercury) and hexachlorobenzene (HCBz). Biomagnification of 2,4,6-trichlorophenol, PCB, 4,5,6-trichloroguaiacol and DDT residues is also clear: if lipid weight basis had been used, levels in pike would be significantly higher than those in roach. Only 2,3,4,6-tetrachlorophenol, pentachlorophenol (PeCP) and tetrachlorocatechol (TeCC) showed no biomagnification from

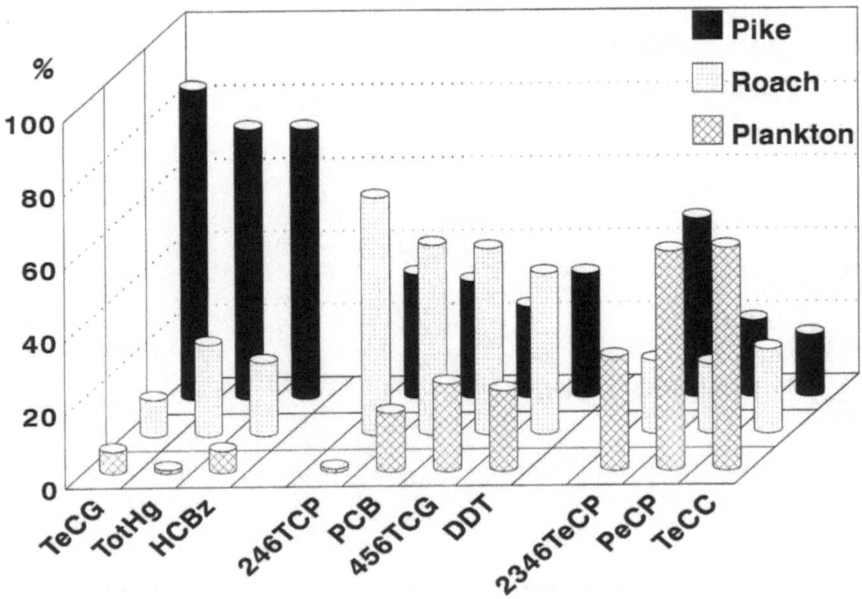

Fig. 6. Distributions of the concentrations in three trophic levels in lakes of central Finland [190, 194]

plankton to roach. In the environment, they were frequently buried in sediment [194].

A four trophic level estimation of biomagnification power can be constructed from concentrations of persistent organic pollutants (POP) analyzed in herring, salmon, seals and eagles in Gulf of Finland and Gulf of Bothnia areas of the Baltic Sea in 1985–1989 [161, 195, 196]. The species in comparison are not exact representatives of the food chain except that Baltic herring is the main food of salmon. However, the averages of the observed concentrations (Table 4) can be used to rank POPs according their environmental hazard. Comparisons of potential hazard as biomagnification power is illustrated for fresh weight (fw) data in Fig. 7 and for lipid weight (lw) data in Fig. 8.

In the above comparisons the PCB, DDT and chlordane (SCHL) residues show high biomagnification. In addition, biomagnification rates of HCBz are relatively high. The dioxin-like toxic POP congeners showed high biomagnification rates in the order PCB169 > PCB126 > 23478PeCDF > PCB77 > 2378TeCDF. They must be considered as serious candidates to cause ecological damages in Baltic biota. The biomagnification of PCB105, HCHs and toxaphene components (PCC) seemed to be low or negligible according to experience of these four species.

Kubiak et al. measured the bioconcentration factors from spottail shriners to Forsters tern as 0.17, 64 and 176 for PBB77, PCB126 and PCB169, respectively [197]. These ratios have very much the same trend as C(eagle)/C(herring) ratios shown in Fig. 8. The high biomagnification of extremely toxic coplanar PCB126 and PCB169 has been demonstrated to form one of the most significant long-

Table 4. Average concentrations ng g^{-1} in Baltic wildlife 1985–1989

Expl. Lipid%	Herring 8.10		Salmon 3.66		Seal 69.8		Eagle 28.9	
	lw	fw	lw	fw	lw	fw	lw	fw
SPCB	1030	83.4	4243	155	254000	177292	848000	245000
SDDT	770	62.4	3254	119	27200	19990	88000	25400
SCHL	43.9	3.56	147	5.38	1100	768	8620	2490
HCBz	57.0	12.4	153	5.60	230	161	2900	838
PCC	545	44.1	2058	75.3	80	55.9	< 10	< 3
α-HCH	111	8.99	70	2.56	90	62.8	< 10	< 3
γ-HCH	71.5	5.79	37	1.35	20	13.9	< 10	< 3
PCB77[a]	1.55	.126	14.7	.538	3.59	2.51	246	71.1
PCB105[a]	17.4	1.41	73.0	2.67	129	90.0	140	40.5
PCB126[a]	.159	.0129	1.75	.0641	2.31	1.61	176	50.9
PCB169[a]	.014	.00113	0.622	.0228	1.30	0.907	68.9	19.9
2378TeCDF	.0041	.000332	0.238	.00871	.199	0.139	< 1	< 0.3
23478PeCDF	.0087	.000705	0.165	.00604	.194	0.135	1.8	0.520

Herring = *Clupea harengus*; Salmon = *Salmo salar*; Seal = grey seal (*Halichoerus grypus*). Eagle = whitetailed eagle (*Haliaetus albicilla*).
SPCB = sum of the main PCB congeners; SDDT = *p,p′*-DDT + *p,p′*-DDE +*p,p′*-DDD.
SCHL = sum of the chlordane residues; PCC = toxaphene.
[a] Ballschmiter numbering [192].

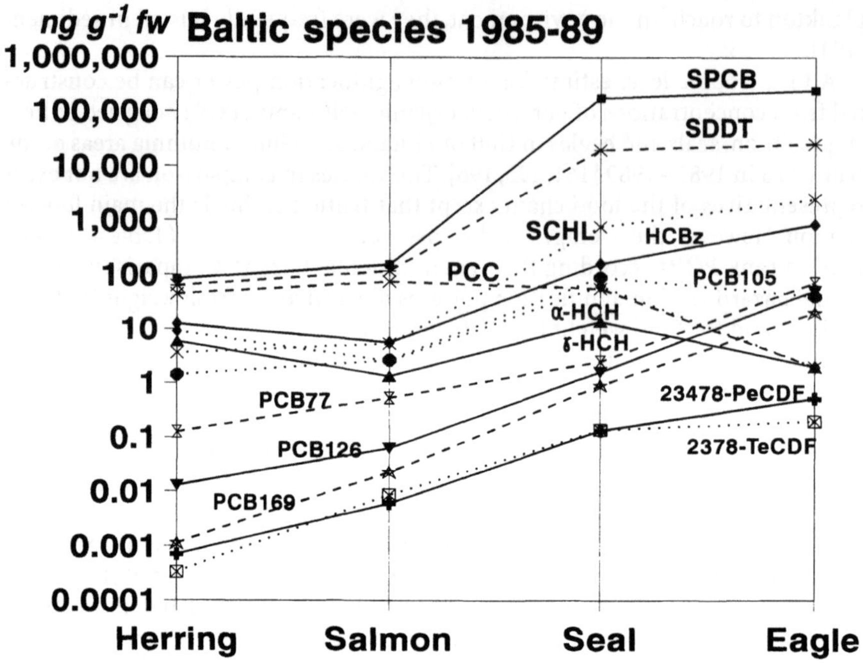

Fig. 7. Average concentrations in fresh muscles of Baltic species shown in logarithmic scale (Table 5)

term hazards to biota globally [198, 199]. They are more persistent than the main PCB congeners and are therefore enriched in the food chain and stay in top consumers longer than PCBs, on average. Although their effect on wildlife cannot fully be isolated from the coaccumulated toxic chloropesticides, PCDDs and PCDFs and other less studied OCCs, there are wide indications of their significant role in reproductive damage, especially among predatory birds [47, 197, 198].

Stable isotope ratios present a novel diagnostic tool to estimate biomagnification, because they are dependent on trophic position of the species. A useful index of trophic level of the organism is the stable nitrogen isotope ratio:

$$\delta^{15}N = [(^{15}N/^{14}N_{sample} \div {}^{15}N/^{14}N_{atmosphere}) - 1] \times 1000 \tag{2}$$

$\delta^{15}N$ is readily measurable by mass spectrometry and it has been found to correlate significantly and positively with food chain length and with concentration of lipophilic xenobiotic. Differences in levels of organochlorine in top predator fish from different waters in the same region can be explained by the different lengths of their food chains measured as $\delta^{15}N$ and supported with observations in the field [177, 200–202].

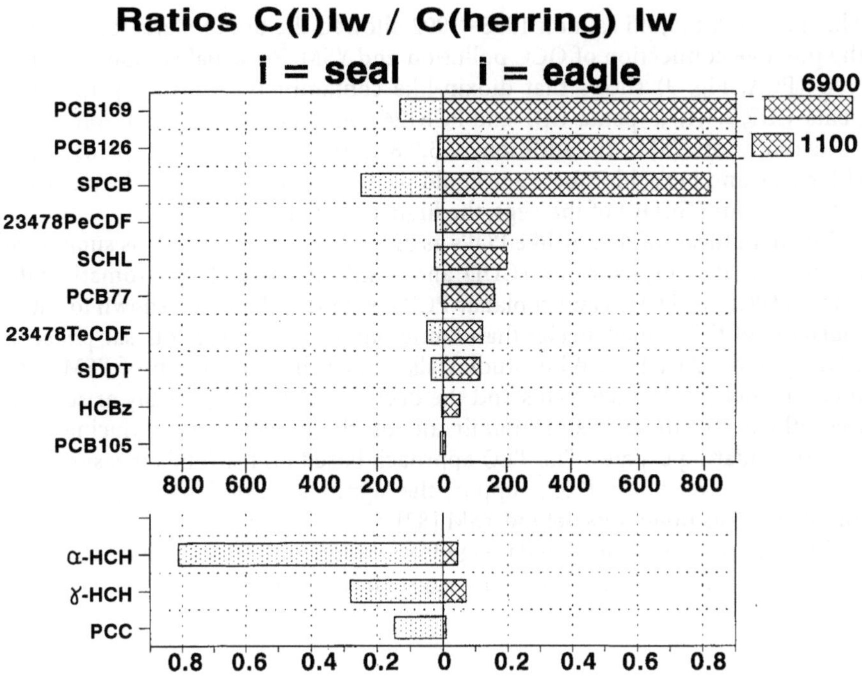

Fig. 8. Biomagnification power estimated as concentrations in lipid of seal and eagle normalized to those in herring (calculated from the data in Table 5)

4.3
Trends of Biomagnified Contaminants and Ecotoxic Effects

Follow-up of time trends is important in assessing possible future hazards of POP pollution. For OCCs in Baltic wildlife, the most useful monitoring data thus far have been obtained from analyses of fish, especially herring and salmon samples. Mother salmon of the natural population breeding in the Simojoki River (Bay of Bothnia, Finland) were analyzed extensively during 1988–1992 [82, 179]. These salmon collect OCCs and other pollutants from their prey during their feeding migration in winter along the Baltic and Gulf of Bothnia. The number of samples was sufficient to detect significant time trends by statistics. Most common OCCs – HCHs, oxychlordane, toxaphene (PCC) and SPCB – showed very significant decreases during 1988–1992. In contrast, dioxin-like toxic PCDDs, PCDFs and coplanar PCBs indicated increasing or no trends.

Colour (GCOL) of the eggs and offspring survival (FERM = fertilizing mortality; ESM = eyed stage mortality and YSM = yolk sac mortality) of the same Simojoki river salmon were also recorded. During the study years, YSM increased dramatically, showing a great upward jump in 1991. Similarly, since 1974 YSM has occurred in many salmon populations spawning in Swedish rivers flowing to the Baltic sea [203], and the epidemics were named M74 syndrome.

The data on Simojoki salmon 1988–1992 allowed the use of statistics to study the possible connection of OCC pollution and YSM. Principal component analysis (PCA; Fig. 9) shows that dioxin-like contaminants, especially 1,2,3,7,8-penta-CDF, 2,3,4,7,8-penta-CDF and PCB126 could have a significant role in the occurrence of YSM. In addition, 2,3,7,8-tetra-CDF, PCB77, PCB169, oxychlordane and hexachlorobenzene could participate in the effect. The paler the orange colour (GCOL) of the eggs, the higher is YSM.

The hypothesis of dioxin-like POPs as an original cause of YSM is supported by the fact that exposure to TCDD and similarly toxic chloroaromatic substances (PCDDs, PCDFs and coplanar PCB congeners) has been shown to cause mortality of fish (other species than Baltic salmon) fry in the yolk sac phase of development [204, 205]. Wide studies also show the connection of YSM with thiamine deficiency [206]. This and the decrease of GCOL possibly arise from the influence of dioxin-like contaminants on the metabolism producing vitamin-degradating ezymes. The TEQ approach based on fish TEFs measured for early juvenile mortality [205] support the significance of 2,3,4,7,8-penta-CDF and PCB126 as pollutants behind YSM [82].

Other known reproductive damage from biomagnification is the extinction of some bird populations due to the eggshell thinning effect, which was asso-

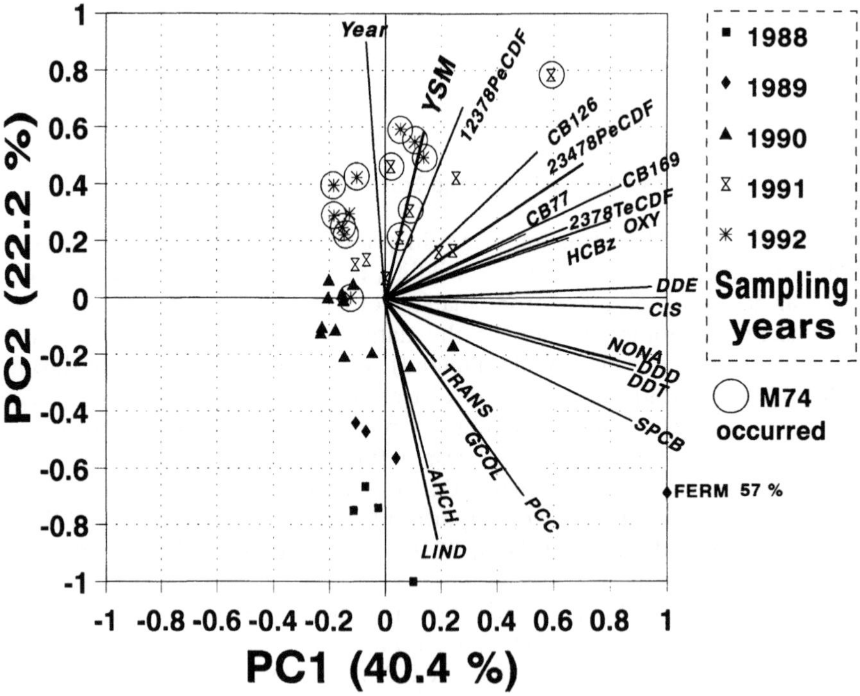

Fig. 9. Biplot from the PCA of concentrations in lipid of 40 samples together with sampling year, egg colour (GCOL) and yolksac mortality. Sample scores are shown as *points*, variable loadings as *vectors*

ciated with DDE [45, 207]. The effect is characterized by an eggshell index (EI) defined by

$$EI = \frac{\text{Weight of the shell (mg)}}{\text{Breadth (mm)} \times \text{Length (mm)}} \tag{3}$$

As an example, DDE contents in the membrane lipids of peregrine falcon (*Falco peregrinus*) eggs collected at different times and from a variety of areas correlate significantly with EI. The linear regression done by this author [6] from data of Peakall and Kiff [208] is illustrated in Fig. 10.

The eggshell thinning effect has been observed as the reason for the low nesting success of Baltic whitetailed eagles in the 1970s [209, 210]. Eggs from the time before DDT usage had EI values of 3.1–3.2, but in the 1970s this was 2.6–2.8. The average DDE concentration in lipid of ten addled eagle eggs collected in 1974–1978 was 452 μg g^{-1} [210]. Further addled eggs (together with whitetailed eagles found dead and their prey) were collected and studied in 1980–1985. These eggs had slighly lower DDE contents and their EIs were a little increased compared to eggs from 1974–1978. Simultaneously, the nesting success of Baltic whitetailed eagles has improved [211, 212].

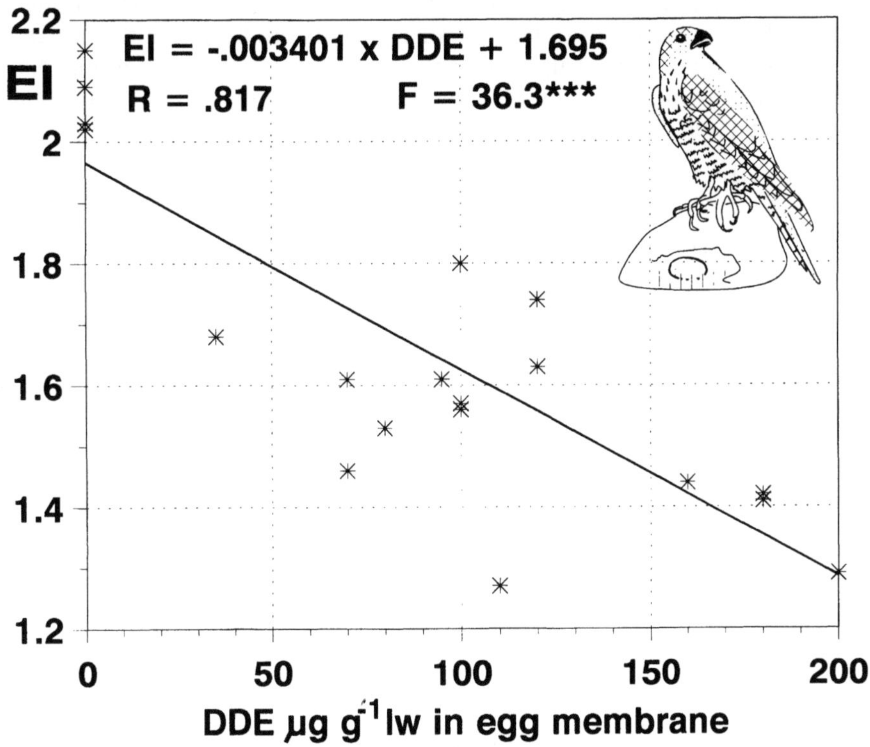

Fig. 10. Dependence of eggshell index (EI) on DDE in membrane lipid of peregrine falcon eggs [6]

Chick oedema disease has been demonstrated in many bird species as result of biomagnification of dioxin-like pollutants [55]. The first symptoms: bloody swelling of embryos and chicks associated with poor hatching success, were observed in herring gull colonies of the lower Great Lakes of North America [54]. Bioaccumulation of chloropesticides, PCBs, PCDDs and PCDFs had led to high concentrations in eggs. Injection and follow-up experiments revealed that 2,3,7,8-tetrachlorodibenzo-*p*-dioxin (TCDD), the most toxic dioxin, was the dominating factor of disease in herring gulls, which were insensitive to chloropesticides and to purified commercial PCB. In eggs collected in the early 1970s from a gull colony in Scotch Bonnett Island, over 1000 ppt (pg g^{-1}) concentrations of TCDD were present. A steep decreasing trend from 1971 to 1981 in TCDD levels, which went down to 50–80 ppt was observed. The chick oedema disease was not seen any more in 1975–1981, when the TCDD level in eggs decreased below 500 ppt [54]. Similar trends of chick oedema disease and levels of TCDD in eggs of the herring gull colonies of Lake Ontario indicate that the bioaccumulation of this single compound was the main cause of breeding damage there.

Analyses of trends and prognosis of future development of bioaccumulated pollutants are important to protect man and the environment against future damage. A predictive method was developed during monitoring of merury in pike of different areas of Lake Päijänne, Finland, after industrial discharge of mercury had been restricted in 1968. From Hg analysis results from different sizes of fish over several years, empirical equations (Eqs.(4)–(6)) were developed by multiple regression for mercury content as function of weight of the fish and year of sampling [213]:

$$\text{TIIRINSELKÄ TotHg} = 0.1094 \times \text{WEIGHT}^{0.327} \times (\text{YEAR} - 1970)^{-0.242} \tag{4}$$

$$\text{RISTINSELKÄ TotHg} = 0.0593 \times \text{WEIGHT}^{0.474} \times (\text{YEAR} - 1970)^{-0.332} \tag{5}$$

$$\text{TEHINSELKÄ TotHg} = 0.1758 \times \text{WEIGHT}^{0.408} \times (\text{YEAR} - 1970)^{-0.525} \tag{6}$$

Validity of the prognosis was investigated in 1981 by collecting ten pike from Ristinselkä, analyzing Hg from their muscles, and calculating by linear regression the observed TotHg for 1-kg pike. The predicted content was 0.70 and the observed was 0.69 mg kg^{-1}. Then the prognosis was revised, adding the new data to previous data to give [190]

$$\text{RISTINSELKÄ TotHg} = 0.0650 \times \text{WEIGHT}^{0.460} \times (\text{YEAR} - 1970)^{-0.331} \tag{7}$$

A second validation was done in 1987 by analyzing another ten pike from Ristinselkä. Predicted TotHg for 1-kg pike then was exactly the same as the observed value, 0.61 mg kg^{-1} [6]. Accordingly, this method seems to be very useful in planning and managing future fishery economics in new reservoirs, which are normally rich in fish but have mercury mobilization and bioaccumulation problems after the first few years [8, 214, 215].

5
Summarizing Conclusions

It is certain that bioaccumulation of persistent chemicals as a consequence of human activities has caused long-term adverse effects in ecosystems. However, in most cases coaccumulation of many toxic substances has made identification of the damaging agents difficult. Cause-effect relationships from bioaccumulation of mercury, lead, cadmium and tin compounds on the local scale have been well demonstrated in numerous cases. Persistent organochlorines (OCC) have obviously caused regional scales of ecological damage, but distinction between the different OCC compounds as major factors behind the damage has not often been achieved. As a result of restriction of release, levels of OCCs in wildlife in many industrialized areas have declined during the 1970s and 1980s. Coinciding with this decline, populations of many wildlife species have been recovered.

References

1. Hutzinger O (ed) (1980) Handbook of environmental chemistry, vol 3 part A. Anthropogenic compounds, Springer, Berlin Heidelberg New York
2. Gavis J, Ferguson JF (1972) Water Res 6:989
3. Johnels AG, Westermark T (1969) Mercury contamination of the environment of Sweden. In: Chemical fallout, chap 10. Thomas, Springfield
4. Lindqvist O, Jernelöv A, Johansson K, Rohde H (1984) Mercury in the Swedish environment, global and local sources. National Swedish Environment Protection Board Report snv pm 1816
5. Environment Agency, Japan (1975) Studies on the Health effects of alkylmercury in Japan
6. Paasivirta J (1991) Chemical ecotoxicology, L-366. Lewis, Chelsea, MI
7. Mitra S (1986) Mercury in the ecosystem. Trans Tech Publications, Aedermannsdorf, Switzerland
8. Potter L, Kidd D, Standiford D (1975) Environ Sci Technol 9:41
9. Brouard D, Demers C, Lalumiere R, Schetagne R, Verdon R (1990) Evolution of mercury levels in fish of the La Grande hydroelectric complex. Quebec. Summary Report, Hydro Quebec (Montreal) and Schooner (Quebec), PQ
10. Rimmer CC (1992) Amer Birds 46:216
11. Nordberg GF, Skerfving S (1972) Metabolism. In: Friberg L, Vostal J (eds) Mercury in the environment. CRC Press, Cleveland, p 29
12. Frank R, Van Hove Holdrinet M, Suda P (1979) Bull Environ Contam Toxicol 22:500
13. Pankakoski E, Koivisto I, Hyvärinen H (1992) Acta Zool Fennica 91:135
14. Wren C, MacCrimmon H, Frank R, Suda P (1980) Bull Environ Contam Toxicol 25:100
15. Wren CD (1986) Environ Res 40:210
16. Hutzinger O (ed) (1980) Handbook of environmental chemistry, vol 1, part A. The natural environment and biogeochemical cycles. Springer, Berlin Heidelberg New York
17. Bull KR, Avery WJ, Freestone P, Hall JR, Osborn D, Cooke AS, Stowe T (1983) Environ Pollut 31A:239
18. Osborn D, Eney WJ, Bull KR (1983) Environ Pollut 31A:261
19. Wilson KW, Head PC, Jones PD (1986) Wat Sci Tech 18:171
20. Ramade F. (1987) Ecotoxicology. Wiley, Chichester, UK
21. Eisler R (1988) Lead hazards to fish, wildlife, and invertebrates: a synoptic review. Biological Report 85(1.14), US Fish and Wildlife Service, Laurel, MD

22. Dieter MP, Finley MT (1979) Environ Res 19:127
23. Blus LJ, Henny CJ, Hoffman DJ, Grove RA (1993) Ecotoxicology 2:139
24. Jörgensen SS, Willems M (1987) Ambio 16:11
25. Ma W (1989) Arc Environ Contam Toxicol 18:617
26. Hodson PV, Whittle DM, Wong PTS, Borgmann U, Thomas RL, Chau YK, Nriagu JO, Hallett DJ (1984) Lead contamination of the Great Lakes and its potential effects on aquatic biota. In: Ngiaru JO, Simmons MS (eds) Toxic contaminants in the Great Lakes. Wiley, New York, p 335
27. Kägi JHR, Nordberg M (eds) (1979) Metallothionein. Birkhäuser, Basel
28. Spehar RL, Anderson RL, Fiandt JT (1978) Environ Pollut 15:195-208
29. Ray S, McLeese DW, Burridge LE (1981)Mar Pollut Bull 12:383
30. Bryan GW (1976) Some aspects of heavy metal tolerance in aquatic organisms. In: Lockwood AP (ed) Effects of pollutants on aquatic organisms. Cambridge University Press, Cambridge, pp 7-34
31. Cadd GM (1993) FEMS Microbiol Rev 11:297
32. Ward GS, Gramm GC, Parrish PR, Trachman H, Schlesinger A (1981) ASTM Spec Tech Publ 737:183
33. Tsuda T, Nakanishi H, Aoki S, Takebayashi J. (1986) Toxicol Environ Chem 12:137
34. Zuolian C, Jensen A (1989) Mar Pollut Bull 20:281
35. Langston WJ, Burt GR (1991) Mar Environ Res 32:61
36. Cosson R, Amiard-Triquet C, Grandier-Vazeille X (1989) Oceanis 15:411
37. Bryan GW, Gibbs PE, Burt GR, Hummerstone LG (1987) J Mar Biol Assoc 67:525
38. Triebskorn R, Koehler H-R, Flemming J, Braunbek T, Negele R-D, Rahmann H (1994) Aquat Toxicol 30:189
39. Saint-Louis R, Pelletier E, Marsot P, Fournier R (1994) Water Res 28:2533
40. Yamada H, Tateishi M, Takayanagi K (1994) Environ Toxicol Chem 13:1415
41. Lewis TE (1989) Environmental chemistry and toxicology of aluminium,.Lewis, Chelsea, MI
42. Gribble GW (1992) J Nat Prod 55:1353
43. Grimwall A, de Leer EBW (eds) (1995) Naturally-produced organohalogens. Kluwer, Dordrecht
44. Paasivirta J (1988) Wat Sci Tech 20:119
45. Ratcliffe DA (1967) Nature 215:208
46. Smith AG (1991) Chlorinated hydrocarbon insecticides. In: Hayes WJ, Laws ER (eds) Handbook of pesticide toxicology classes of pesticides 2:731
47. Detzell E, Doull J, Giesy J, Mackay D, Munro I, Williams G (1994) Interpretative review of the potential adverse effects of chlorinated organic chemicals on human health and the environment. Regul Toxicol Pharmacol 20, part 2, pp 1-1056
48. Courtney KD (1979) Environmental Research 20:225
49. Ahling B, Brjørseth (1978) Chemosphere 7:799
50. Jensen S (1966) New Scientist 32:612
51. Bowes GW, Simoneit BR, Burlingame AL, de Lappe BW, Risebrough RW (1973) Environ Health Persp 5:191-198
52 Helle EM, Olsson M, Jensen S (1976) Ambio 5:261
53 Kihlström JE, Olsson M, Jensen S (1976) NORDFORSK Milövårdsekreteriatet Publikation 2:567
54 Gilbertsson M (1983) Chemosphere 12:357
55. Firestone D (1973) Environ Health Persp 5:59
56. Vos JG, Koeman JH, van der Maas LH, ten Noever de Brauw MC, Vos RH (1970) Food Cosmet Toxicol 8:625-633
57. Buser RH, Bosshardt HP, Rappe C (1978) Chemosphere 7:109
58. Paasivirta J, Herzschuh R, Humppi T, Kantolahti E, Knuutinen J, Lahtiperä M, Laitinen R, Salovaara J, Tarhanen J, Virkki L (1985) Envir Health Persp 60:269
59. Sandermann W, Stockmann H, Casten R (1957) Chem Ber 90:690
60. Kimmig J, Schulz KH (1957) Dermatologia 115:540

61. Hutzinger O, Fiedler H (1989) Chemosphere 18:23
62. Rappe C, Andersson R, Bergqvist P-A, Brohede C, Hansson M, Kjeller L-O, Lindström G, Marklund S, Nygren M, Swanson SE, Tysklind M, Wiberg K (1987) Chemosphere 16:1603
63. Hutzinger O, Frei RW, Merian E, Pocchiari F (eds) (1982), Chlorinated dioxins and related compounds. Impact on the environment. Pergamon, Oxford
64. Poland A, Clover E, Kende AS (1976) J. Biol. Chem. 251:4936
65. Conney AH, Kuntzman R (1971) Metabolism of normal constituents by drug-metabolizing enzymes in liver microsomes, In: Brodie BB, Gillette J (eds) Concepts in biochemical pharmacology. Springer, Berlin Heidelberg New York, p 401
66. McKinney JD (1981) Environmental health chemistry. Ann Arbor Science Publ, Ann Arbor, MI
67. Tanabe S (1988) PCB problems in the future: foresight from current knowledge. Environ Pollut 50:5-28
68. Kimbrough RD, Jensen AA (eds) (1989) Halogenated biphenyls, terphenyls, naphthalenes, dibenzodioxins and related products, 2nd edn. Elsevier, Amsterdam
69. Ueda S, Fujiwara A, Umemura T, Tokuda H (1995) Organohalogen Compounds 21:209
70. Safe S (1987) Chemosphere 16:791
71. Gillner M, Fernström B, Gustafsson J-Å, Bergman J, Cambillau C (1986) Chemosphere 15:1673
72. Barnes DG, Bellin J, Cleverly D (1986) Chemosphere 15:1895
73. Nordisk Expertgrupp (1988) Nordisk Dioxinriskbedömning. Nordisk Ministerråd. NORD 49
74. McKinney JD, Singh P (1981) Chem-Biochem Interact 33:271
75. McKinney JD, Chae K, McConnell EE, Birnbaum LS (1985) Ann NY Acad Sci 320:179
76. McKinney JD, Chae K, McConnell EE, Birnbaum LS (1985) Environ Health Persp 20:57
77. Safe S (1984) CRC Crit Rev Toxicol 13:319
78. Safe S (1992) Chemosphere 25:61
79. Ahlborg UG, Brouwer A, Fingerhut MA, Jacobson JL, Jacobson SW, Kennedy SW, Kettrup AAF, Koeman JH, Poiger H, Rappe C, Safe SH, Seegal RF, Tuomisto J, van der Berg M (1992) Eur J Pharmacol - Envir Toxicol Pharmacol Section 228:179
80. Ahlborg U, Becking GC, Birnbaum RS, Brouwer A, Derks HJGM, Feeley M, Golor G, Hanberg A, Larsen JC, Liem AKD, Safe SH, Schlatter C, Waern F, Younes M, Yrjänheikki E (1994) Chemosphere 28:1049
81. Bergqvist P-A, Bergek S (1989) Chemosphere 19:513
82. Paasivirta J, Vuorinen PJ, Vuorinen M, Koistinen J, Rantio T, Hyötyläinen T, Welling L (1995) Organohalogen Compounds 25:355
83. Koistinen J, Koivusaari J, Nuuja I, Paasivirta J, (1995) Chemosphere 30:1671
84. Jakobsson E (1994) Synthesis and analysis of chlorinated naphthalenes. Biological and environmental implications. PhD thesis, Stockholm University
85. Hanberg A, Waern F, Asplund L, Haglund E, Safe S (1990) Chemosphere 20:1161
86. Engwall M, Jakobsson E, Brunström B (1994) Arch Toxicol 68:37
87. De Boer J (1995). Analysis and biomonitoring of complex mixtures of persistent halogenated micro-contaminants. PhD thesis, Amsterdam University. DLO-Netherlands Institute for Fisheries Research, The Netherlands
88. Hale RC, Graeves J, Gallagher K, Vadas GG (1990) Environ Sci Technol 24:1727
89. Freudenthal J, Greve PA (1973) Bull Environ Contam Toxicol 10:108
90. Zitko V, Hutzinger O, Jamieson WD, Choi PMK (1973) Bull Environ Contam Toxicol 10:200
91. Renberg L, Sundström G (1978) Chemosphere 6:477
92. Jensen AA, Jörgensen KF (1983) Sci Total Environ 27:231
93. Kaminsky R, Hites RA (1984) Environ Sci Technol 18:275
94. Griffiths RW (1991) Hydrobiologia 219:143
95. Chu I, Villeneuve DC, Secours VE, Valli VE, Leeson S, Shen SY (1986) Fundam Appl Toxicol 6:69
96. Oliver BG (1987) Environ Sci Technol 21:785

97. Oliver BG, Niimi AJ (1988) Environ Sci Technol 22:388
98. De Boer J (1989) Chemosphere 18:2131
99. Pereira WE, Rostad CE, Chiou CT, Brinton TI, Barber LB II, Demcheck DK, Demas CR (1988) Environ Sci Technol 22:772
100. USEPA (1989) Hexachlorobutadiene: drinking water health advisory. NTIS Report PB91-160663
101. Bjørseth A, Lunde G, Gjs N (1977) Acta Chem Scand B 31:797
102. Kuokkanen T (1989) Chlorocymenes and chlorocymenenes: persistent chlorocompounds in spent bleach liquors of kraft pulp mills, Deparment of Chemistry, University of Jyväskylä, Research Report 32
103. Rantio T (1992) Chemosphere 25:505
104. Paasivirta J, Koistinen J, Kuokkanen T, Maatela P, Mäntykoski K, Paukku R, Rantalainen A-L, Rantio T, Sinkkonen S, Welling L (1993) Chemosphere 27:447
105. Rantio T, Koistinen J, Paasivirta J (1996) Bioaccumulation of pulp chlorobleaching-originated aromatic chlorohydrocarbons in recipient watercourses. In: Servos MR, Munkittrick KR, Carey JH, Van der Kraak GJ (eds) Environmental fate and effects of pulp and paper mill effluents. Sr. Lucie Press, Delray Beach, FL, p 341
106. Nevalainen T, Koistinen J (1990) Organohalogen componds 4:361
107. Nevalainen T (1995) Synthesis of chlorinated alkylbibenzyls and 9-chlororetene, Chemosphere 30:847
108. Koistinen J, Nevalainen T, Tarhanen J (1992) Environ Sci Technol 26:2499
109. Koistinen J (1992) Chemosphere 24:559
110. Koistinen J, Paasivirta J, Lahtiperä M (1993) Chemosphere 27:149–156
111. Koistinen J, Paasivirta J, Nevalainen T, Lahtiperä M (1994) Chemosphere 28:1261–1277
112. Hodson PV (1996) Mixed function oxygenase induction by pulp mill effluents: advances since 1991. In: Servos MR, Munkittrick KR, Carey JH, Van der Kraak GJ (eds), Environmental fate and effects of pulp and paper mill effluents, Sr. Lucie Press, Delray Beach, FL, p 349
113. Bjørseth A, Carlberg GE, Møller M (1979) Sci Total Environ. 11:197
114. Koistinen J, Paasivirta J, Nevalainen T, Lahtiperä M (1994) Chemosphere 28:2139
115. Paasivirta J, Tarhanen J, Soikkeli J (1986) Chemosphere 15:1429
116. Paasivirta J, Koistinen J (1994) Chlorinated ethers. In: Kiceniuk JW, Ray S (eds) Analysis of contaminants in edible aquatic resources, chap 6. Book II: Organics. VCH Publ, New York, p 411
117. Curtis RF, Land DG, Griffiths NM, Gee M, Robinson D, Peel JL, Dennis C, Gee JM (1972) Nature 235:223–224
118. Paasivirta J, Klein P, Knuutila M, Knuutinen J, Lahtiperä M, Paukku R, Veijanen A, Welling L, Vuorinen M, Vuorinen PJ (1987) Chemosphere 16:1231
119. Paasivirta J, Rantalainen A-L, Welling L, Herve S, Heinonen P (1992) Wat Sci Tech 25:105–113
120. Schreitmüller J, Ballschmiter K (1995) Environ Sci Technol 28:207
121. Firestone D, Ress J, Brown NL, Barron RP, Damico JN (1972) J Assoc Offic Anal Chem 55:85
122. Jensen S, Renberg L (1972) Ambio 1:1–4
123. Paasivirta J, Lahtiperä M, Leskijärvi T (1982) Experiences of structure analyses of chlorophenol dimers and trimers found in different samples. In: Hutzinger O, Frei RW, Merian E, Pocchiari F (eds) (1982) Chlorinated dioxins & related compounds. Impact on the environment. Pergamon, Oxford, p 191
124. Humppi T (1985) Synthesis, identification and analysis of dimeric impurities of chlorophenols. Department of Chemistry, University of Jyväskylä, Research Report 23
125. Humppi T, Knuutinen J, Paasivirta J (1984) Chemosphere 13:1235
126. Humppi T (1985) Chemosphere 14:523
127. Paasivirta J, Tarhanen J, Juvonen B (1987) Chemosphere 16:1787
128. Paasivirta J, Heinola K, Humppi T, Karjalainen A, Knuutinen J, Mäntykoski K, Paukku R, Piilola T, Surma-Aho K, Tarhanen J, Welling L, Vihonen H, Särkkä, J (1985) Chemosphere 14:469

129. Humppi T, Heinola K (1985) J.Chromatogr 331:410
130. Sundström G, Hutzinger O, Safe S (1976) Chemosphere 5:267
131. Nevalainen.T (1995) Polychlorinated diphenyl ethers: synthesis, NMR spectroscopy, structural properties, and estimated toxicity. Department of Chemistry, University of Jyväskylä, Research Report 50
132. Nevalainen T, Koistinen J, Nurmela, P (1994) Environ Sci Technol 28:1341
133. Hu J, Kolehmainen E, Nevalainen T, Kauppinen R (1994) Chemosphere 28:1069
134. Kurz J, Ballschmiter K (1995) Fresenius J Anal Chem 351:98
135. Becker M, Phillips T, Safe S (1991) Toxicol Environ Chem 33:189
136. Koistinen J, Vuorinen PJ, Paasivirta J (1993) Chemosphere 27:2365
137. Koistinen J, Paasivirta J, Suonperä M, Hyvärinen H (1995) Environ Sci Technol 29:2541
138. Miyazaki T, Kaneko S, Horii S, Yamagishi T (1981) Bull Environm Contamin Toxicol 26:577
139. Atlas E, Sullivan K, Giam CS (1986) Atm. Environ. 20:1217
140. Grimwall A, Borén H (1995) Preprints of papers presented at the 210th ACS National Meeting, Chicago, IL, August 20–24, 1995, 35(2):294
141. Anonymous (1975) Feed contaminant in farmers' blood. Chem Eng News 24:7
142. Sundström G, Hutzinger O, Safe S (1976) Chemosphere 5:11
143. Andersson Ö, Blomkvist B (1981) Chemosphere 10:1051
144. Suter GW II.(ed) (1993) Ecological risk assessment. L785. Lewis, Chelsea, MI
145. Calamari D (ed) (1993) Chemical exposure prediction. Lewis Publ, Roca Baton, FL
146. Mackay D (1991) Multimedia environmental models. The fugacity approach. L-242. Lewis, Chelsea, MI
147. Paasivirta J, Sinkkonen S, Rantio T, Calamari D, Di Guardo A, Matthies M, Trapp S (1995) Modelled and observed fate of selected organochlorines in the Nordic environment. In: Munawar M, Luotola M (eds) The contaminants in the Nordic ecosystem: the dynamics, processes and fate. SPB Academic Publ, Amsterdam, pp 11–24
148. OECD (1981) Guidelines for testing of chemicals. OECD, Paris
149. Lyman WJ, Reehl WF, Rosenblatt DH (1990) Handbook of chemical property estimation methods. American Chemical Society, Washington, DC
150. Hermens JLM, Oppenhuizer A (eds) (1991) QSAR in environmental toxicology, Elsevier, Amsterdam
151. ECETOC (1992) Estimating environmental concentrations of chemicals using fate and exposure models. European Centre for Ecotoxicology and Toxicology of Chemicals, Technical Report 50, Brussels
152. Trapp S, Matthies M (1996) Dynamik von Schadstoffen-Umweltmodellierung mit CemoS, Springer, Berlin Heidelberg New York
153. Baumgarten G, Reiter B, Scheil S, Schwartz S, Wagner J-O (1995) CemoS Handbuch für ProgrammeVersion 1.0. University of Osnabrück, Germany
154. Thomann, RV (1989) Environ Sci Technol 23:699
155. Clark KE, Gobas FAPC, Mackay D (1990) Environ Sci Technol 24:1203
156. Neely WB, Mackay D (1982) Evaluative model for estimating environmental fate. In: Dickson KL, Maki AW, Cairns J (eds) Modeling the fate of chemicals in the aquatic environment. Ann Arbor Science Publ, Ann Arbor, MI
157. Paasivirta J (1990) Organohalogen Compounds 1:367
158. Gobas FAPC, McCorquodalle JR, Haffner GD (1993) Environ Toxicol Chem 12:567–576
159. Bidleman TF, Wideqvist U, Jansson B, Söderlund R (1987) Atmos Environ 21:641–654
160. Gaul H (1992) ICES Mar Sc Symp 195:110
161. Paasivirta J, Rantio T (1991) Chemosphere 22:47
162. Paasivirta J, Rantio T, Koistinen J, Vuorinen PJ (1993) Chemosphere 27:2011
163. Brüggemann R, Münzer B (1987) EXWAT Multicompartiment Modell für den Transport von Stoffen in Oberflächenwässern, GSF-Bericht 33/87, Neuherberg
164. Matthies M, Trapp S (1988). Environmental models for exposure and hazard assessment of chemicals. In: Lokke H, Tyle H, Bro-Rasmussen F (eds) 1st European Conference on Ecotoxicology, October 17–19, 1988, Copenhagen, Denmark, p 420

165. Mackay D, Southwood JM (1992) Modelling the fate of organochlorine chemicals in pulp mill effluents. Report of the Pulp and Paper Centre of the University of Toronto
166. Vetter RD, Carey MC, Patton JS (1985) J Lipid Res 26:428
167. Masuda Y, Kagawa R, Kuroki H, Kuratsune M, Yoshimura T, Take I, Kusuda M, Yamashita F, Hayashi M (1978) Food Cosmet Toxicol 16:543
168. Kodama H, Ota H (1980) Arch Environ Health 35:95
169. Morris RJ, Law RJ, Allchin CR, Kelly CA, Fileman CF (1989) Mar Pollut Bull 20:512
170. Ballschmiter K (1992) Angew Chem Int Edn Engl 31:487
171. Pacyna J, Oehme M (1988) Atmos Environ 22:243.
172. Patton GW, Hinkley DA, Walla MD, Bidleman TF (1989) Tellus 41B:243
173. Patton GW, Walla MD, Bidleman TF, Barrie LA (1991) J Geophys Res 96D:10867
174. Hargrave BT, Vass WP, Erickson PE, Fowler BR (1988) Tellus 40B:480
175. Gregor DJ, Gummer W (1989) Environ Sci Technol 23:561
176. Bidleman TF, Patton GW, Walla MD, Hargrave BT, Vass WP, Erickson PE, Fowler BR, Scott V, Gregor DJ (1989) Arctic 42:307
177. Muir DCG, Norstrom RJ, Simon M (1988) Environ Sci Technol 22:1071
178. Muir DCG, Wagemann R, Hargrave BT, Thomas DJ, Peakall DB, Norstrom RJ (1992) Sci Total Environ 122:75
179. Paasivirta J, Koistinen J, Rantio T, Vuorinen PJ (1994) Organohalogen Compounds 20:529
180. Bright DA, Duschenko WT, Grundy S, Reimer KJ (1995) Sci Tot Environ 160/161:265
181. Wania F, Mackay D (1993) Ambio 22:10
182. Wania F, Mackay D (1993) Chemosphere 27:2079
183. Paasivirta J, Rantio T, Koistinen J, Vuorinen PJ (1993) Chemosphere 27:2011
184. Särkkä J, Hattula M-L, Paasivirta J, Janatuinen J (1978) Holarctic Ecol 1:326
185. Hattula M-L, Janatuinen J, Särkkä J, Paasivirta J (1978) Environ Pollut 15:121
186. Hattula M-L, Särkkä J, Janatuinen J, Paasivirta J, Roos A (1978) Environ Pollut 17:19
187. Särkkä J, Hattula M-L, Janatuinen J, Paasivirta J, Palokangas R (1978) Pestis Monit J 12:26
188. Paasivirta J, Särkkä J, Pellinen J, Humppi T (1981) Chemosphere 10:787
189. Paasivirta J, Särkkä J, Aho M, Surma-Aho K, Tarhanen J, Roos A (1981) Chemosphere 10:405
190. Paasivirta J, Särkkä J, Surma-Aho K, Humppi T, Kuokkanen T, Marttinen M (1983) Chemosphere 12:239
191. Winter S, Streit B (1992) Chemosphere 24:1765
192. Ballschmiter K, Bacher R, Mennel A, Fischer R, Riehle U, Swerev M (1992) J High Resolut Chromatogr 15:260–270
193. Bush B, Bennett AH, Snow JT (1986) Arch Environ Contam Toxicol 15:333
194. Paasivirta J, Särkkä J, Leskijärvi T, Roos A (1980) Chemosphere 9:441
195. Bergqvist P-A, Bergek S (1989) Chemosphere 19:513
196. Paasivirta J, Paukku R (1989) Chemosphere 19:1551
197. Kubiak TJ, Harris HJ, Smith LM, Scwartz TR, Stalling DL, Trick JA, Silfo L, Docherty DE, Erdman TC (1989) Arch Environ Contam Toxicol 18:706
198. Tanabe S, Kannan N, Subramanian An, Watanabe S, Tatsukawa R (1987) Environ Pollut 47:147–163
199. Tarhanen J, Koistinen J, Paasivirta J, Vuorinen PJ, Koivusaari J, Nuuja I, Kannan N, Tatsukawa R (1989) Chemosphere 18:1067
200. Schindler DW, Kidd KA, Muir DCG, Loclkhart WL (1995) Sci Tot Environ 160/161:1
201. Kidd KA, Schindler DW, Muir DCG, Lockhart WL, Hesslein RH (1995) Science 269:240
202. Rasmussen JB, Rowan DJ, Lean DRS, Carey JH (1990) Can J Fish Aquat Sci 47:2030
203. Norrgren L, Andersson T, Bergqvist P-A, Björkelund I (1993) Environ Toxicol Chem 12:2065
204. Helder Th (1982) Effects of 2,3,7,8-tetrachlorodibenzo-*p*-dioxin (TCDD) on early life stages of two fresh-water fish species. In: Hutzinger O, Frei RW, Merian E, Pocchiari F (eds) Chlorinated dioxins and related compounds, impact on the environment. Pergamon, Oxford, pp 455–462

205. Walker MK, Peterson RE (1991) Aquat Toxicol. 21:219
206. Naturvårdsverket (1995) Second Workshop on Reproduction Disturbances in Fish, 20–23 November 1995, Stockholm, Sweden, Mimeograph
207. Ratcliffe DA (1970) J Appl Ecol 7:67
208. Peakall DB, Kiff LF (1979) Ibis 121:200–204
209. Odsjö T, Johnels AG (1972) Äggskalförtunning hos några svenska fågelarter. Report to the National Swedish Environmental Protection Board 1972-2-14, Stockholm (mimeographed)
210. Koivusaari J, Nuuja I, Palokangas R, Finnlund M (1980) Environ Pollut 23:41
211. Paasivirta J, Kääriäinen H, Paukku R, Surma-Aho K, Koivusaari J, Nuuja I, Palokangas R (1985) Chlorohydrocarbons and mercury in whitetailed eagles, their eggs and food. In: Paasivirta O (ed) Ministry of Environment, Finland, Division of Environmental Protection, Publication A3, p 27 (in Finnish)
212. Koivusaari J, Nuusa I, Paasivirta J, unpublished data
213. Paasivirta J, Särkkä J, Aho M, Surma-Aho K, Tarhanen J, Roos A (1981) Chemosphere 10:405
214. Abernathy AR, Cumbie PM (1977) Bull Environ Contam Toxicol 17:595
215. Surma-Aho K, Paasivirta J, Rekolainen S, Verta M (1986) Chemosphere 15:353–372

The Assessment of Bioaccumulation

Bernd Beek, Stella Böhling, Ursula Bruckmann, Christian Franke, Ulrich Jöhncke, Gabriele Studinger

Federal Environmental Agency, Seecktstraße 6–10, D-13581 Berlin, Germany

Bioaccumulation and biomagnification of chemicals in biota may be a prerequisite for adverse effects in individuals, species, and ecosystems. From disastrous events posed by xenobiotic chemicals, e.g. PCBs, Dioxins, DDT etc. it must be concluded retrospectively that such impacts cannot be avoided and predicted sufficiently with existing hazard and risk assessment strategies. Even sophisticated testing for chronic effects cannot rule out a possible risk of retarded effects completely. Since adverse effects as a consequence of bioaccumulation may become obvious long after a chemical's release and recovery may be retarded if not even hampered, authorities concerned with notification and registration of chemicals need a conceptual approach how to minimise risks posed by dangerous substances. Different concepts for the assessment of bioaccumulation (USA, Canada, Japan, Netherlands, ECETOC and EU) are critically discussed and compared regarding their precautionary principles. The risk assessment for bioaccumulation presented here is more comprehensive than the EU Technical Guidance Document (TGD) for new and existing substances. It gives guidance how to proceed stepwise from testing bioaccumulation, ranking of results, decision-making on the basis of triggered ecotoxicological tests and finally to an assessment of risks for bioaccumulation and biomagnification. Going beyond the scope of existing concepts this approach takes into account the complexity of bioaccumulation processes including uptake and depuration kinetics, bioconcentration factor, metabolism, and bound residues, relating these data to critical body burden concentrations. The risk assessment of biomagnification is driven by the outcome of the bioaccumulation assessment. If following the refined risk assessment recommended by the TGD an uncertain risk of biomagnification in ecosystems cannot be ruled out, the application of an unsafety factor of 10 on the final PEC/PNEC is proposed for discussion.

Keywords: Assessment, Bioaccumulation, Biomagnification, Concepts, Secondary Poisoning.

The Handbook of Environmental Chemistry, Vol. 2 Part J
Bioaccumulation (ed. by B. Beek)

Abbreviations

APEO	alkylphenolethoxylate; alkylphenol polyglycol ether
BAF	bioaccumulation factor
BAP	bioaccumulation potential; indicator of a risk of bioaccumulation in living organisms due to the physico-chemical and structural properties of a substance
BCF	bioconcentration factor
BM	biomagnification
BMP	biomagnification potential
CBB_{food}	critical body burden concentration for food in the organisms
CEPA	Canadian Environmental Protection Act
CMC	critical micelle concentration
ct_{50}	half-life clearance time, i.e. the time needed to reach 50% removal
DDD	a main metabolite of DDT
DDE	a main metabolite of DDT
DDT	dichlorodiphenyltrichloroethan
ECETOC	European Chemical Industry Ecology & Toxicology Centre
EU	European Union
HCB	hexachlorobenzene
HCH	hexachlorocyclohexane
α, β, γ and δ-HCH	isomeres of hexachlorocyclohexane
γ-HCH	Lindane
K_{OC}	partion coefficient organic carbon/water
K_{OW}	n-octanol/water partition coefficient; synonym of P_{OW}
LAS	linear alkylbenzene sulfonate
LC_{50}	lethal concentration for 50% of a population
log K_{OC}	logarithmic form of K_{OC}
log K_{OW}	K_{OW} in its logarithmic form
log K_P	logarithmic form of partition coefficient for a compartment, e.g. sediment. K_P is the product of K_{OC} and the weight fractions of organic carbon solids for the respective compartment
MITI	Ministry of International Trade and Industry Japan
MW	molecular weight
NOEC	no observed effect concentration
NOEL	no observed effect level
OECD	Organisation of Economic Co-operation and Development
OTS	Office of Toxic Substances
PAH	polycyclic aromatic hydrocarbons
PCB	polychlorinated biphenyls
PCDD	polychlorinated dibenzo-*p*-dioxins
PCDF	polychlorinated dibenzofurans
PCP	pentachlorophenol
PEC	predicted environmental concentration

PEC/PNEC	ratio triggering tests and risk reduction measures, respectively
PEC_{oral}	predicted environmental concentration of the prey; product of a NOEL and tiered safety factors and conversion factors
PEC_{reg}	predicted regional environmental concentration
PEC_{water}	predicted concentration of a substance in water, exposure concentration
pH	negative common logarithm of the hydrogen activity
pK	negative common logarithm of the constant for a chemical reaction at equilibrium
pKa	acid exponent; negative common logarithm of the acid constant
PNEC	predicted no effect concentration
$PNEC_{oral}$	predicted no effect concentration of the predator
P_{OW}	n-octanol/water partition coefficient; synonym of K_{OW}
QAT	quarternary ammonium compounds
QSAR	quantitative structure activity relationship
R48	danger of serious damage to health by prolonged exposure
R60	may impair fertility
R61	may cause harm to the unborn child
R62	possible risk of impaired fertility
R63	possible risk of harm to the unborn child
R64	may cause harm to breastfed babies
SAR	structure activity relationship
T	toxic
T+	very toxic
T_{95}	time to reach 95% of the steady state concentration
TBT	tributyltin
TCDF	tetrachlorodibenzofuran
TGD	Technical Guidance Document
TSCA	Toxic Substances Control Act
UBA	Umweltbundesamt (German Federal Environmental Agency)
US EPA	Environmental Protection Agency of the United States
Xn	harmful

1
Bioaccumulation

1.1
Significance of Bioaccumulation for Risk Assessment of Chemicals in the Environment

Enrichment of chemical compounds in organisms, i.e. the bioaccumulation, is a fundamental strategy along evolutionary processes which may act as a driving force towards a selective advantage among competing species, e.g. in situations of limited resources.

Whereas species developed selective uptake mechanisms for naturally occurring beneficial substances, avoiding strategies for the uptake of unwanted substances causing detrimental effects often do not exist. This is particularly true for xenobiotics.

It may be assumed that during evolution the time for adaptation towards naturally existing substances was long enough at least for creating avoidance strategies empirically, but not sufficient to adapt to xenobiotics, e.g. halogenated organics enter food webs and ecosystems as well as any other chemical substance.

Bioaccumulation in organisms may have different consequences:

- selective advantage for species
- building up of a depot and neutral behaviour without causing adverse effects
- reversible, transitory effects, e.g. activation of detoxification systems such as enzyme induction, metabolism, biotransformation, inactivation, depuration
- bioaccumulation in organs/tissues inducing adverse acute, subacute, chronic or unknown long-term effects in individuals, populations, species and ecosystems

The latter phenomena – bioaccumulation and biomagnification of xenobiotics leading to irreversible adverse effects in biota and ecosystems – are subjects of concern and integral parts in legislative and administrative regulations.

Since there is not always conformity about the definitions of the terms bioaccumulation and biomagnification, they are briefly defined in the following:

Bioaccumulation is the uptake of chemicals in organisms from the surrounding medium (water, pore water) by gills, skin, etc. or by ingestion of particle-bound chemicals. However, the distinction between the exclusive uptake of the truly dissolved phase and other fractions (colloidal, dispersed, emulgated) is not clearly definable.

Bioaccumulation is quantitatively expressed by the **bioaccumulation factor** (BAF), the ratio of the concentration reached in the organism under steady state condition and the concentration of the surrounding medium. This factor can be related to the whole organism or tissues and organs thereof on a wet, dry or lipid weight basis depending on the context.

The terms **bioconcentration** and **bioconcentration factor** (BCF) as defined by OECD guidelines should be limited to laboratory test systems (e.g. OECD Guideline No. 305 [1]), where the uptake of a chemical is nearly exclusively restricted to the soluble fraction and any other uptake routes can be neglected e.g. by minimizing the particle-bound fraction of suspended matter.

But even under controlled laboratory conditions a certain uptake of adsorbed fractions onto food may occur.

Generally all fractions must be considered potentially bioavailable regardless of the route of uptake.

Biomagnification is generally defined as the process of bioaccumulation along food chains or more precisely – within food webs – following various pathways on different trophic levels. However, this process must not necessarily end up in a magnification, leading to a stepwise increase with highest concentrations in organisms being in terminal positions of food webs e.g. whales, crocodiles, humans.

More frequently, there is a transfer of a chemical or its metabolites over several trophic levels which, although often not on a spectacularly high concentration level may cause long-term effects e.g. DDE.

Hence, the term biomagnification should express the transfer of a chemical or its metabolites within several trophic levels which may lead to a stepwise increase of the concentration level, if no metabolisation and depuration exist.

After the detrimental toxification events a few decades ago caused by the magnification of inorganic/metallo-organic chemicals, e.g. mercury compounds, there was increasing scientific interest to examine principles and extent of such processes for all chemicals released into the environment.

Whereas fate and effects of the most important metallo-organic and inorganic chemicals – the number and volume of which are smaller compared to organic chemicals – is relatively well known, we feel that there is still a considerable lack of knowledge about the risk of bioaccumulation/biomagnification processes for the bulk of organic chemicals.

Hence, predominantly organic xenobiotic chemicals are focused on in the following, inorganic/metallo-organic chemicals only in cases where relevant risk aspects are of concern.

Until now most risk assessment approaches for bioaccumulation and ecotoxicological processes relate to aquatic systems due to easier test performance and test systems available. However, risk assessment schemes and risk management concepts for all environmental compartments are urgently necessary.

1.2 Overview on Bioaccumulation Processes in Ecosystems

1.2.1 *Predictability Versus Reality*

From chemical structure, partitioning behaviour, fate and exposure of a chemical many bioaccumulation processes may be predicted with sufficient exactness and confirmed by monitoring data.

However, considerable events have been experienced in the last decades due to unknown intrinsic properties of chemicals, physiological responses of species and ecosystemically complex interdependencies which were beyond any imaginative power, e.g. the reduced eggshell-thickness of bird eggs caused by DDE [2] or the estrogenic effects mediated by organochlorines [3] or APEOs [4, 5] leading to endocrine disruption and reproductive impairment in organisms.

There is common agreement that lipophilicity of a chemical is the major prerequisite for bioaccumulation in organisms. Hence most of the current **Quantitative Structure Activity Relationships** (QSARs) are based on the **octanol/water partitioning coefficient** (P_{OW}) whereby octanol is used as a surrogate for the compartment fat in an organism. It is further generally assumed that the application of such correlations allows for a prediction of the presumptive BCF. P_{OW} is also called $\mathbf{K_{OW}}$ and since K_{OW} has become more common it is referred to in the following.

However, such correlations are applicable only under the premises, that lipophilicity and hydrophilicity are inversely proportional, but not in cases, where chemicals are either insoluble both in octanol and water or soluble in any ratios, both possibilities resulting eventually in a low K_{OW} and thus not predicting a bioaccumulation potential.

Also not considered by these models are bipolar chemicals, e.g. surface active chemicals like detergents and chemicals with certain nitrogen structures like water soluble bipyridinium compounds and quarternary ammonium compounds which inspite of a low K_{OW} and high water solubility may be bioaccumulated considerably and therefore incorrectly assessed.

Prediction of biomagnification potentials are also doubtful. After disastrous intoxication events arising from bioaccumulation of metallo-organic compounds across food-webs in the sixties and seventies it was expected that organics show the same behaviour. First investigations of simple food-chain relationships and compilation of monitoring data on concentrations and effects in ecosystems led to the premature conclusion that biomagnification of organic chemicals is overestimated and plays only a role for a few highly lipophilic compounds [6].

However, meanwhile it is evident, that not only substances like DDT, HCB, PCBs and PCDD, but also less lipophilic substances like Lindane (γ-HCH) with a log K_{OW} of 3.63 are candidates for biomagnification, although laboratory results indicate no biomagnification potential, proven by complete depuration [7]. Also the occurrence of synthetic musk derivatives in humans and biota [8, 9, 10] was surprising and far beyond any expectation in the light of the relatively small amounts placed on the market.

In the following a short selection of paradigmatic bioaccumulation and biomagnification processes compiled from literature and supplemented by results of research and development projects of Umweltbundesamt (UBA) is presented demonstrating the complexity of bioaccumulation processes in ecosystems and the difficulties of predicting accumulation and long-term effects from simple generic test and risk assessment strategies.

1.2.2
Bioaccumulation, Biomagnification, and Long-term Effects of Organochlorines

Apart from well documented bioaccumulation and biomagnification processes of highly lipophilic chemicals, the capacity of enrichment of the rather moderate lipophilic γ-HCH (log K_{OW} 3.63) in ecosystems is remarkable demonstrating that not only the degree of lipophilicity but also the degree and position of chlorination and particularly the elimination pathways determine the potential of biomagnification.

During the 1987/88 mass mortality of bottlenose dolphins along the Atlantic coast γ-HCH was among the nine of the most frequently detected pesticides [11].

However, comparing biomagnification efficiencies and residues of top predators with the same diet, e.g. fish eating tuna fish and dolphins, there are data suggesting that not the predator status per se, but the lack of branchial elimination pathways of mammals as compared to the elimination potential of gill-breathing fish may explain the higher residues and bioaccumulation/magnification potential in marine mammals [12].

This is in conformity with investigations on the bioaccumulation and transfer of γ-HCH, PCBs and DDTs in pike (*Esox lucius*) [13]. Lipids and concentrations of contaminants in hard roe were 10 times higher as compared to muscle suggesting that the transfer via roe is an important elimination pathway for the individual and a prerequisite for persisting residues in the offspring.

γ-HCH residues have also been found in water, sediments, eggs of pelicans and eels, the main pelican prey. Data suggest a biomagnification with a factor of 1.8 between eel and pelican eggs. The log BCFs/BAFs for eel and pelican eggs were 3.33 and 3.58, respectively related to water, i.e. nearly as high as the log K_{OW} for γ-HCH, making evident the risk of underestimating the bioaccumulation from laboratory investigations [14].

But also in terrestrial food-webs γ-HCH is often present when residues of organochlorines in biota are reported. Systemic impact of pesticides was investigated in a terrestrial food-chain based on plant (cabbage) – host (*Pieris brassicae, Lepidoptera*) – endoparasitic beneficials (*Apanteles glomeratus, Pteromalus puparum, Hymenoptera*). Compared to Parathion which was metabolized and excreted along the food-chain, Lindane despite of a relatively low acute toxicity revealed a high chronic food-chain toxicity mainly by prevention of metamorphosis in the endoparasitic wasp population. Although depuration amounted up to 80%, a high pupal mortality occurred [15].

Evidence for the decline of the cattle egret (*Bubulcus ibis*), feeding predominantly on insects in agricultural areas, caused by DDE and γ-HCH is reported by Mullié et al. 1992 [16].

Reduction of breeding success, eggshell strength, and of migration and breeding behaviour of the great tit (*Pares major*) was evidenced by laboratory experiments in a three-step food-chain based on oak-leaf, caterpillar and great tit. Apart from PCB 153 which was detected in all samples, a remarkable amount of γ-HCH was detected in 86% of all samples [17]. Results on eggshell thickness and population dynamics are in agreement with the assumption of negative population effects in the great tit suggesting effects during the early stages of the developing bird.

Summarizing it can be stated that not only highly lipophilic but also moderate lipophilic organochlorines like Lindane exhibit a considerable potential for bioaccumulation/biomagnification in all environmental compartments.

1.2.3
Bioaccumulation of Non-lipophilic Chemicals

Predictions of bioaccumulation potentials for water soluble non-lipophilic chemicals applying generic QSARs based on log K_{OW} may underestimate the true bioaccumulation capacity in certain cases:

Listed in the so-called Japanese MITI – list [18] are several chemicals which despite of a high water solubility, relatively low log K_{OW}, show a considerable bioaccumulation. Three N-containing chemicals may exemplify this (Table 1).

A high tendency for adsorption onto organic carbon (humus) which may be bioavailable for soil/sediment ingesting terrestrial organisms e.g. earthworm, was demonstrated also for the N-containing pesticides [19] (Table 2).

The herbicide Paraquat is bioaccumulated and adsorbed in snails more than 200 fold [21] and induces significant tadpole mortality resulting from tadpole feeding on Paraquat-contaminated plant material [22].

A quarternary ammonium compound used as reference substance in electrophotographic toners with a log K_{OW} between 2 and 3, a water solubility >100 mg/l and a surface tension <40 mN/m was accumulated up to a BCF >300. Steady state was not reached before 6–8 weeks. QSARs would have predicted a BCF <50 [23].

Obviously molecules containing reductive nitrogen tend to bind to negatively charged sites of molecules, e.g. mucopolysaccharides, due to free electrons resulting in positive loadings of N independent of the log K_{OW} of the substance. Equilibrium for such substances is reached late and depuration is often retarded.

Table 1. BCFs of well water-soluble chemicals

Chemical name	CAS No	Water solubility in mg/l	48 h LC_{50} fish in mg/l	BCF
Basic green-4	569-64-2	>1000 (log K_{OW} -0.17)	0.32	36–91 (20 µg/l) 44–75 (2 µg/l)
4-Vinylpyridine	100-43-6	>10,000	1.57	58–96 (20 µg/l) 48–96 (2 µg/l)
4-(N,N-Dimethyl- amino-1,2 dithio-lan	1631-58-9 1631-58-9	>2000 >2000	0.207 0.207	29–59 (1.56 µg/l) 40–64 (0.156 µg/l)

Table 2. Examples for N-containing and well water-soluble pesticides with the tendency of high adsorption

Group	Trade name	log K_{OW} [20]	Water solubility in g/l [20]
QATs	Chlormequat (chloride)	-1.58	950
Diazines	Chloridazon	1.2	0.34
Bipyridylium	Diquat	-4.6	7–8
	Paraquat	-4.5 – -4,7	620

Also anionic surfactants, e.g. the well water-soluble LAS, are accumulated reaching BCFs >100. Long-chain homologues are accumulated more than 1000 fold. The BCFs are an order of magnitude higher than expected from the log K_{OW} [24].

Concluding from these results it can be stated that the log K_{OW} in certain cases is an inadequate descriptor predicting the BCF or BAF, respectively. Surface activity and structural properties together with the intended use category of a chemical which may give indications on a bioaccumulation potential must also be considered when applying QSARs.

1.2.4
Bioavailability of Chemicals for Bioaccumulation

In many publications released on bioavailability during the last years there are still assumptions to be found that a chemical can only be accumulated either by uptake of the truly dissolved fraction or by ingestion of contaminated food, and that sediment-bound fractions are not longer bioavailable.

In a study on sediment-associated hydrophobic organic contaminants from the Great Lakes it was shown that the contaminants were accumulated by benthic organisms exposed to whole sediment, pore water, elutriates and aqueous medium making use of different uptake strategies whereby the BAFs for aqueous extracts of sediment-associated chemicals indicated a much lower bioaccumulation as compared to whole sediment [25].

Bioavailability of sediment-associated hydrocarbons is also demonstrated in a five-compartment steady-state food-web model including fish and a benthic amphipod. Uptake by ingestion of sediment-associated chlorinated hydrocarbons with log $K_{OW} > 5$ was more significant than the uptake via interstitial and overlying water, respectively, in this amphipod-sculpin food-web of Lake Ontario [26].

Adsorption and bioaccumulation of PAHs and pesticides were investigated in sediment and the benthic-feeding bivalve *Corbicula fluminea*.

Bioaccumulation factors of DDT, DDD, and particularly of DDE in *Corbicula* were greater than predicted values from the K_{OW}. The bioaccumulation factors for the hydrophobic pesticides were one order of magnitude higher than values generally obtained in laboratory studies under equilibrium conditions [27].

In a 10-days bioassay the earthworm (*Lumbricus terrestris*) and fathead minnow (*Pimephales promelas*) accumulated significant amounts of PCBs when exposed to Great Lake sediments [28].

Tubificids (*Tubifex tubifex*, *Limnodrilus hoffmeisteri*) accumulated sediment-associated γ-HCH and HCB in a laboratory test system up to a factor of 4 and 7, respectively, related to sediment concentrations [29].

The oligochaete *Lumbriculus variegatus* accumulated sediment-associated pyrene rapidly [30]. Although not significantly accumulated itself, sediment bound polydimethylsiloxane influenced the uptake kinetics of benzo(*a*)pyrene, resulting in a lower bioaccumulation factor as compared to the uptake of benzo(*a*)pyrene alone [31].

Even after 5 years PCDDs remained bioavailable to freshwater mussel and crayfish exposed to contaminated sediments [32].

A large fish kill observed in the river Tajo in Spain was caused by the lipophilic resin dehydroabietic acid which was associated to suspended matter. Toxicity could be dropped and regained by filtration and resuspension, respectively. The toxicity front moved downstream more slowly than the water body in conformity with the retarded distribution of suspended matter [33].

Although a sharp distinction between uptake routes via water, pore water, colloids, suspended solids and sediment is not always possible, these few examples clearly demonstrate the general bioavailability of sediment-associated fractions. However, a prediction on the extent of bioavailability is limited.

1.2.5
Overestimation and Underestimation of Bioaccumulation

Metabolisation, distribution, and excretion are major detoxification processes. Hence, BCFs may be lower than expected from log K_{OW} as exemplified for benzo(*a*)pyrene [34].

However, enzyme induction may be hampered by high exposure concentrations e.g. of the insecticide Chlorpyrifos resulting in a retarded depuration kinetic [35]. Consequently bioaccumulation would be underestimated when applying laboratory derived low-exposure depuration kinetic constants in high exposure scenarios.

Bioaccumulation of superlipophilic substances may be overestimated. Experiments with PCB congeners revealed that obviously not the molecular weight but size and steric factors of molecules may reduce the bioaccumulation of very hydrophobic compounds [36]. Log K_{OW}/log BCF correlations could be described by a 2nd order polynom showing maximal BCFs dependent on the degree of chlorination and log K_{OW} and decreasing BCFs at further increasing log K_{OW}, hydrophobicity and degree of chlorination.

Also disperse dyestuffs with low water solubilities show no or a bioaccumulation lower than expected mainly due to their large molecular size and reduced bioavailability owing to their very low water solubilities [37].

Conversely, methodological shortcomings such as testing bioaccumulation of superlipophilic chemicals in concentrations far above their true water solubility by means of solubilisers may result in low BCFs from the ratio of concentrations in fish/water and insufficient time to gain a steady state, respectively, thus underestimating the bioaccumulation. Testing within the true water solubility without solvent carriers and calculating the BCF on the basis of kinetic rate constants result in values in agreement with current QSARs [38, 39].

Significantly different BCFs for chemicals existing in isomeric structures were reported for α, β, γ, and δ-HCH [7] and for insecticidal pyrethroids with higher BCFs up to a factor of 8 for the cis-isomers [40]. Whereas the BCFs of HCH were dependent on different depuration rate constants, the higher BCFs of the cis-isomers of the pyrethroids could only be explained by greater uptake rate constants, since the depuration rates were similar.

Among PCBs, PCDDs, and PCDFs the degree of chlorination and the chlorine position of the molecule will greatly influence the bioaccumulation behaviour, e.g. the BCFs between the coplanar tetrachlorobiphenyl congener No. 77 and the ortho-substituted congener No. 54 differ by a factor of 32 [41].

Beside isomeric differences causing varying BCFs also enantioselectivity and chiral discrimination of optically active chemicals may influence the degree of bioaccumulation. Organ-specific ratios of enantiomers of α-HCH and α-, β-, and γ-HCH isomers were detected in brain and other tissues of neonatal northern fur seals (*Callorhinus ursinus*) revealing surprisingly high ratios of the two α-HCH enantiomers (1.8 to 28) which were discussed in context with the different health status of the seals [42]. Existence of enantioselectivity and specific transport systems point out that bioaccumulation processes may be decisively governed by small submolecular differences leading to results far from predictability.

From residues in biota and surface waters monitored in the field, BAFs can be estimated and compared with laboratory-generated data. Field BCFs were higher by a factor of 50 for α-chlordane and 220 for DDE [43].

Due to the presumption that only undissociated molecules can penetrate membranes and that uptake through aqueous pores is limited, dissociating substances are generally considered to have no essential bioaccumulation potential. However, the pH may influence the bioaccumulation patterns decisively.

Bioaccumulation of dissociating pentachlorophenol in northern pike in acidified lakes (pH ~ 5.8) was nearly twice as high as in alkaline lakes (pH average 8.1) [44]. This may be relevant when assessing the risk of bioaccumulation processes in areas with serious acidification, e.g. Southwest Sweden.

Bioaccumulation studies with 5-chloro-2-(2,4-dichlorophenoxy)-phenol dissociating within a pH range of 5.8 to 8.8 demonstrated that at pH 8.8, where a high degree of dissociation (~88% dissociated molecules) is present, body concentrations and BCFs measured in zebra fish were similar compared to those at pH 5.8 even though uptake and depuration rates were considerably lower. Beside the uptake of undissociated molecules by diffusion through the membrane the permeation of dissociated molecules through gap-junctions is discussed [36].

Predictions of bioaccumulation in plants according to models based on log K_{OC} are doubtful considering the different uptake routes, types of plants and soils, lipid content and translocation processes in plants.

Investigations on the soil-plant relationships for root crops and the soil-borne part of foliar contamination revealed different uptake and translocation processes in plants which only in part can be explained by the physico-chemical properties of the chemicals [45]. Bioconcentration factors were in several cases much higher than predicted from the K_{OW}.

Bioaccumulation in plants by foliar uptake resulting from partitioning between soil-air-plant may be the main uptake route also for more lipophilic substances with theoretically low vapour pressure and high degree of chlorination. BCFs for PCDDs in plants were 2–3 orders of magnitude higher than could be expected from their log K_{OW} [46]. Hence air to leaf transfer of gaseous organics may be a key process for bioaccumulation in plants and the primary step towards a magnification in ecosystems.

Also temperature can influence bioaccumulation and sorption processes significantly resulting in increased bioaccumulation with raising temperature as demonstrated for green-algae [47].

Concluding from these selected examples, overestimation of bioaccumulation potentials due to methodological shortcomings and lacking scientific knowledge may be embarrassing but so far without consequences, underestimation, however, may imply a serious risk when applying wrong prediction in risk assessment approaches.

Since it can be assumed that even with sophisticated scientific work wrong predictions of bioaccumulation potentials cannot be avoided, risk potentials must be countered by precautionary principles, e.g. safety factors.

1.2.6
Sublethal and Indirect Effects by Bioaccumulation

Surface active substances already in low concentrations i.e. in the range of µg/l may cause sublethal effects with a broad spectrum of actions.

Although controversially discussed whether a lowered surface tension is responsible for toxic effects, tensides may have an impact on chemoreceptors leading e.g. to disturbed orientation of food-searching fish, on functional disruption of cell membranes, on enzyme induction, and embryogenesis [48].

Bioaccumulation of tributyltin (TBT) compounds which have a broad biocidal action and are used as antifoulants is by far underestimated when estimated using the log K_{OW} varying between 3.2 and 3.8 for the different compounds. BCFs as high as 133,000 for mussels (*Mya arenaria*) and 100,000 for snails (*Nucella lapillus*) have been reported [49, 50].

Clear evidence exists between bioaccumulation of TBT compounds in very low concentrations and the imposition of male sexual characters on female snails (imposex) which is a worldwide observed phenomenon and already used as bioindicator. Sublethal concentrations in the range of ng/l are discussed inducing histopathological malformation in the female gonadal system and leading to complete sterility of the marine mollusks *Littorina littorea* and *Hydrobia ulvae* [51].

A correlation between planar PCB concentrations in eggs, enzyme activities, occurrence of deformities and reproductive success in double-crested cormorants (*Phalacrocorax auritus*) is reported as a consequence of environmental contamination [52]. Bill deformities (> 50% of investigated chicks) were significantly greater at Lake Michigan than in other nesting colonies in the other less contaminated Great Lakes or Canada.

Sublethal effects such as cytological alterations in the liver ascribable to the primary acute toxic mechanism of acetylcholin esterase inhibition were observed in rainbow trout exposed to the insecticide Disulfoton in concentrations well below such producing any macroscopically visible effect [53]. Disulfoton has a short half-life time in water and a moderate BCF of about 400. Even if this indistinct mode of action is interpreted as an adaptive/compensative rather than degenerative phenomenon, this example may reveal basic mechanisms on an ultrastructural level demonstrating potential long-term effects also by substances with an acute toxic mode of action.

1.2.7
Compartment-crossing Transfer of Accumulated Chemicals

Apart from the transfer of sediment-associated chemicals via benthic organisms to benthos-feeding fish, there exist further transfer routes enhancing mobility and distribution of contaminants and leading to a compartment-crossing transfer from sediments to other food-webs.

By diurnal migrations of the epibenthic freshwater shrimp *Mysis relicta* substantial amounts of accumulated sediment-associated PCB congeners were transferred into the pelagic food-web thus coupling the benthic and pelagic zones [54].

A transboundary transport of contaminants from sediments to air and terrestrial ecosystems occurs by the emergence of insects, mainly diptera.

Laboratory experiments showed that 0.2% to 2.1% of total sediment contaminant content are exported annually by emerging insects which had accumulated sediment-sorbed *2,3,7,8*,-TCDF [55].

Midge larvae (*Chironomus decorus*) which accumulated the pesticide trans-chlordane in a whole life cycle laboratory exposure assay over the course of a 50 day study, transferred 82.6% of the contaminant during metamorphosis to the adult insects, whereas 11.4% was left behind in the shed exuviae [56].

Since emergence events often occur synchronically over a short time interval due to the season, high quantities of contaminants may be available e.g. for midge-eating birds thus enhancing the risk of quickly reaching a critical body burden.

1.2.8
Bioaccumulation, Critical Body Burden and Effects

The bioaccumulation, although a risk factor per se, cannot be assessed without consideration of effects, since enrichment of chemicals in or on organisms or tissues thereof is an necessary prerequisite independent of the mode of action.

With regard to the amount of chemicals accumulated, not the relative amount of accumulated substance, expressed as BCF or BAF, is decisive, but the internal concentration level may cause effects after reaching a critical threshold, either unspecific (e.g. narcotic) or specific (e.g. neurotoxic).

The relationship between bioaccumulation and effects has first been described by Kobayashi et al. 1979 [57], further investigated and confirmed by Mc Carty, 1986 [58], and formulated as the concept of "lethal body burden" as a toxicological endpoint by Sijm et al. 1993 [59]. This internal whole-body concentration in millimoles per kilogram at time of death or immobilization is the product of BCF and steady state LC_{50} which has a constant value for certain groups of closely related compounds, e.g. phenols, with respect to a certain end point and the mode of action. This concept was first verified for narcotic substances with an unspecific mode of action, but probably seems to be applicable also for substances with other modes of action.

Since it is evidenced that the BCF is not a characteristic property of a chemical, respectively an organism, but may depend on the concentration tested and other factors influencing the uptake and depuration kinetics, a more complex strategy for the assessment of bioaccumulation is suggested [60, 61]. Combining this approach with the lethal body burden concept allows for the decision whether an already reached body concentration is of concern and how far it is away from becoming critical for an organism at a given exposure concentration if no depuration system exists avoiding a further increase.

Considering longterm effects the depuration has a direct influence on the time-dependent toxicity. Species with the ability of elimination will reach an equilibrium for the internal concentration and also an ultimate LC_{50}, whereas the LC_{50} in species that are not capable to eliminate e.g. cadmium, may reach values near to zero. For these species the time to reach the lethal body burden is decisive. Taxonomically related species appear to have comparable accumulation patterns, but lethal body burdens may differ. The authors conclude, that knowledge of the accumulation pattern is indispensable for the evaluation of a species' sensitivities to toxicants [62].

Lethal body burdens were also used to estimate the toxicological susceptibility of a species [63]. As an alternative to the LC_{50}, which expresses both the bioaccumulation potential and its intrinsic toxicity, the lethal body burden is more appropriate to reflect the intrinsic properties of a chemical and to explain species susceptibility to toxicants.

Moreover, beside the time-dependent toxicity for an individual organism there is the risk of a transfer of not eliminated body burden from females to the offspring via roe [44], bird eggs [14, 16] and lactation [12].

Incomplete depuration and non-eliminated residues of pentachlorophenol (PCP) were also observed in a bioaccumulation study with the benthic oligochaete *Tubifex tubifex*. Although the body burden concentration of approximately 9 µmol/l was low, residue concentration of parent PCP during the depuration phase remained on a plateau of approximately 3.7 µmol/l [60].

Also in fish (*Leuciscus idus*) a retarded depuration of PCP has been observed resulting in residues on a low, but detectable concentration level [64].

It is a reasonable assumption that non-eliminated body burdens are the main prerequisites for biomagnification in food-webs.

1.3 Scope of Risk Assessment of Bioaccumulation

Drawing conclusions from the cited examples revealing unexpected and non-predicted effects one might assume that with our current risk assessment schemes we are doing the mistakes today which we will become aware of tomorrow.

As experienced and demonstrated for certain chemicals, e.g. PCBs, it must be recognized that bioaccumulation/magnification processes may be phenomena lasting over decades and inducing effects even after release into the environment had been stopped years before and residues in almost all compartments of the environment have declined [65]. Remediation measures are limited to curative activities only. This is particularly true for such highly bioaccumulating and persistent substances unknown as yet.

Hence, for providing a better protection of environment and man we need approaches for a future-oriented risk assessment covering that part of risk which obviously never can be determined ultimately.

To gain more insight into the causal relationships interdisciplinary investigations including food-biology, physiology, biochemistry, immunobiology, pharmatoxicology, neurotoxicology, genetics etc. should be performed. To encounter non-predictable effects by risk assessment strategies, precautionary principles such as the use of appropriate uncertainty factors should be included and measures of risk management and risk reduction implemented.

2 Assessment Concepts of Bioaccumulation

2.1 Criteria for a Bioaccumulation Assessment Concept

In contrast to the assessment of bioaccumulation potentials based on QSARs or specific indications, the measurement of bioaccumulation has to consider all relevant criteria described in the following. Existing concepts for the assessment of bioaccumulation should be critically judged with regard to the consideration of these criteria.

2.1.1 *Test Organisms*

With the choice of test organisms a far-reaching decision is made concerning the test design and the assessment of data gained. Because of the intra- and interspecies variations it is not possible to transfer the results from one species to another. Therefore it is not only necessary to have representative species for at least all relevant environmental compartments such as fresh/marine water, sediment and soil, but also adequate assessment approaches when uptake routes are different e.g. fish and sludge-worm, respectively.

2.1.2 *Uptake Routes*

Principally substances can be taken up from the surrounding medium (water, sediment, soil, air), via food or through body surfaces. For an adequate assessment of bioaccumulation it has to be considered which uptake routes or which combination of them are relevant for a specific substance **and** species. All uptake routes mentioned are possible e.g. for fish, but in combination with substance specific properties like molecular size and shape, charge or surface activity some routes may be excluded in favour of others.

2.1.3
Metabolism

Depending on species and chemical accumulated, metabolism may differ in specificity and extent leading ideally to complete depuration. However, this mechanism cannot be regarded as a mitigating property in general, since uptake may be faster than metabolism and metabolites may be stable and not being eliminated still causing adverse effects. Therefore metabolites should be identified and their quantity measured.

2.1.4
Persistence

Another important factor for an integrated approach of assessing bioaccumulation is the persistence/degradation of a substance in environmental compartments. Like metabolism, degradation cannot be regarded generally as a mitigating property because uptake may be faster than degradation. Therefore persistence/degradation have to be integrated in an appropriate way into an assessment concept.

2.1.5
Precautionary Principles and Trigger Values

Although bioaccumulation is not necessarily a prerequisite for adverse effects, unpredictable risk potentials must be encountered by adequate risk management strategies. Therefore, when assessing the risk of bioaccumulation, two aspects have to be considered:

- the **qualitative** assessment of bioaccumulation defining precautionary principles and characterizing risk potentials,
- the **quantitative** evaluation of data on bioaccumulation defining trigger- or cut off-values on the basis of bioaccumulation categories for further testing or administrative measures according to the respective environmental legislations.

2.1.6
Monitoring Data

Monitoring data on biota indicating adverse effects or alterations in food-webs resulting from bioaccumulation/biomagnification are of utmost value and should be integrated in an overall risk assessment scheme with highest priority supporting and refining the final risk assessment.

2.2
Key Parameters for the Assessment of Bioaccumulation

Parameters for the assessment of bioaccumulation are:

- the BCF in the whole fish and in parts thereof, such as fillet, viscera or carcass.

The BCF alone should be considered critically. It does not reflect the complexity of the bioaccumulation process. Which BCF is reached depends largely on the test organism and the test method, so that the height of the BCF is relative and consequences for individuals are not directly related to this value. For example a relatively low BCF may be harmful when a lethal body burden is already reached or the substance is not eliminated and hence transferred into the food-web.

- organ specific accumulation, reversible as well as irreversible.

This may give rise to special effects (for example behavioural alterations possibly adverse to an individual), which cannot be related easily to a relative low BCF in the whole organism. These effects may also be expressed in a later phase of life or in the following generation.

- the elimination or depuration expressed as half-life clearance time (ct_{50} i.e. the time needed to reach 50% removal).

From the half-life clearance time it can be seen how long a substance remains in an organism no longer exposed to this substance. A short half-life clearance time may be a real mitigating property for even a high BCF, a long half-life clearance time may be, however, an incriminating factor for a low BCF.

- uptake routes and elimination kinetics.

The uptake/elimination may be bi- or multiphasic, i.e. with different velocities at the beginning and the end. Therefore a kinetic description of uptake/elimination is needed to reflect the complete uptake and depuration process.

- incomplete elimination/ plateau formation.

An incomplete elimination of a substance or its metabolites gives rise to bound residues, which may form a plateau in tissues or organs over time and raise the risk of adverse effects (if the substance or its metabolites show a low **No Observed Effect Concentration** (NOEC)) or of biomagnification.

- information on metabolism especially with regard to stable metabolites.

Likewise, stable metabolites may remain in the organism possibly causing adverse effects or may be transferred to higher trophic levels, hence raising the risk of biomagnification.

2.3 Indications of Bioaccumulation Potential

Bioaccumulation studies are laborious and require animal testing. Therefore, as an initial step of a testing strategy it was internationally agreed to use a simple screening method for assessing the hazard that a substance might accumulate in organisms, from a minimal set of (physico-chemical) data and the knowledge of its chemical structure: the determination of **bioaccumulation potential** (BAP). Bioaccumulation potential may serve as a qualitative, or to a limited extent as a quantitative, indicator of a risk of bioaccumulation in living organisms due to the physico-chemical and structural properties of a substance.

There are several possible indications of a bioaccumulation potential which are discussed below. Most of the criteria can only be applied to organic, hydrophobic substances. Only a few can also be used for polar organic or for inorganic substances.

2.3.1
n-Octanol-water Partition Coefficient

Bioaccumulation potentials are generally estimated on the basis of the n-octanol/water partition coefficient in its logarithmic form (log K_{OW}). It is easily available and does not require expensive animal testing. If measured values are not available, log K_{OW} can be calculated from the chemical structure of a substance as a first approach.

This approach assumes that accumulating organic substances are hydrophobic, can freely diffuse through cell membranes, and are only enriched in the lipid-fraction of organisms. Therefore, partition equilibrium of a substance between n-octanol and water is regarded as a model of bioaccumulation.

On the other hand, the correlation between n-octanol/water partition coefficient (calculated as log K_{OW}) and the bioconcentration factor (calculated as log BCF) has been proved to be poor for some types of chemicals. It cannot be expected that the n-octanol/water partition coefficient generally is a sufficient model of bioaccumulation behaviour of organic chemicals because it does not take into consideration factors influencing bioaccumulation in organisms, including e.g.:

- phenomena of active transport,
- the influence on the diffusion behaviour through cell membranes,
- metabolism in organisms and accumulation behaviour of metabolites,
- accumulation in specific organs and tissues (also by adsorption onto biological surfaces like gills, skins),
- special structural properties (e.g. amphiphilic substances, dissociating substances leading to multiple equilibrium processes),
- uptake and depuration kinetics, residue plateau of the substance or of metabolites after depuration.

A subtle problem is the log K_{OW} measurement of ionisable substances because this may lead to multiple partition equilibria. The new test guidelines for log K_{OW} measurement (cf. e.g. Annex to Commission Directive 92/69/EEC of the European Communities No. A.8 [66] or OECD Guideline for Testing of Chemicals No. 107 [67]) suggest that log K_{OW} measurements should be performed with ionisable substances only in their non-ionized form (free acid or free base), thus allowing to determine maximum lipophilicity of a tested substance. Therefore, the pH-value of an appropriate buffer chosen for log K_{OW} measurement must be at least one pH unit below (free acid) or above (free base) pK-value. Other measurements of log K_{OW} are not valid with regard to assessment of bioaccumulation potentials.

Despite of these limitations it is internationally accepted that log K_{OW} values greater than or equal to 3 indicate that the substance has the potential to bioaccumulate.

2.3.2
Fat Solubility

Fat solubility may also give an indication of a bioaccumulation potential assuming the same partitioning equilibrium and membrane diffusion processes for bioaccumulation as derived from the octanol solubility. Since fish lipids cannot be considered as a uniform compartment, the partitioning between water and the different lipid fractions have to be taken into account [68]. For polar membrane lipids (phospholipids) octanol is an appropriate surrogate, whereas non-polar storage lipids are better represented by hexane. Hence, high fat solubility signalizes a high probability of a bioaccumulation potential, particularly with respect to storage fat, e.g. in adult fish.

2.3.3
Surface Activity

Surface active substances, like tensides and many pesticides, may also have the potential to bioaccumulate even if their log K_{OW} values are < 3. Surface activity is measured as surface tension of a solution of a substance in water [69]. If a substance has a surface tension of ≤ 50 mN/m at a concentration ≤ 1 g/l, i.e. is surface active, it may be bioaccumulated itself or enhance the bioaccumulation of other chemicals present. A low **Critical Micelle Concentration** (CMC) also may indicate facilitated uptake and alteration of membrane fluidity.

2.3.4
Adsorption

Adsorption onto biological surfaces (e.g. gills, skin) may also lead to bioaccumulation and uptake of substances via food chain (see section 3 Biomagnification). Therefore, high adsorptive capacity (log $K_P \geq 3$) can be regarded as an additional indication of a bioaccumulation potential. This aspect may be of relevance for metallo-organic, organic or polar compounds, e.g. dye-stuffs.

2.3.5
Structural Features

A further indication of bioaccumulation potential is given for analogues of organic or inorganic substances known to have the potential to bioaccumulate in organisms. The same is true for substances which contain nitrogen, e.g. amines, pyridinium compounds, which accumulate higher than expected from their log K_{OW} (e.g. herbicide Paraquat, log K_{OW} -4.6, BCF > 200).

2.3.6
Mitigating Aspects

Certain physico-chemical, biological, and structural criteria might exclude a bioaccumulation potential for a distinct substance even if it exhibits an indication for a bioaccumulation potential, i.e. log $K_{OW} \geq 3$. Because uptake of a chemical may be very fast as is exemplified in figure1, uptake rate and hydrolysis half-life time have to be related for substances which are predominantly emitted directly into aquatic compartments. If the half-life time of hydrolysis for such a substance is less than 1 h, it is assumed that hydrolysis proceeds quicker than the uptake by organisms. No indication of bioaccumulation potential is assumed in this case. However, it may be necessary to check the hydrolysis products for their bioaccumulation potential.

Ready biodegradability of a chemical is commonly considered as a mitigating aspect, however, uptake rates of bioaccumulation may be significantly faster than biodegradation as is shown in Fig. 1. Hence bioaccumulation might occur even though the substance is readily biodegradable. This has to be assessed carefully on a case-by-case basis considering kinetic information on both processes.

It may be necessary to check the products of abiotic and biotic degradation for their bioaccumulation potential.

2.4 Existing Assessment Concepts of Bioaccumulation

Bioaccumulation potential is assessed differently in national and international regulations. The assessment category may even differ in a certain country for

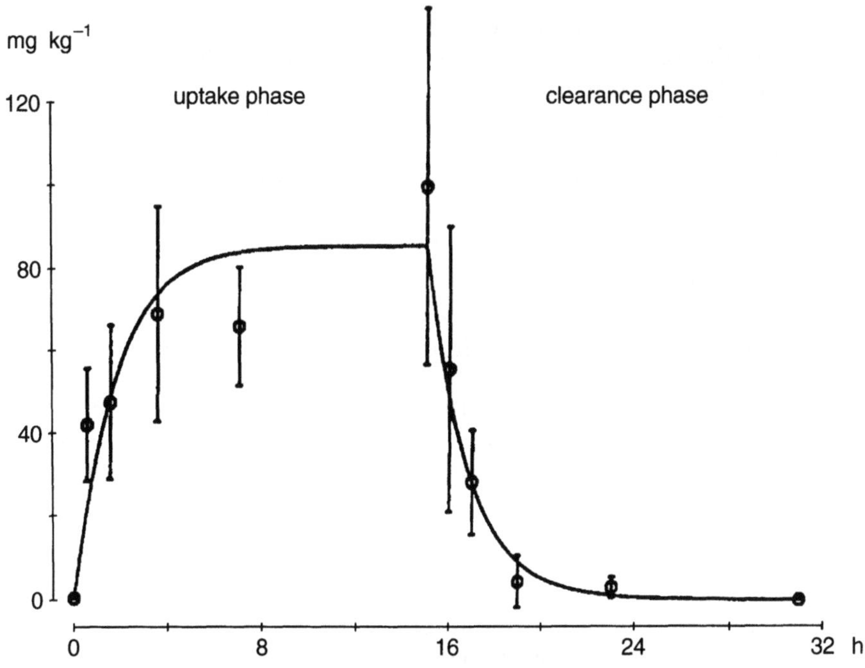

Fig. 1. Rapid uptake of 2-t-butylphenol by zebra fish (*Brachydanio rerio*): steady-state concentration within 5 h (taken from [64]) (wet weight basis)

different groups of chemicals. Furthermore, the indication of bioaccumulation potential for a certain substance may be used for different purposes, e.g.:

- for classification and labelling,
- for deciding on the test duration of ecotoxicological tests, with respect to the steady-state concentration,
- as trigger for bioaccumulation testing.

Table 3 and the following text present a short overview of the regulations and proposals for criteria and trigger for indications of bioaccumulation potential in different countries so far available or known, respectively.

2.4.1
USA

legal scope:	**Environmental Protection Agency** (EPA), Toxic Substances Control Act (TSCA), new and existing chemicals [78]
criteria:	"sufficient" toxicity in the lower TIERS, or indications of chronic effects, or uptake and effective persistence
trigger:	log K_{OW} ≥3.5
consequences:	bioaccumulation study

Within the EPA the **Office of Toxic Substances** (OTS) is responsible for implementing the **Toxic Substance Control Act** (TSCA). The OTS has developed the following approach:

1) identification of appropriate ecological endpoints,
2) a tier-testing scheme for estimating impacts on such endpoints,
3) ecotoxicological testguidelines,
4) models and techniques for estimating ecotoxicity from chemical structure (SAR/QSAR),
5) hazard assessment factors for establishing chemicals concentration of environmental concern,
6) risk assessment methodologies characterizing the risk by including hazard (ecotoxicity) exposure data.

The tier-testing scheme has four tiers (I-IV) of toxicity testing with aquatic and terrestrial organisms. On TIER III the bioaccumulation is included gaining importance for further decisions.

Bioaccumulation testing at TIER III is conducted if there is "sufficient" toxicity in the lower TIERS, or indications of chronic effects or uptake and effective persistence (based on half-lives in water, soil and plants) could be shown.

A degradation half-life in water ≥ 4 d and log K_{OW} ≥ 3.5 would trigger a bioaccumulation study at this TIER.

Other indications e.g. surface tension are not mentioned, no further explanations are given concerning derivation of the half-lives and no guidance is given concerning the decision of testing bioaccumulation in fish and/or oyster.

QSAR is only described/used for estimating toxicity.

In the evaluation of bioaccumulation data the BCF is the only criterion used in the assessment resulting in three categories:

Table 3. Different National and International Regulations and Proposals for Indication of Bioaccumulation Potential

State/union	Regulation	Criterion and Trigger	Result, Consequence	Reference
Germany	Chemicals Act (new and existing chemicals)	cf. EU (new and existing chemicals)	see EU	[70]
	Plant Protection Act (pesticides)	$\log K_{OW} \geq 3$	bioconcentration study in fish	[71]
EU	67/548/EEC (all chemicals)	$\log K_{OW} \geq 3$	Classification and labelling (indication of bioaccumulation potential)	[72]
	93/67/EEC (new and existing chemicals)	$\log K_{OW} \geq 3$ or highly adsorptive or belongs to a class of substances known to have a potential to accumulate in living organisms or indications from structural features and no mitigating properties (cf. 2.4.4)	Indication of bioaccumulation potential SAR estimation of BCF for assessment of secondary poisoning which may trigger a bioaccumulation study Trigger of aquatic long-term ecotoxicological tests	[73]
	91/414/EEC (pesticides)	$\log K_{OW} \geq 3$	Bioconcentration study in fish	[74]
Canada	Toxic Substances Management Policy (new and existing chemicals)	$\log K_{OW} \geq 5$	Bioaccumulation study (cf. 2.4.3)	[75]
	Pesticides	$\log K_{OW} \geq 3$ og K_{OW}: 2–6	Bioconcentration study in fish Raises concern about potential bioaccumulation	[76]
Japan	Chemicals Substance Control Law (new and existing chemicals)	Not readily biodegradable and $\log K_{OW} \geq 3$	Bioconcentration study in fish (cf. 2.4.2)	[18, 77]
The Netherlands	Pesticides	$\log K_{OW} \geq 4.3$	Bioconcentration study	[76]
USA	Toxic Substances Control Act (new and existing chemicals)	$8 > \log K_{OW} \geq 3.5$ and $MW \leq 1000$, effective persistence	Bioconcentration study (cf. 2.4.1)	[78]
	Pesticides	$\log K_{OW} \geq 3$	Bioconcentration evidence	[76]

high: BCF ≥1000
medium: BCF ≥100<1000
low: BCF <100

Elimination behaviour, the formation of a plateau of residues or persistent metabolites are not mentioned.

In the assessment process of new substances so called assessment factors are used, ranging from 10, or 100 to 1000. They have to be understood as "uncertainty factors" and they are used only with toxicity test results.

No factor for a bioaccumulation risk is mentioned.

2.4.2
Japan

legal scope: Chemical Substances Control Law, new and existing chemicals [18]
criteria: non-biodegradability
trigger: log $K_{OW} \geq 3$
consequences: bioaccumulation study

Biodegradability, bioaccumulation and toxicity are basic criteria for regulating chemical substances in Japan under the Chemical Substances Control Law. Chemical substances are not subject to regulation when they have high biodegradability, low bioaccumulation and low toxicity in general.

In any case it is generally assumed that chemicals have a low bioaccumulation when their BCFs are less than 1000 in the bioaccumulation test OECD 305 C [79]. However, the final decision is drawn after a review in the Judgement Committee considering also additional factors. In principle, the bioaccumulation test is applied to non-biodegradable chemicals. If a test substance is altered to another chemical substance in the biodegradation test, the bioaccumulation test is conducted with the altered substance until the BCF reaches equilibrium.

Two concentration levels are tested; if the BCF is shown to be concentration-dependent, more than two levels are tested.

Generally non-polar substances with a log $K_{OW} \geq 3$ have to be tested. Dissociating substances also have to be tested if their log K_{OW} is < 3 and the criteria for weak acids (pKa-pH < 1.7) and weak bases (pH-pKa < 1.7) have to be applied.

Testing is also indicated if the substance reveals other properties, e.g. hydrolysis and QSARs are not applicable. If rapid transformation occurs, the transformation products have to be considered.

2.4.3
Canada

legal scope: **Canadian Environmental Protection Act** (CEPA), new and existing chemicals [75]
criteria: persistence
trigger: log $K_{OW} \geq 5$
consequences: bioaccumulation study

Key objectives of the Canadian Toxic Substance Management Policy are virtual elimination from the environment of toxic substances that result from human activity and that are persistent and bioaccumulative (referred to as Track 1 substances).

Criteria for selection of Track 1 substances are:

- Persistence given as half-life for air (≥ 2 d), water and soil (≥ 182 d), sediment (≥ 365 d),
- Bioaccumulation with BAF or BCF ≥ 5000 or log K_{OW} ≥ 5.0
- Toxicity[1],
- Predominantly anthropogenic[2].

Expert judgement and the weight of scientific evidence will be used in determining whether these criteria are met.

Substances not meeting all four criteria are so-called track 2 substances for which a life-cycle management is demanded to prevent or minimize release into the environment.

Concerning bioaccumulation lipid content of the organisms should be considered for a better comparability of data.

Only BCF or BAF and log K_{OW} are used, no other parameters, e.g. elimination from the organisms, are used. However, BCF and BAF are considered as environmentally more relevant than K_{OW}, and bioavailability of the substance has to be considered particularly when BAF is determined. Field data (i.e. BAF) are preferred over laboratory data (e.g. BCF).

There is no guidance on test guidelines and no concept on testing strategies for other compartments than water.

2.4.4 *European Union*

legal scope: 1) Commission Directive 93/67/EEC for new and existing substances [80], Technical Guidance Document [73]
2) Council Directive 91//414/EEC for pesticides [74]
Criteria ad 1): log K_{OW} ≥ 3 **or** highly sorptive **or** belongs to a class of substances known to have a potential to accumulate in organisms **or** indications from structural features **and** no miti-

[1] Defined in the Canadian Environmental Protection Act, Sec. 11: "a substance is toxic if it is entering or may enter the environment in a quantity or concentration or under conditions (a) having or that may have an immediate or long-term harmful effect on the environment; (b) constituting or that may constitute a danger to the environment on which human life depends; or (c) constituting or that may constitute a danger in Canada to human life or health."

[2] On the basis of expert judgement, the concentration of the substance in any environmental medium is due largely to the quantities of the substance used or released as a result of human activity relative to contribution from natural sources.
Elements and naturally occurring inorganic compounds are not candidates for virtual elimination.

gating properties such as hydrolysis (dt_{50}< 12 h) or molecular mass > 700
Criteria ad 2): log $K_{OW} \geq 3$

Concerning new notified chemicals risk potentials and testing requirements depend on the quantity placed on the market. On the base set level (> 1 t/a < 100 t/a) the bioaccumulation potential is assessed according to the criteria listed above. If one of the criteria is met, the substance will be classified with "indication of bioaccumulation potential" in a first approach. If a potential to bioaccumulate can definitely be excluded, it will be classified as showing "no indication of bioaccumulation potential".

Stable transformation products from abiotic (hydrolysis, photolysis, photooxidation) or biotic degradation processes (biodegradation, metabolisation) have also to be checked for their possible bioaccumulation potential.

The bioaccumulation potential is used for three purposes:

- as indicator of a risk for possible adverse long-term effects in ecosystems caused by bioaccumulation,
- as trigger for a bioaccumulation study according to Council Directive 92/32/EEC [81] and German Chemicals Act [70], respectively. A bioaccumulation potential defined by a log $K_{OW} \geq 3$ or the other criteria in combination with mammals/bird toxicity indicate a risk of secondary poisoning and may trigger a bioaccumulation study already on the base set level. On level 1 (> 100 t/a < 1000 t/a) or at 10 t/a depending on ecotoxicological data a bioaccumulation study is mandatory if an indication of a bioaccumulation potential was assessed.

Depending on the result of the study and the risk assessment taking into account exposure and ecotoxicological data (PEC/PNEC), further tests, e.g. bioaccumulation with other organisms may be required on level 2 (> 1000 t/a).

- classification and labelling according to Council Directive 92/32/EEC.

Concerning existing chemicals all available data on bioaccumulation in biota are considered and assessed by expert judgement case by case.

Deficiencies of the risk assessment of the TGD are:

- only log K_{OW} and BCF are considered,
- bioaccumulation is not integrated into the risk assessment scheme and ecotoxicological testing strategy.

Concerning pesticides a bioaccumulation study is mandatory if the active ingredient of a pesticide has a log $K_{OW} \geq 3$.

If the BCF is > 1000 or > 100, respectively, depending on biodegradation and ecotoxicological data, no registration may be granted. However, the complexity of bioaccumulation, particularly the formation of possible bound residues is generally not considered for those hazardous pesticides passing cut-off values.

2.4.5
European Chemical Industry Ecology & Toxicology Centre (ECETOC)

legal scope:	ECETOC-Concept [82] based on Commission Directive 93/67/EEC, new and existing chemicals
criteria:	high persistence, toxicity, negligible metabolism
trigger:	log K_{OW} between 5 and 8
consequences:	bioaccumulation study

In this concept a distinction is made between bioconcentration and bioaccumulation. Bioconcentration is defined as the net result of uptake, distribution, and elimination of a substance in an organism due to water-borne exposure, whereas bioaccumulation includes all routes of exposure including food.

Bioaccumulation is not regarded as an adverse effect or hazard in itself. Bioconcentration and bioaccumulation may lead to an increase in body burden which may cause toxic effects due to direct (water) and/or indirect (dietary) exposure. If no measured BCF data are available, SAR relationships are recommended. K_{OW} is preferred but other relationships based on water solubility and molecular connectivity indices may also be applied. Bioaccumulating substances are characterized by high persistence, toxicity, negligible metabolism and a log K_{OW} between 5 and 8. However, they are only of concern when widely distributed in the environment. Hence the bioaccumulation potential is regarded as an exposure-related parameter in risk assessment. Molecular volume (molecular weight well above 700), low lipid solubility, low bioavailability, rapid biotransformation and structural features are considered as mitigating aspects.

The risk assessment is driven by the key criteria:

- environmental exposure,
- possible uptake.

For substances which reach a steady-state body burden within the organism during the toxicity test, direct effects of bioconcentration are included. Hence a PNEC derived under this condition is regarded as appropriate for use in risk assessment. However, for substances which are taken up and depurated very slowly by fish, the steady-state body burden concentration may not be reached during the toxicity test. Hence, it is recommended to consider the time to reach steady-state (recommended is T_{95}, i.e. time to reach 95% of the steady state concentration) when calculating the **Predicted No Effect Concentrations** (PNECs) for such substances.

If exposure and uptake is possible, bioaccumulation potential is integrated in the ECETOC assessment concept in two ways:

- T_{95}, calculated from K_{OW} is used to select an appropriate duration of aquatic ecotoxicological tests.
- A calculated BCF greater then 1000 is used as trigger for an assessment of secondary poisoning applying the value of this BCF then to estimate a PEC_{oral} (cf. 3.2.3).

If exposure and uptake are not assumed, further assessment is not necessary. Deficiencies are summarized in the context of biomagnification (cf. 3.2.3).

2.5
Proposal for a Comprehensive Assessment Concept of Bioaccumulation

Resuming the criteria, trigger values, and deficiencies in the fore-mentioned concepts, an attempt is made for a comprehensive concept of risk assessment of bioaccumulation as follows:

Criteria for the assessment of bioaccumulation should be the BCF in the whole fish and the elimination or depuration expressed as **half-life clearance time** (ct_{50}) i.e. the time needed to reach 50% removal, as well as organ specific accumulation and incomplete elimination leading to bound residues.

Information on the course of elimination kinetics, however, can only be obtained from a dynamic test based on a two- or more compartment fish model.

The BCF is calculated from the steady-state concentrations in fish and water or from the quotient of the uptake and elimination rate constants, k_1 and k_2. Ct_{50} is calculated from the elimination curve in substance free water after a certain time of exposure.

The complexity of bioaccumulation processes makes it necessary to take into account all measurable processes influencing bioaccumulation, such as

- metabolism, transformation, conjugation,
- organ-specific accumulation (reversible/irreversible),
- incomplete elimination (bound residues),
- bioavailability of the chemical (binding to particulate and dissolved fractions),
- uptake routes,

as well as criteria which are difficult to quantify, such as

- intra- and interspecies variance,
- conditioning factors,
- developmental stages.

Since the degree of elimination of an accumulated chemical is decisive with regard to a possible transfer to higher trophic levels, BCF and half-life time of depuration, ct_{50}, are equally taken into account resulting in 4 respective assessment categories covering the whole range of experimental results, as is shown in Table 4.

The combination of the BCF and ct_{50} will lead to 4 averaged overall assessment categories characterizing the degree of concern.

A more restrictive classification may result in the overall assessment if e.g. there is an indication of organ specific bioaccumulation or of incomplete elimination leading to bound residues forming a plateau, thus raising the risk of biomagnification significantly.

In this case-by-case assessment various aspects have to be considered two of them pointed out below:

- bi- or multiphasic elimination kinetics

  Ct_{50} usually is determined from the elimination curve of the first few days assuming a first order kinetic. Therefore, bioaccumulation risk will be underestimated for substances showing an elimination kinetic with an order higher than 1 if ct_{50} is regarded only.

Table 4. Classification of Bioconcentration Factor and Elimination and Overall Assessment of Bioaccumulation

Bioconcentration Factor (BCF)

BCF range	Assessment Category	Comment
< 30	I	low BCF
30–100	II	moderate BCF
100–1000	III	high BCF
> 1000	IV	very high BCF

Elimination

ct_{50} range	Assessment Category	Comment
< 3 days	I	Rapid elimination
3–10 days	II	Delayed elimination: short term bioaccumulation
10–30 days	III	Slow elimination: medium term bioaccumulation
> 30 days	IV	Insignificant elimination:long term bioaccumulation

Overall Assessment of Bioaccumulation

The categories of the bioaccumulation criteria BCF and ct_{50} are equally taken into account in the overall assessment of bioaccumulation as follows:

$$\frac{\text{BCF category} + ct_{50}\text{ category}}{2}$$

The result of this calculation will lead to one of four bioaccumulation assessment categories. If the resulting quotient lies between two categories, the higher is taken. If elimination data are not available, then only the BCF category can be used.

Overall Assessment Category	Comment
I	no concern
II	indication of risk potential
III	cause for concern
IV	high risk (recommendation for risk reduction)

In the overall assessment a more negative classification may be made if there is an indication of organ specific bioaccumulation or of uncompleted elimination leading to bound residues forming a plateau which would raise the risk of biomagnification significantly.

- plateau formation
  This aspect can also only be assessed case-by-case. If residues of a chemical or its metabolites remain in tissues or organs over a time period which exceeds the duration of long-term ecotoxicity tests, even a plateau as low as 10% of the total ammount of accumulated substance raises the risk of biomagnification.

These examples stress the necessity of an overall assessment of bioaccumulation behaviour which may lead to a more relevant classification than indicated by the BCF and ct_{50} alone.

The different bioaccumulation assessment categories reflect various degrees of concern. The flow-scheme in figure 2 gives guidance how these categories

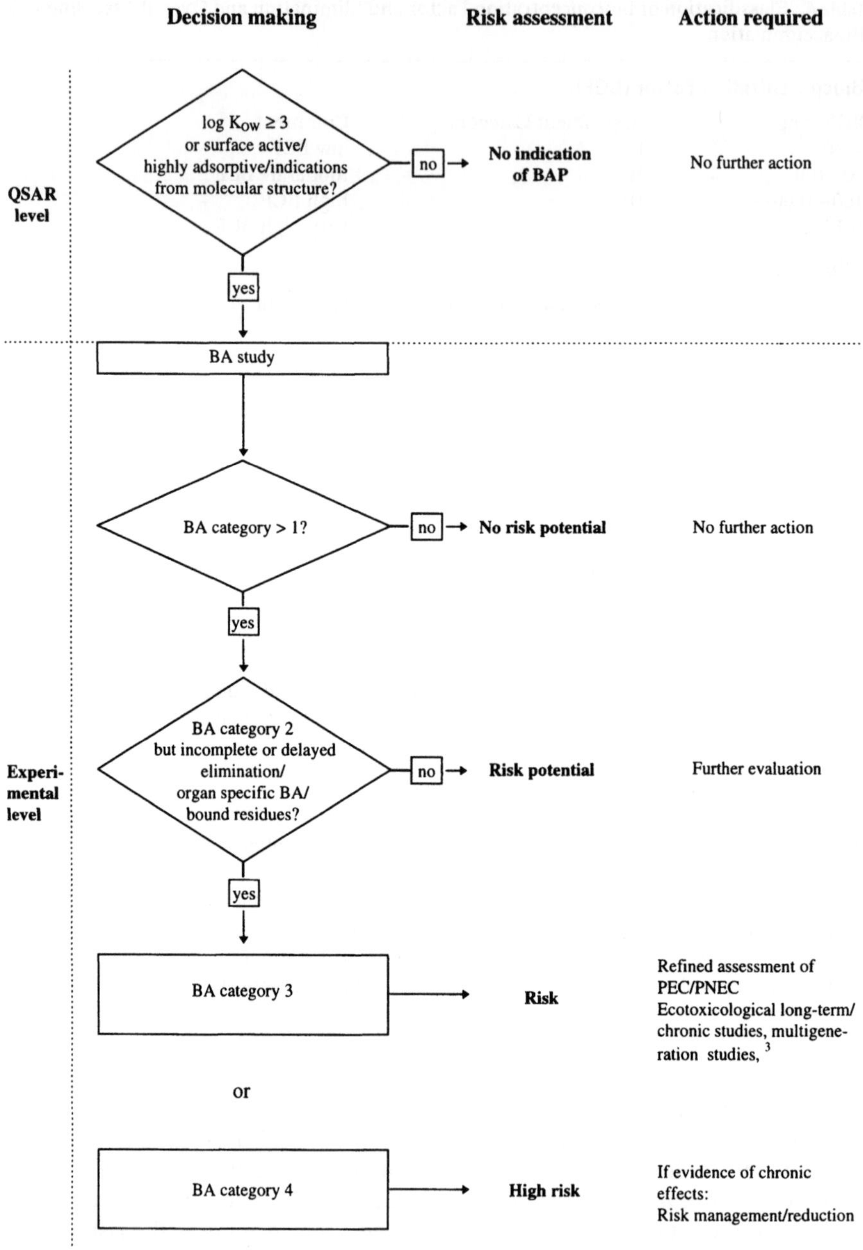

[3] testing for complex i.e. genetic, physiological, histopathological endpoints, endocrine disruptions etc., field studies.

Fig. 2. Risk assessment strategy of bioaccumulation

could be used to assess the risk of bioaccumulation and trigger more conclusive ecotoxicological tests.

With regard to testing and assessment strategy the bioaccumulation categories should lead to the following consequences:

Category I:

No immediate concern with regard to bioaccumulation.

Category II or III:

For chemicals in these categories the risk of biomagnification and secondary poisoning becomes important. On a case-by-case basis it has to be decided whether immediate further testing may be necessary or whether a higher production volume or changes in the use patterns can be awaited. In this decision the category of bioaccumulation, the calculated risk from the indirect effects assessment, data from prolonged (eco)toxicity tests, and exposure data have to be taken into account. Further testing should include tests for chronic effects, e.g. full life cycle tests, preferably together with residue analysis, and testing for other more complex (e.g. genetic, physiological, histopathological) endpoints and multi-generation tests.

To obtain a more comprehensive picture of bioaccumulation, biosorption and biomagnification as well as further aspects such as the impact of highly adsorptive substances on terrestrial and benthic organisms have to be considered. Therefore, bioaccumulation studies with these species may become necessary at this stage.

Category IV:

Chemicals in this bioaccumulation category possess a very high risk of bioaccumulation and biomagnification under environmental conditions. For these chemicals it may be necessary to propose specific recommendations for risk reduction.

3 Biomagnification

3.1 Significance of Biomagnification for Risk Assessment of Chemicals in the Environment

Biomagnification (BM) is the transfer of chemical substances via food-webs passing different trophic levels and resulting in residues which may be detrimental for organisms in terminal positions within food-webs, e.g. dolphins, seals, crocodiles, humans (cf. [6, 17]).

A biomagnification potential (BMP) is indicated if within a food-web the concentration of a chemical or its metabolites in an organism is higher than in its food as major source of uptake.

A special aspect of biomagnification is the concept of "secondary poisoning" which is concerned with toxic effects on higher members of a food chain. Secondary poisoning results from ingestion of organisms at different trophic levels that contain accumulated substances (indirect exposure). A strategy for the assessment of the potential for secondary poisoning has been developed e.g. by Romijn et al. [83, 84] and has become part of the assessment of New and Existing Chemicals in the European Communities [73, 80] (see Fig. 3). In this concept the predicted chemical concentration in food of higher organisms is compared with the mammalian toxicity of the chemical as an indication of possible effects on birds and mammals.

Prerequisite for biomagnification is the bioaccumulation/biosorption of chemicals either by direct uptake from the aquatic or terrestrial environment (via water, pore-water) or by the uptake of particle-bound chemicals and concentration in the organisms respectively (e.g. micro-organisms, algae, invertebrates, vertebrates). Furthermore, there is convincing evidence (cf. e.g. examples mentioned above) that non-metabolized or metabolized residues, which are not excreted completely, may be transferred to the next trophic stage. A part from the BCF the consideration of bound residues are of main concern when conducting and evaluating a bioaccumulation study.

Biomagnification of a substance can hardly be measured in laboratory testing systems existing so far. Therefore, the possibility that a chemical might bioaccumulate – the biomagnification potential (BMP) – has to be considered as an initial step. The flow scheme in Fig. 4 (cf. 3.3) gives guidance on how to conduct assessment of biomagnification in a tiered system taking exposure scenarios and toxicological as well as ecotoxicological effects into consideration. Generally, accumulation, depuration kinetics, and bound residues are the key criteria for a biomagnification potential. If there are strong indications of such residues, further tests including more sophisticated investigations, e.g. of organ-specific concentrations, may become mandatory.

Prior to the final environmental risk assessment of biomagnification, adverse toxicological/ecotoxicological chronic effects and refined exposure assessment must be considered.

3.2 Existing Assessment Concepts of Biomagnification

3.2.1 *USA*

There is no special concept for biomagnification, but US-EPA applies so-called "food chain multipliers" which account for bioaccumulation starting at log K_{OW} of 4.0 [78].

3.2.2 *European Union*

The EU risk assessment approach involves bioaccumulation, biomagnification and secondary poisoning, i.e. the indirect intoxication along a short food-chain water → fish → fish-eating bird or mammal (see Fig. 3).

Secondary poisoning is indicated if the concentration in fish (PEC_{oral} of the prey) reaches a level exceeding the threshold for adverse effects in most sen-

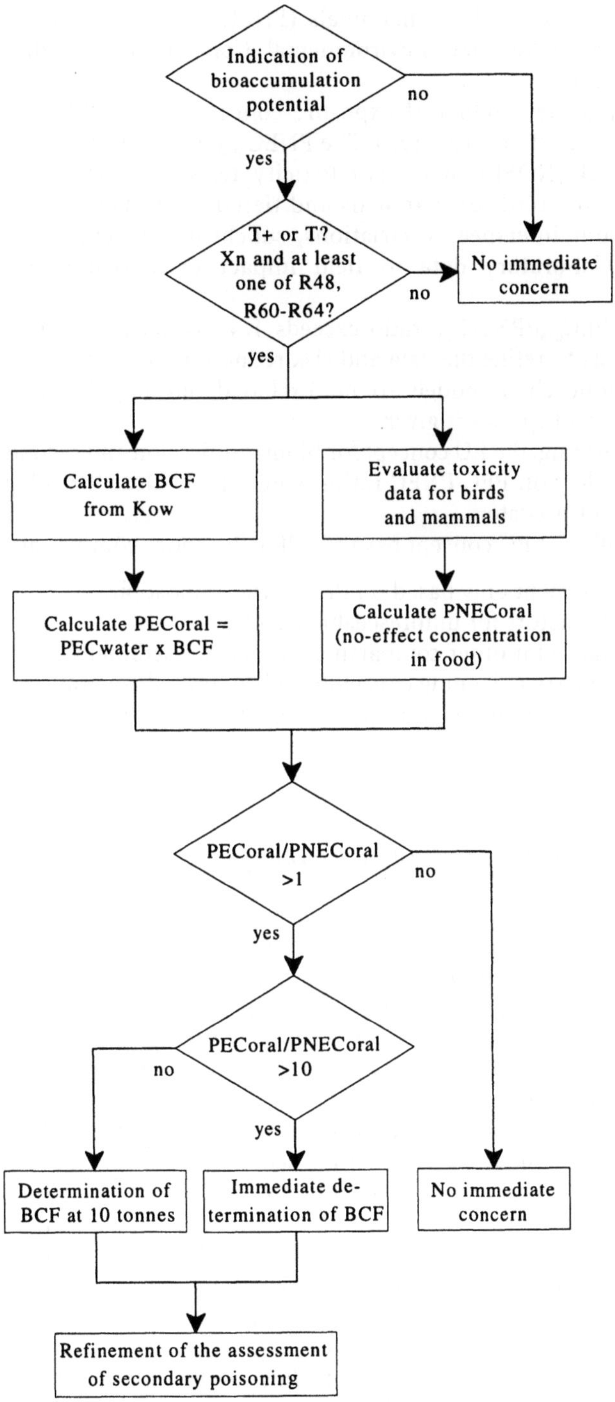

Fig. 3. Strategy for the risk assessment of secondary poisoning

sitive fish-eating birds or mammals ($PNEC_{oral}$ of the predator), i.e. **PEC_{oral}/$PNEC_{oral}$** >1 (Predicted Environmental Concentration/Predicted No Effect Concentration).

PEC_{oral} is the product of exposure concentration (PEC_{water}) and the BCF in fish (modeled or measured). The $PNEC_{oral}$ is the product of a **No Observed Effect Level** (NOEL) in dietary toxicity tests with animals representative for fish-eating birds or mammals and tiered safety factors. The latter taking into account interspecies variations, subchronic to chronic toxicity extrapolation, laboratory data to field impact extrapolation and conversion factors.

When PEC_{oral}/$PNEC_{oral}$ ratio exceeds 1, a concern is signalized triggering further tests to refine the data and risk reduction measures, respectively.

Other food-chain models are not excluded, but no guidance for other compartments and species is given.

Summarizing, the EU concept for biomagnification and secondary poisoning is based solely on PEC/PNEC ratios which imply clear toxicologically defined endpoints of predators.

Generally the EU concept has the following short-comings and deficits:

- risk assessment only based on PEC/PNEC philosophy,
- no safety factors for unforeseeable effects,
- no guidance for other compartments (marine, sediment, terrestrial),
- the secondary poisoning concept is only a limited aquatic food-chain model,
- no consideration of aquatic, sediment associated and terrestrial food-webs.

3.2.3
European Chemical Industry Ecology & Toxicology Centre (ECETOC)

In this concept the biomagnification is integrated in an overall risk assessment scheme. If the calculated BCF is >1000 as the outcome of an initial assessment of bioaccumulation potential (cf. 2.4.5), the risk assessment of secondary poisoning is triggered. Dietary uptake by aquatic organisms is considered only if the BCF of prey organisms is >1000 corresponding to a log K_{OW} of 4.3. If this criterion is met, a PEC_{oral}/$PNEC_{oral}$ assessment for predators is conducted and refined if considered necessary.

Referring to the EU Technical Guidance Document the ECETOC concept criticizes that the risk assessment for the secondary poisoning concept is initiated at log $K_{OW} > 3$. This approach would thus overestimate the risk of chemicals of „little" relevance (i.e. already with log $K_{OW} \geq 3$) and underestimate the risk of chemicals in the log K_{OW} range of 4.5–8 with higher lipophilicity where dietary uptake is more significant for biomagnification.

Further it is suggested that the PEC_{oral} should be based rather on actual body burden concentration than on log K_{OW} regression, since elimination may significantly reduce the body burden thus overestimating the magnification. Unfortunately other potential end points are not discussed as well as consequences resulting from body burden concentration.

Deficiencies of ECETOC concept:

- underestimation of biomagnification for substances with lower log K_{OW} values,
- risk assessment solely based on the PEC/PNEC approach,
- no consideration of bound residues,
- only consideration of known (acute toxic) effects,
- no safety factors to counter unforeseeable effects,
- persistent, lipophilic and toxic substances (fish/mammals) are not considered in the three case studies presented.

3.2.4
Van Leeuwen and Hermens

The biomagnification has been modelled by Van Leeuwen and Hermens (1995) [85] taking into account uptake of food and ingestion of sediment, e.g. by sediment dwelling organisms.

The BMP does not consider the BCF alone but also information regarding kinetics of uptake, metabolism and elimination in the organism.

Risk characterization of BM is based on exposure and effect assessment (PEC/PNEC) defining risk quotients. Effect assessment is mostly based on acute, less frequently on subacute or chronic tests, i.e. well defined ecotoxicological end points.

After performance of an uncertainty analysis the probability of the occurrence of defined and known effects is identified and quantified.

Secondary poisoning, i.e. the indirect intoxication via a short food chain, e.g. fish → fish-eating bird or mammal is not only related to increased mortality, but may consider also fitness parameters and more subtle effects, e.g. impact on eggshell thickness.

Although unforeseeable effects can never be ruled out completely by any risk assessment strategy, the PEC/PNEC approach, in our opinion, does not sufficiently counter the risk of unexpected effects, even by application of uncertainty factors on the final PNEC.

3.2.5
Cowan et al.

An integrated approach for environmental assessment of new and existing substances is presented by Cowan et al (1995) [86] which specifically evaluates persistence and bioaccumulation of a substance in order to assess the potential for direct and indirect effects on species in aquatic, sediment and terrestrial environments.

This concept is based on the assessment of bioconcentration, dietary pathways, potential for bioaccumulation and effects resulting in 4 tiers of concern.

A measured BCF > 1000 signalizes the potential of dietary exposure for aquatic organisms and the ratio $PEC_{oral}/PNEC_{oral}$ is calculated. If the ratio exceeds 1, long term ecotoxicological studies are demanded. Risk management is indicated, if after a refined assessment which includes monitoring data the ratio is still > 1.

This approach exceeds conventional risk assessment concepts, but aims obviously only at mortality rather than at long-term/chronic effect in ecosystems.

3.3 Proposal for a Comprehensive Assessment Concept of Biomagnification

The main principles of a risk assessment strategy for biomagnification are shown in the flow-scheme in Fig. 4.

Basic prerequisite of a risk assessment for biomagnification is a valid comprehensive bioaccumulation study taking into account uptake and depuration kinetics, organ-specific distribution and accumulation, metabolic activities and conjugation products, bound residues and, preferably, critical body burden concentrations for subtle end points e.g. chromosomal aberrations.

However, there is a considerable variability of these parameters among species and transmission of data from one species to another is generally not possible.

Not or only partially metabolized chemicals, classified in a bioaccumulation category ≥ III (cf. 2.5), which persist in individuals and may be transferred to further generations or trophic levels, respectively, signalize an indication of biomagnification potential.

In turn, lack of these incriminating criteria may as a first approach lead to the conclusion that there is no immediate concern for a risk so far and no further action is required for the moment.

Monitoring data on the environmental fate of pollutants in terrestrial and aquatic compartments and the occurrence in biota may give decisive indications on biomagnification processes.

Such indications may also be derived from laboratory scale food-chain-studies. Additionally, if permanent exposure is anticipated or proven or if the predicted regional environmental concentration (PEC_{reg}) is in the same order of magnitude as the **critical body burden** (CBB_{food}) concentration for food in the organisms, a risk of biomagnification exists and the probability for the occurrence of effects must be assessed, provided data are sufficient.

If such indications are not recognizable on this level the risk of a biomagnification potential is low. Immediate action is not required but further data should be gathered for a refined assessment.

Is a risk characterization not possible on this level due to lacking data more sophisticated laboratory studies have to be performed comprising chronic multigeneration studies and investigating, e.g. genetic, physiological, histopathological and endocrinological endpoints.

The outcome of such studies may confirm the suspicion revealing adverse effects or – in case of non-visible effects – may lead to the conclusion that an uncertain and unforeseeable risk remains, which should be countered by the application of a risk factor of 10 on the final PEC/PNEC.

A clear evidence of adverse effects in food-webs means an unacceptable risk for ecosystems and should result in risk management and reduction/regulation measurements.

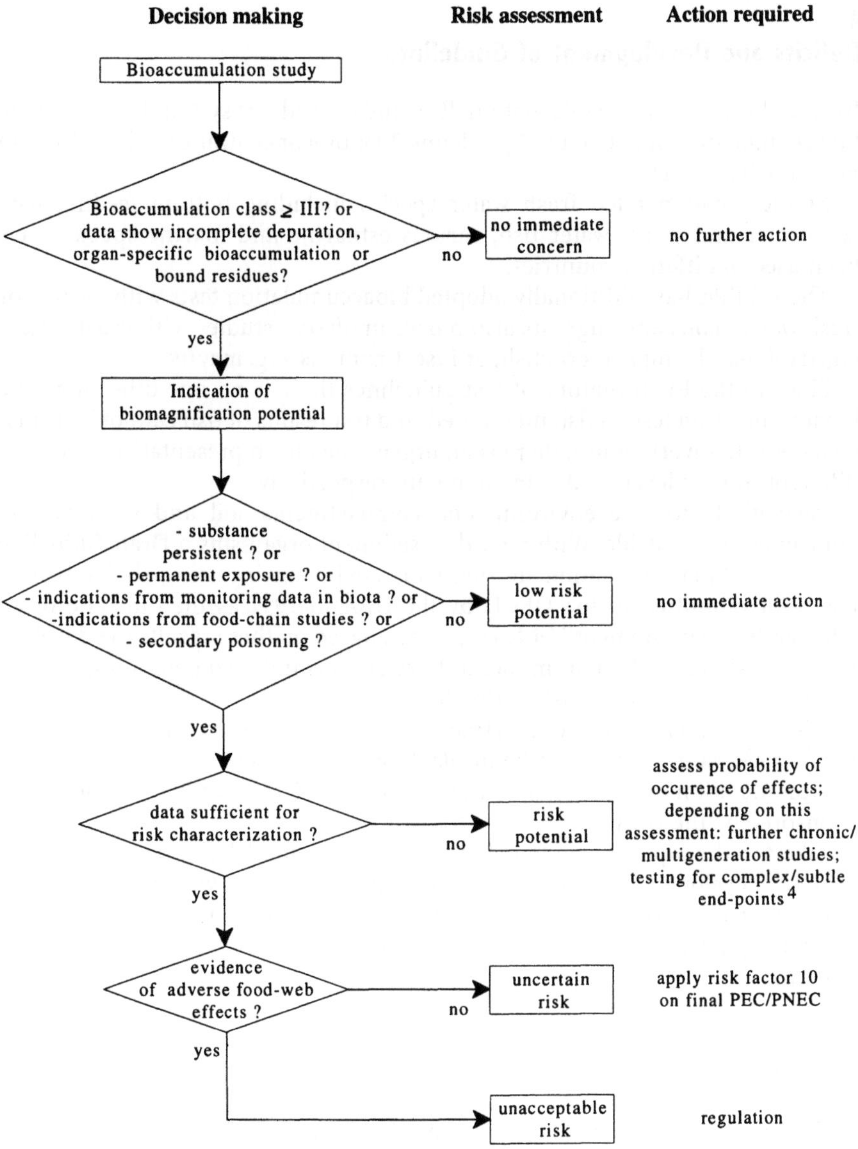

[4] genetic, physiological, histopathological, endocrine disruptions etc.

Fig. 4. Risk assessment strategy of biomagnification

4
Deficits and Development of Guidelines

By now there is only one internationally standardized test system for testing the bioaccumulation in fish: OECD guideline 305: Bioconcentration: Flow-through Fish Test (1996) [1].

Beside recommended fresh water species including bottom feeding fish, cold-water and warm-water fish, various estuarine and marine species have been used in different countries.

The US EPA has additionally adopted bioaccumulation tests with oysters or fresh water clams and suggests also bioaccumulation studies with crustaceans, e.g. daphnia, shrimps or crayfish, or insect nymphs, e.g. mayfly.

Since in the EU inventory of test guidelines there is no one other than the bioaccumulation test on fish mentioned, and test results of fish cannot be transferred e.g. to invertebrates, there is an urgent need for representative species of different trophic levels and compartments, respectively.

Particularly for the environmental compartments soil and sediment no guidelines are available. With regard to sediment organisms a Draft Guideline for Testing Bioaccumulation in Tubificids (sediment ingesting sludge-worm) has been submitted to the OECD by the UBA in 1997 as the outcome of an "**Research & Development**" (R & D) project. Based on these results the development of a short food-chain model test system was initiated consisting of sediment – Tubifex → fish → fish-eating bird.

Also for the terrestrial compartment no test system exists until now. Within the framework of the OECD Chemicals Program a test system investigating the bioaccumulation in earthworms was scheduled in 1997 by an R & D project also sponsored by the UBA.

Like for the aquatic compartment, food-chain model test systems should also be developed for the terrestrial compartment as well as for marine and estuarine environments thus considering the most important environmental compartments for an overall comprehensive risk assessment for ecosystems.

5
Conclusions

- The risk assessment of bioaccumulation by environmental authorities should not be based on QSARs alone since many chemicals do not obey commonly applied correlations. Whereas overestimation of bioaccumulation may be irrelevant to real environmental conditions, underestimation of risks may have serious consequences for ecosystems.
- Risk assessment of bioaccumulation solely based on the BCF is insufficient and may be misleading. Instead, a tiered risk assessment strategy of bioaccumulation and biomagnification is proposed taking into account the complexity of bioaccumulation processes integrating equally the key parameters BCF/BAF and depuration half-life times for deriving four classes which characterize the risk of bioaccumulation and if necessary trigger further ecotoxi-

cological tests. Incomplete depuration and the occurrence of bound residues are additionally considered as incrimination factors.

- The critical body burden concentration (i.e. the internal concentration in tissues or organs above which effects may be induced) is finally the decisive parameter of bioaccumulation with regard to effects. This ecotoxicological endpoint is more meaningful than conventional EC_x-values defining external concentrations. Having knowledge of the internal threshold concentration for a specific endpoint, the safety margin for the risk resulting from the difference between the concentration already reached and the concentration inducing effects, can be defined (see contribution of Sijm and Hermens, this volume).
- BCFs and BAFs as numerical values should not equally be used in risk assessment approaches. BAFs of organisms related to sediment or soil concentrations are of course usually lower than BCFs of aquatic organisms for which the main uptake route for moderately lipophilic substances is predominantly via water. However, regardless which uptake routes are involved a terrestrial or sedimental BAF >1 is considered as relevant regarding a significant bioaccumulation.
- Even exonerating results from sophisticated chronic bioaccumulation and biomagnification studies, if ever conducted for each relevant chemical released into the environment, cannot completely rule out the risk for adverse long-term effects. Hence, if for the risk assessment of biomagnification a risk for ecosystems is indicated, the application of an additional safety factor of 10 on the final PEC/PNEC according to the TGD is proposed. If as a consequence of the risk assessment strategy an unacceptable risk for ecosystems is shown, restrictions and bans, respectively, should be considered.

References

1. Organisation for Economic Co-operation and Development (OECD). OECD Guideline for Testing of Chemicals No. 305: Bioconcentration: Flow-through Fish Test. 14 June 1996, Paris
2. Henny CJ, Bennett JK (1990) Environ Tox Chem 9:797
3. Bitmann J, Cecil HC (1970) J Agric Food Chem 18:1108
4. Routledge EJ, Sumpter JP (1996) Environ Tox Chem 15 (3):241
5. Jobling S, Sheahan D, Osborne JA, Matthiessen P, Sumpter JP (1996) Environ Tox Chem 15 (2):194
6. Elster H-J (1987) Bioakkumulation in Nahrungsketten: Zusammenfassende Bewertung der Befunde. In: Lillelund K, de Haar U, Elster H-J, Karbe L, Schwoerbel I, Simonis W (eds) Bioakkumulation in Nahrungsketten. DFG, Deutsche Forschungsgemeinschaft. VCH, Weinheim
7. Butte W, Fox K, Zauke G-P (1991) Sci Total Environ 109/110:377
8. Liebl B, Ehrenstorfer S (1993) Chemosphere 27 (11):2253
9. Rimkus G, Rimkus B, Wolf M (1994) Chemosphere 28 (2):421
10. Yamagishi T, Miyayuki T, Horii S, Akiyama K (1983) Arch Environ Contam Toxicol 12:83
11. Kuehl DW, Haebler R, Potter C (1991) Chemosphere 22 (11):1071
12. Marsili L, Gaggi C, Bortolotto A, Stanzani L, Franchi A, Renzoni A, Bacci E (1995) Chemosphere 31 (8):3919
13. Larsson P, Lennart O, Collvin L (1993) Environ Tox Chem 12:855

14. Albanis TA, Hela DG, Hatzilakos D (1995) Chemosphere 31 (11/12):4341
15. Wolf-Roskosch F, Schuphan I (1990) A model of testing the food chain toxicity of low-level concentrations of pesticides in host-parasitoid-systems. In: Casida JE (ed) Pesticides and alternatives. Elsevier BV (Biomedical Division)
16. Mullié WC, Massi A, Focardi S, Renzoni A (1992) Egypt Bull Environ Contam Toxicol 48:739
17. Winter S, Streit B (1992) Chemosphere 24 (12):1765
18. Chemicals Inspection & Testing Institute (CITI), Japan, (ed) (October 1992) Biodegradation and Bioaccumulation. Data of Existing Chemicals based on the Chemical Substances Control Law Japan
19. Blume H-P, Ahlsdorf B (1993) Ecotoxicol Environ Saf 26:313
20. UBA files in context with registration procedures acc. to the German Plant Protection Act
21. Streit B (1990) Chemikalien im Wasser: Experimente und Modelle zur Bioakkumulation bei Süßwassertieren. In: Kinzelbach/Friedrich (eds): Biologie des Rheins. Gustav Fischer Verlag, Stuttgart New York (Limnologie aktuell, vol 1)
22. Dial CA, Dial N A (1995) Bull Environ Contam Toxicol 55 (6):870
23. UBA files in context with the implementation of the German Chemicals Act (specific data confidential)
24. Woltering DM, Larson RJ, Robert J, Hopping WD, Jamieson RA, De Oude NT, (1987) Tenside, Surfactants, Deterg, 24 (5):286
25. Harkey GA, Landrum PF, Klaine SJ (1994) Environ Tox Chem 13 (8):1315
26. Thomann RV, Conolly JP, Parkerton TF (1992) Environ Tox Chem 11:615
27. Pereira WE, Domagalski JL, Hostettler FD, Brown LR, Rapp JB (1996) Environ Tox Chem 15 (2):172
28. Mac MJ, Noguchi GE, Hesselberg RJ, Edsall CC, Shoesmith JA, Bowker JD (1990) Environ Tox Chem 9:1405
29. Egeler PT, Römbke J, Knacker Th, Meller M, Franke C, Studinger G, Nagel R (1997) Chemosphere 35 (4):835
30. Kukkonen J, Landrum PF (1994) Environ Tox Chem 13 (9):1457
31. Kukkonen J, Landrum PF (1995) Environ Tox Chem 14 (3):523
32. Segstro MD, Muir DCG, Servos MR, Webster GRB (1995) Environ Tox Chem 14 (10):1799
33. Munoz MJ, Castano A, Blazquez T, Vega M, Carbonell G, Ortiz JA, Carballo M, Tarazona JV (1994) Chemosphere 29 (1):55
34. Balk L, Meijer J, DePierre J W, Appelgren L-E (1984) Toxicol Appl Pharmacol 74:430
35. Deneer JW (1994) Chemosphere 29 (7):1561
36. Butte W, Fox K, Schmidt A, Zauke G-P (1993) Bioaccumulation at the borderline of lipophilicity and molecular structures. R & D project, Umweltbundesamt (Federal Environmental Agency), Berlin
37. Anliker R, Moser P, Poppinger D (1988) Chemosphere 17 (8):1631
38. Geyer H-J, Steinberg CEW, Kettrup A (1992) Umweltchem Ökotox 4 (2):74
39. Schmieder P, Lothenbach D, Tietge J, Erickson R, Johnson R (1995) Environ Tox Chem 14 (10):1735
40. Muir DCG, Rawn GP, Townsend BE, Lockhart WL, Greenhalgh R (1985) Environ Tox Chem 4 (1):51
41. Fox K, Zauke G-P, Butte W (1994) Ecotoxicol Environ Saf 28:99
42. Mössner S, Spraker TR, Becker PR, Ballschmiter K (1992) Chemosphere 24 (9):1171
43. Oliver BG, Niimi AJ (1985) Environ Sci Technol 19:842
44. Larsson P, Bremle G, Okla L (1993) Environ Contam Toxicol 50:653
45. Schroll R, Bierling B, Cao G, Dörfler U, Lahaniati M, Langenbach T, Scheunert I, Winkler R (1994) Chemosphere 28 (2):297
46. Bacci E, Cerejeira MJ, Gaggi C, Chemello G, Calamari D, Vighi M (1990) Chemosphere 21 (4/5):525
47. Koelmanns AA, Jiménez CS (1994) Chemosphere 28 (12):2041
48. Stahlschmidt-Allner P (1992) Toxikokinetik und Toxikodynamik von grenzflächenaktiven Stoffen in aquatischen Organismen. Expert report on behalf of the Umweltbundesamt, reg. no. I 4.1–97 316/24

49. Rippen G (1990) in: Handbuch Umweltchemikalien. Stoffdaten, Prüfverfahren, Vorschriften. Ecomed, Landsberg am Lech. (3. Auflage)
50. Kalbfus W, Zellner A, Frey S, Stanner E (1990) Risks for surface waters caused by tinorganic compounds in antifouling paints. R & D project, Umweltbundesamt (Federal Environmental Agency), Berlin
51. Oehlmann J, Ide I, Bauer B, Watermann B, Schulte-Oehlmann U, Fioroni A (1996) Erfassung histopathologischer Effekte von Organozinnverbindungen auf marine Mollusken und Prüfung ihrer Verwendbarkeit für ein zukünftiges biologisches Effektmonitoring. R & D project, Umweltbundesamt (Federal Environmental Agency), Berlin
52. Larson JM, Karasov WH, Sileo L, Stromborg KL, Hanbidge BA, Giesy JP, Jones PD, Tillitt DE, Verbrugge DA (1996) Environ Tox Chem 15 (4):553
53. Arnold H, Pluta H-J, Braunbeck T (1996) Ecotoxicol Environ Saf 34:43
54. Lester DC, McIntosh A (1994) Environ Tox Chem 13 (11):1825
55. Fairchild WL, Muir DCG, Currie RS, Yarechewski AL (1992) Environ Tox Chem 11:867
56. Harkey GA, Klaine SJ (1992) Chemosphere 24 (12):1911
57. Kobayashi K, Akitake H, Hanabe K (1979) Bull Japan Soc Sci Fish 45 (2):173
58. Mc Carty LS (1986) Environ Toxicol Chem 5:1071
59. Sijm DTHM, Schipper M, Opperhuizen A (1993) Environ Toxicol Chem 12:1117
60. Franke C, Studinger G, Berger G, Böhling S, Bruckmann U, Cohors-Fresenborg D and Jöhncke U (1994) Chemosphere 29 (7):1501
61. Egeler PT, Knacker T, Römbke J, Schallnaß H-J, Studinger G, Franke C, Nagel R (1996) Bioaccumulation: Sediment test using benthic oligochaetes. Draft guideline acc. to OECD format. R & D project, Umweltbundesamt (Federal Environmental Agency), Berlin
62. Crommentuijn T, Doodeman CJAM, Doornekamp A (1994) Environ Toxicol Chem 13 (11):1781
63. Wezel AP van, Sijm DTHM, Seinen W, Opperhuizen A (1995) Chemosphere 31 (5):3201
64. Butte W, Willig A, Zauke G-P (1988) Bioaccumulation of phenols using a flow-through fish test (OECD 305 E) and its correlation to structural parameters. R & D project, Umweltbundesamt (Federal Environmental Agency), Berlin
65. Tillitt DE, Ankley GT, Giesy JP, Ludwig JP, Kurita-Matsuba H, Weseloh DV, Ross PS, Bishop CA, Sileo L, Stromborg KL, Larson J, Kubiak TJ (1992) Environ Toxicol Chem 11:1281
66. Annex to Commission Directive 92/69/EEC of 31 July 1992 adapting to technical progress for the seventeenth time Council Directive 67/548/EEC on the approximation of laws, regulations and administrative provisions relating to the classification, packaging and labelling of dangerous substances. Official Journal of the European Communities L 383 A, 35, 29 December 1992. Part A: Methods for the determination of physico-chemical properties, A.8. Partion coefficient
67. OECD Guideline for Testing of Chemicals No. 107 Partion coefficient (n-Octanol/Water): Shake Flask Method
68. Wezel AP van, Opperhuizen A (1995) Chemosphere 31 (7):3605
69. Annex to Commission Directive 92/69/EEC of 31 July 1992 adapting to technical progress for the seventeenth time Council Directive 67/548/EEC on the approximation of laws, regulations and administrative provisions relating to the classification, packaging and labelling of dangerous substances. Official Journal of the European Communities L 383 A, 35, 29 December 1992. Part A: Methods for the determination of physico-chemical properties, A.5. Surface tension
70. German Chemicals Act. Gesetz zum Schutz vor gefährlichen Stoffen (Chemikaliengesetz – ChemG) vom 14. März 1990, BGBl. I, 22. März 1990
71. German Plant Protection Act. Erstes Gesetz zur Änderung des Pflanzenschutzgesetzes vom 14. Mai 1998, BGBl I, 27. Mai 1998
72. Annex I, II, III and IV to Commission Directive 93/21/EEC of 27 April 1993 adapting to technical progress for the eighteenth time Council Directive 67/548/EEC on the approximation of the laws, regulations and administrative provisions relating to the classification, packaging and labelling of dangerous substances. Official Journal of the European Communities L 110 A, 36, 4 May 1993

73. Technical Guidance Documents (TGD) in support of the Commission Directive 93/67 EEC on Risk Assessment for new notified substances and the Commission Regulation (EC) 1488/94 on Risk Assessment for existing substances. ECB, 1996
74. Council Directive 91/414/EEC of 15 July 1991 concerning the placing of plant protection products on the market. Official Journal of the European Communities L 220, 34 Annex II
75. Toxic Substances Management Policy. Persistence and Bioaccumulation Criteria. Final Report of the ad hoc Science Group on Criteria, Government of Canada, Environment of Canada, June 1995
76. International workshop on approaches to a harmonization of environmental assessment of plant protecting products for authorization purposes, OECD Meeting, 27.–30. November 1995, Fraunhofer IRB Verlag
77. Wellenreuther G (1990) Melliand Textilberichte 71:466
78. Landis WG, Hughes JS, Lewis MA (eds) Environmental Toxicology and Risk Assessment, ASTM STP 1179, American Society for Testing and Materials, Philadelphia (1997) pers. comm.
79. Organisation for Economic Co-operation and Development (OECD). OECD Guideline for Testing of Chemicals No. 305 A-E: Bioaccumulation. 12 May 1981, Paris
80. Commission Directive 93/67/EEC of 20 July 1993 laying down the principles for assessment of risk to man and the environment of substances notified in accordance with the requirements of Council Directive 67/548/EEC. Official Journal of the European Communities L 227, 36, 18 September 1993
81. Council Directive 92/32/EEC of 30 April 1992 amending for the seventh time Directive 67/548/EEC on the approximation of the laws, regulations and administrative provisions relating to the classification, packaging and labelling of dangerous substances. Official Journal of the European Communities L 154, 35, 5 June 1992
82. The Role of Bioaccumulation in Environmental Risk Assessment: The Aquatic Environment and Related Food Web (Oct. 1995) ECETOC Technical Report No.67
83. Romijn CFA, Luttik R, Meet D v.d., Slooff W, Canton JH (1993) Ecotoxicol Environ Saf 26:61
84. Romijn CFA, Luttik R, Meet D v.d., Slooff W, Canton JH (1998) Presentation of a general algorithm to include effect assessment on secondary poisoning in the derivation of environmental quality criteria. Part 2. Terrestrial food chains. Ecotoxicol Environ Saf In press
85. Van Leeuwen CJ, Hermens JLM (1995) Risk assessment of chemicals: An introduction, Kluwer Academic Publishers, Dordrecht/Boston/London
86. Cowan CE, Versteeg DJ, Larson RJ, Kloepper-Sams PJ (1995) Regulatory Toxicology and Pharmacology 21:3

Subject Index

MIX
Papier aus verantwortungsvollen Quellen
Paper from responsible sources
FSC® C105338
FSC
www.fsc.org

If you have any concerns about our products,
you can contact us on
ProductSafety@springernature.com

In case Publisher is established outside the EU,
the EU authorized representative is:
Springer Nature Customer Service Center GmbH
Europaplatz 3, 69115 Heidelberg, Germany

Printed by Libri Plureos GmbH
in Hamburg, Germany